Many of the fundamental theories of modern physics can be considered as descriptions of dynamical systems subjected to constraints. The study of these constrained dynamical systems, in particular the problems encountered in formulating them as quantum systems, has many profound links with geometry, and these were explored in the Symposium on Geometry and Gravity held at the Newton Institute in 1994. This book arose from a conference held during that symposium, and is a collection of papers devoted to problems such as Chern-Simons theory, sigma-models, gauge invariance and loop quantization, general relativity and the notion of time, and quantum gravity. They present a lively, varied and topical perspective on this important branch of theoretical physics from some of the leading authorities in the subject.

PUBLICATIONS OF THE NEWTON INSTITUTE

Geometry of Constrained Dynamical Systems

Publications of the Newton Institute

Edited by P. Goddard
Deputy Director, Isaac Newton Institute for Mathematical Sciences

The Isaac Newton Institute of Mathematical Sciences of the University of Cambridge exists to stimulate research in all branches of the mathematical sciences, including pure mathematics, statistics, applied mathematics, theoretical physics, theoretical computer science, mathematical biology and economics. The four six-month long research programmes it runs each year bring together leading mathematical scientists from all over the world to exchange ideas through seminars, teaching and informal interaction.

Associated with the programmes are two types of publication. The first contains lecture courses, aimed at making the latest developments accessible to a wider audience and providing an entry to the area. The second contains proceedings of workshops and conferences focusing on the most topical aspects of the subjects.

GEOMETRY OF CONSTRAINED DYNAMICAL SYSTEMS

Proceedings of a Conference held at the
Isaac Newton Institute, Cambridge, June 1994

edited by
John M. Charap
Queen Mary and Westfield College
University of London

CAMBRIDGE UNIVERSITY PRESS
Cambridge, New York, Melbourne, Madrid, Cape Town, Singapore, São Paulo

Cambridge University Press
The Edinburgh Building, Cambridge CB2 8RU, UK

Published in the United States of America by Cambridge University Press, New York

www.cambridge.org
Information on this title: www.cambridge.org/9780521482714

First published 1995

A catalogue record for this publication is available from the British Library

ISBN 978-0-521-48271-4 hardback

Transferred to digital printing 2008

Contents*

* Where a paper has more than one author, an asterisk is used to indicate who presented the paper at the Conference.

The network

The thirteen participating institutions in the HCM contract which established the "Constraints Club" network are:

Belgium:	Université Libre de Bruxelles
France:	Université Pierre et Marie Curie, Paris
Germany:	Universität Kaiserslautern
Ireland:	St Patrick's College, Maynooth
Italy:	INFN Sezione di Firenze, (with Università di Firenze, Parma and Pisa)
	INFN Sezione di Napoli, (with Università di Napoli, Bologna, Camerino, Trieste)
	INFN Sezione di Pavia, (with Università di Pavia)
	Università di Torino
The Netherlands:	NIKHEF-H, Amsterdam
Portugal:	Instituto Superior Técnico, Lisboa
Spain:	Universitat de Barcelona
United Kingdom:	University of Lancaster
	Queen Mary and Westfield College, London

The five members from the Former Soviet Union who joined the network supported by a contract with INTAS are:

Armenia:	Yerevan Physics Institute
Russia:	PN Lebedev Physical Institute, Moscow
	ITEP, Moscow
	Tomsk State University
Ukraine:	Bogolyubov Institute for Theoretical Physics, Kiev

And the five members it is proposed to add with support from the European Communities' PECO programme are:

Bulgaria:	Institute of Nuclear Research and Nuclear Energy, Sofia
Poland:	University of Łódź
	University of Wrocław
Romania:	University of Cluj-Napoca
Russia:	Steclov Mathematical Institute, Moscow

Preface

In the hills above Florence lies Arcetri, where Galileo Galilei spent his years of house arrest after his trial in 1633. Galileo's statement of the law of inertia launched the study of dynamics on the voyage of discovery and invention on which most if not all of modern physics depends. We also remember Galileo when we speak of Galilean relativity, and it was therefore fitting that it was in Arcetri that the workshop on 'Constraint's theory and relativistic dynamics' was held in 1986. This workshop was organised by Giorgio Longhi, Luca Lusanna and Giuseppe Marmo "to examine the current situation of relativistic dynamics", and I had the good fortune to be able to attend, meeting there many of those who were later to become the "Constraints Club".

A few years later there arose an opportunity to establish this more formally as an association of researchers active in the area of constrained dynamical systems. Under the European Communities' SCIENCE programme funds were available for "networks", and an application was duly prepared to support young postdoctoral fellows to work within the "Constraints Club", which at that time was a rather loosely coordinated group of five laboratories. This application was unsuccessful, but in 1992 an enlarged (and improved!) application was submitted to the Human Capital and Mobility programme, the successor to SCIENCE. It too failed, but on resubmission in 1993 was at last accepted*, and the network in "Constrained Dynamical Systems" came officially into being on 1 January 1994. There were by now 13 laboratories included in the network, and many of them had been represented at the second workshop to be organised by the Florence group, on 'Constraint theory and quantization methods', held this time at Montepulciano (Siena) in the summer of 1993.

Our network has grown, and the 13 original HCM members have now been joined by five laboratories in the former Soviet Union with support from the INTAS programme. Applications are at the time of writing pending for a further enlargement by adding five more from Central and Eastern Europe through the PECO programme of the European Union.

With financial support derived from the HCM contract, the Conference on 'Geometry of constrained dynamical systems' was held on 15–18 June 1994 at the Isaac Newton Institute for Mathematical Sciences in Cambridge. Again

*As Contract ERBCHRXCT930362

the choice of venue was apt; if Galileo launched the study of dynamics on its voyage of discovery, it was Newton who steered it to prosperity. The Newton Institute conducts a series of 6-month research programmes, and in the first half of 1994 Gary Gibbons, Steven Hawking and Chris Isham had organised a programme on 'Geometry and Gravity'. I am grateful to them, and to Peter Goddard the Deputy Director of the Institute, for their hospitality which allowed the Conference to be included as part of that research programme. To quote Galileo: 'He who undertakes to deal with questions of natural sciences without the help of geometry is attempting the unfeasible'. The Conference was attended by around 100 physicists and mathematicians, about equally divided between those from within the network and those from elsewhere. Cambridge was at its best, with clear blue skies which rivalled those of Arcetri and Montepulciano. Its success augurs well for the future of the "Constraints Club".

The 36 papers presented at the Conference are the basis for this volume. Two papers which have been published elsewhere have been omitted, and one paper included here was not presented because its author was unable to be in Cambridge. The order of papers in this volume is the same as their presentation at the Conference.

I am grateful to all the contributors for their care in preparing the TeX or LaTeX files from which this volume has been compiled. I am also grateful to the staff at the Isaac Newton Institute for their invaluable support and assistance which ensured that the meeting ran smoothly and successfully. Finally, my thanks also to David Tranah at Cambridge University Press and to the Syndicate of the Press for including this volume in the Publications of the Newton Institute.

1994 John M Charap

The constraint algebra of higher dimensional Chern-Simons theories

Máximo Bañados

Chern-Simons theories in three dimensions have received much attention in the last years. The aim of this work is to investigate whether or not some of the properties of the three-dimensional Chern-Simons theory can be carried over to higher dimensions. In particular, we shall be interested in the Hamiltonian formulation of the theory and its constraint algebra.

Higher dimensional Chern-Simons theories are defined by the relation

$$Tr(\underbrace{F \wedge F \wedge \cdots \wedge F}_{n+1}) = d \wedge L_{2n+1} \tag{1}$$

where $F = dA + A \wedge A$ is the curvature 2-form for a non-Abelian connection 1-form $A = A^a J_a$. The J_a's form a basis for a Lie-algebra that we assume to have a non-degenerate Killing metric normalised by $g_{ab} = Tr(J_a J_b)$.

The left hand side of (1) is a gauge invariant quantity. Thus, L_{2n+1} is gauge-invariant up to a total derivative. Let M be a $2n+1$ dimensional manifold without a boundary. The functional

$$I = \int_M L_{2n+1} \tag{2}$$

is then a gauge invariant quantity. We shall call it the Chern-Simons action in $2n+1$ dimensions. [For manifolds with a boundary one has to add a surface term to the action (2) in order to achieve gauge invariance. We shall not discuss this point here.]

If the manifold has the topology $M = \Re \times \Sigma$, the gauge field can be globally decomposed as $A^a_\mu dx^\mu = A^a_0 dt + A^a_i dx^i$ where t runs along $\Re$ and the x^i (i=1,...,$2n$) are coordinates on Σ. It is easy to see that the action does not contain time derivatives of A^a_0 and that it depends linearly on $\dot{A}^a_i$. In fact, the action (2) has the form

$$I = \int_{\Re} dt \int_{\Sigma} d^{2n}x [L^i_a(A) \dot{A}^a_i + A^a_0 G_a], \tag{3}$$

where G_a is given by

$$G_a = \frac{1}{2n} g_{aa_1\ldots a_n} \epsilon^{i_1\ldots i_{2n}} F^{a_1}_{i_1 i_2} \cdots F^{a_n}_{i_{2n-1} i_{2n}} \tag{4}$$

and the "metric" $g_{aa_1\ldots a_n}$ [the trace of a symmetrised product of generators] satisfies the identity

$$f^a{}_{ba_1} g_{a_2\ldots a_p} + f^a{}_{ba_2} g_{a_3\ldots a_p a_1} + \cdots + f^a{}_{ba_p} g_{a_1\ldots a_{p-1}} = 0, \tag{5}$$

which will be used latter. The potential $L^i_a(A)$ is a complicated function of the gauge field and its spatial derivatives. However, for our purposes here we only need its exterior derivative in phase space which has the remarkably simple form

$$\Omega^{ij}_{ab}(x,y) \equiv \frac{\partial L^i_a(x)}{\partial A^b_j(y)} - \frac{\partial L^j_b(y)}{\partial A^a_i(x)} \tag{6}$$

$$= g_{aba_1\ldots a_{n-1}} \epsilon^{iji_1\ldots i_{2(n-1)}} F^{a_1}_{i_1 i_2} \ldots F^{a_{n-1}}_{i_{2n-3} i_{2n-2}} \delta(x,y). \tag{7}$$

Note that (7) has a delta function at the right hand side. This is a non-trivial property of the theory. Indeed, even though L^i_a depends on the derivatives of the gauge field, all the terms that are proportional to derivatives of the delta function cancel out when the antisymmetrised derivative (6) is calculated. [We shall hereafter refer to Ω as the function in (7) without the delta function.]

The matrix Ω satisfies the identity

$$\Omega^{ij}_{ab} F^b_{ik} = \delta^j_k G_a \tag{8}$$

which is a consequence of the combinatorial properties of Ω and F only.

If Ω were invertible, it would define a symplectic structure on phase space and the Poisson bracket of the gauge fields would have been given by its inverse[1, 2]. However, from (8) one can see that, on the constraint surface, $(F^a_i)_j$ are $2n$ eigenvectors of Ω with zero eigenvalue. Of course not all of them are independent because the constraint has been imposed. Looking at simple examples, say, the abelian case in five dimensions, one can convince oneself that there are at least two zero eigenvalues. This makes rather difficult to extract the symplectic structure of the theory and it seems to be impossible to do so without breaking general covariance. What is behind the non-invertibility of Ω is that some of the fields A^a_i are not fixed by the equations of motion, but rather are completly arbitrary functions of time. Indeed, the equations of motion derived from the action (3) are,

$$\Omega^{ij}_{ab} \dot{A}^b_j = \Omega^{ij}_{ab} D_j A^b_0 \tag{9}$$

plus the constraints $G_a = 0$. Therefore, the velocities along the null directions of Ω cannot be solved. We remark that no further constraints arise because

Ω appears in both sides of (9). The only consequence of the non-invertibility of Ω is that some of the equations of motion are identically satisfied. This introduces a large gauge group that drastically reduces the number of degrees of freedom[3].

Here we would like to point out that even without giving an explicit formula for the Poisson bracket of the dynamical variables one can prove, first, that the constraints G_a are preserved in time, that is, their time derivatives are linear combinations of themselves

$$\frac{dG_a}{dt} = \Lambda^b_a \, G_b \tag{10}$$

and, second, they obey the first class algebra

$$\{G_a(x), G_b(y)\} = h^c{}_{ab} G_c(x)\delta(x,y). \tag{11}$$

The proof of (10) and (11) is rather simple and relies crucially in the formula

$$\frac{\delta G(\lambda)}{\delta A^a_i(x)} = \Omega^{ij}_{ab} D_j \lambda^b \tag{12}$$

satisfied by the smeared constraint $G(\lambda) = \int \lambda^a G_a$.

Consider the time derivative of the smeared constraint

$$\frac{dG(\lambda)}{dt} = \int \frac{\delta G(\lambda)}{\delta A^a_i(x)} \dot{A}^a_i(x) \tag{13}$$

where, for simplicity, we assume that the parameter λ^a does not depend on time, but it is otherwise arbitrary. [If not, there is an extra term proportional to the constraint which is irrelevant for our calculations.] In this step one usually uses the equations of motion to express the time derivative of the gauge field in terms of a Poisson bracket with the Hamiltonian. But, in our case, due to the non-invertibility of Ω, the velocities can not be solved from the equations of motion.

However, replacing (12) in (13) we see that in this calculation one does not need the velocities but rather their product with Ω which can be read from (9). One can check, with the help of (5), that the time derivative of $G(\lambda)$ is given by

$$\begin{aligned} \frac{dG(\lambda)}{dt} &= \int f^a{}_{bc} A^b_0 \lambda^c G_a \\ &= G(f^a{}_{bc} \lambda^b A^c_0). \end{aligned} \tag{14}$$

Eq. (14) is nothing but the smeared form of (10). We find that $\Lambda^a_b = f^a{}_{bc} A^c_0$. The proof of (11) goes as follows. From (3) we see that the Hamiltonian of the theory is given by the constraint G_a smeared with A_0^a. On the other hand, the Poisson bracket is defined in such a way that the time derivative of a phase space function is given by its Poisson bracket with the Hamiltonian. One has to have in mind, however, that the Poisson bracket only sees the time dependence induced by the true canonical variables. In our case, the constraints G_a depend on the gauge fields A_i^a and some of its components are not dynamical, but arbitrary functions of time. Hence, from (14) we have

$$G(f^a{}_{bc}\lambda^b A_0^c) = \{G(\lambda^a), G(A_0^b)\} + \int \frac{\delta G(\lambda)}{\delta z^m(x)} \dot{z}^m(x) \tag{15}$$

where the z^m denotes collectively all the non-dynamical variables. The important point is that the second term in the right hand side of (15) is a combination of the constraints. Indeed, from (12) we know that the derivative of the constraints with respect of any of the variables is proportional to Ω. Now, by definition, the z^m are those variables whose associated component of Ω vanishes on the constraint surface, and therefore, by a theorem proved in [4], it must be a combination of the constraints. The velocities $\dot{z}^m$ are, on the other hand, arbitrary. Thus, we find that the G_a satisfy (11) with $h^a{}_{bc} = f^a{}_{bc} - \Gamma^a{}_{bc}$ where Γ is the proportional function between the second term in (15) and the constraints.

I would like to thank L.J. Garay for many discussions and observations that have strongly contributed to the final form of this presentation. It is also a pleasure to thank M. Henneaux for his sharp comments of a previous version of this work. Conversations with C. Isham, C. Martínez, C. Teitelboim and J. Zanelli are also acknowledged. This work was partially supported by grants 193.0910-93 and 194.0203-94 from FONDECYT (Chile) and by institutional support to the Centro de Estudios Científicos de Santiago provided by SAREC (Sweden) and a group of Chilean private companies (COPEC,CMPC,ENERSIS). The author holds a Jack Ewer/Fundación Andes Fellowship.

References

[1] S. Hojman and L.F. Urrutia (1981), *J. Math. Phys.*, **22** 1896.

[2] L. Faddeev and R. Jackiw (1988), *Phys. Rev. Lett.*, **60** 1692.

[3] M. Bañados, L. J. Garay and M. Henneaux, in preparation.

[4] M. Henneaux and C. Teitelboim (1992), *Quantization of Gauge Systems*, Princeton University Press.

Nonrelativistic Chern-Simons vortices from the constrained Hamiltonian formalism

Igor V. Barashenkov and
Alexander O. Harin

Abstract

The Jackiw-Pi model of the self-gravitating gas of nonrelativistic bosons coupled to the Chern-Simons gauge field is known to exhibit asymptotically vanishing, lump-like soliton solutions. Here we discuss a recently proposed generalisation of this theory, which is applicable to systems of repulsive particles and allows to incorporate asymptotically nonvanishing fields, in particular topological vortices. We demonstrate the absence of the condensate state in the Jackiw-Pi model, relate this fact to a particular Lagrangian formulation of its nongauged precursor and derive the new model by modifying this Lagrangian appropriately and using it as a basis for the gauge theory. Reformulating the modified model as a constrained Hamiltonian system allows us to find the self-duality limit in the pure Chern-Simons and in the mixed Chern-Simons-Maxwell cases. These self-duality equations are shown to exhibit both asymptotically nonvanishing topological vortices and lump solitons.

1 Introduction

Vortices, topologically nontrivial localized structures, lie at the heart of all theories of particles with fractional statistics. It is these collective excitations of the field quanta that are considered as candidates for anyonic objects in the quasi-planar condensed matter physics. More precisely, in the case of the charged matter interacting with the Maxwell field, the anyon is a bound state of an (electrically neutral) vortex and a field quantum, a "flux" and a "charge". If the gauge field is of the Chern-Simons type, the vortex is no more electrically neutral and behaves as an anyon itself.

Both the Maxwell and Chern-Simons vortices were discussed in literature, for both the relativistic and nonrelativistic matter fields. So far the relativistic model has been amenable for a more thorough analysis. It admits a nontrivial ground state, over which topological vortices can be superimposed [1], [2], [3]. In addition to these asymptotically nonvanishing solutions with quantized flux, the relativistic model exhibits nontopological solitons for which the matter field ϕ vanishes at infinity [3]. (We will call such bell-shaped solitons "lumps" in this paper.) Although the two sets of solutions are supported by the same scalar potential $V(\phi) = (|\phi|^2 - \rho_0)^2(|\phi|^2 - \alpha)$, they obviously pertain to different regimes of the self-interaction, vortices to repulsion, lumps to attraction.

The *nonrelativistic* model, i.e. the gauged nonlinear Schrödinger equation [4], is presumably more relevant for condensed matter applications. However, it was found to exhibit only asymptotically vanishing solitons [4], [5]. Although the soliton's flux is quantized and so these collective excitations are also referred to as topological vortices, these are clearly distinct from the relativistic topological vortices which have the form of defects interpolating between topologically distinct vacua. Solutions of the latter type do not arise in the standard nonrelativistic model as formulated by Jackiw and Pi [4]. There is even no condensate, i.e. nontrivial vacuum solution in this model.

As in the nongauged, ordinary nonlinear Schrödinger equation, the abovesaid bell-like solitons arise in the case of the *self-attractive* boson gas (coupled to the Chern-Simons field). By analogy with the nongauged equation, one could expect the condensate state to emerge in the case of the nonlinear Schrödinger equation with repulsion. However, as will be shown below, the condensate can be incorporated into the "standard" model by no choice of the scalar self-interaction. It turns out that this model is applicable only for the description of asymptotically vanishing fields. The aim of the present work [6] is to modify it so as to incorporate the condensate (and the topological defects.)

We demonstrate that the necessary modification can be attained by the revision of the Lagrangian formulation of the nongauged precursor of the standard model. Using the revised Lagrangian as a basis for the gauge theory, we arrive at a new version of the gauged nonlinear Schrödinger equation which is completely compatible with the nonvanishing ("condensate") boundary conditions.

2 The standard model and asymptotically nonvanishing fields

The gauged nonlinear Schrödinger equation was formulated by Jackiw and Pi [4], [5]:

$$i\psi_t - eA_0\psi + \mathbf{D}^2\psi - U'(\rho)\psi = 0, \tag{2.1}$$

where $\rho = |\psi|^2$, $U'(\rho) = dU/d\rho$, $\mathbf{D} = \nabla - ie\mathbf{A}$, A_0 is scalar and $\mathbf{A}$ vector potential. The gauge field $A^\alpha = (A_0, \mathbf{A})$ satisfies its own equation for which the conserved matter current $J^\alpha = (J_0, \mathbf{J})$ serves as a source. (Note that there are no *external* gauge fields.) The most general linear equation for A^α

in (2+1) dimensions comprises both Maxwell and Chern-Simons terms:

$$\mu\partial_\beta F^{\beta\alpha} + \frac{\kappa}{2}\epsilon^{\alpha\beta\gamma}F_{\beta\gamma} = eJ^\alpha. \tag{2.2}$$

Here $F_{\alpha\beta} = \partial_\alpha A_\beta - \partial_\beta A_\alpha$; Greek and Latin indices run over 0,1,2 and 1,2, respectively; and the metric signature is $(+,-,-)$. The matter current of Jackiw and Pi has the following form:

$$J_0 = \rho = |\psi|^2, \quad \mathbf{J} = \frac{1}{i}(\overline{\psi}\mathbf{D}\psi - \psi\overline{\mathbf{D}\psi}). \tag{2.3}$$

The parameters $\mu > 0$ and $\kappa > 0$ control the relative contributions of the Maxwell and Chern-Simons terms in the corresponding Lagrangian:

$$\begin{aligned}\mathcal{L} &= \frac{i}{2}(\overline{\psi}D_0\psi - \psi\overline{D_0\psi}) - \overline{D_k\psi}D_k\psi \\ &\quad -\frac{\mu}{4}F_{\alpha\beta}F^{\alpha\beta} + \frac{\kappa}{4}\epsilon^{\gamma\alpha\beta}A_\gamma F_{\alpha\beta} - U(\rho).\end{aligned} \tag{2.4}$$

In eq. (2.4), $D_0 = \partial_0 + ieA_0$. We will demonstrate that, no matter what the scalar self-interaction $U(\rho)$ is, the system (2.1)–(2.4), which we will call the standard model, does not have the condensate solution.

2.1 No condensate in the standard model

Componentwise, the gauge field equation (2.2) can be written as

$$\mu\,\mathrm{div}\,\mathbf{E} - \kappa B = eJ_0, \tag{2.5}$$

$$-\mu\mathbf{E}_t + \mu\,\mathrm{curl}\,B - \kappa\widetilde{\mathbf{E}} = e\mathbf{J}, \tag{2.6}$$

where $E^i = F^{i0}$ and $B = F^{21}$ are the electric and magnetic fields, respectively, and $\widetilde{E}^i = \epsilon_{ij}E^j$. We will call *condensate* any solution to the equations (2.1), (2.5), (2.6) with the following properties: (i) ψ is a time-independent matter field distribution with a uniform density, i.e. $\psi = \sqrt{\rho_0}e^{i\chi(\mathbf{x})}$, $\rho_0 = \text{const}$; (ii) the electric and magnetic fields are also static, $\mathbf{E} = \mathbf{E}(\mathbf{x})$, $B = B(\mathbf{x})$, and bounded.

In the pure Maxwell case ($\kappa = 0$), the nonexistence of the solution with the above properties is straightforward. Eq. (2.5) becomes simply $\mathrm{div}\,\mathbf{E} \sim \rho_0$ and so the electric field has to grow infinitely. In the pure Chern-Simons case ($\mu = 0$), eq. (2.5) becomes $B \sim \rho_0$ and the nonexistence of the condensate is not so obvious. Below, we prove this fact for the most general situation when both the Maxwell and Chern-Simons terms are present.

We choose the real gauge, $\chi(\mathbf{x}) = \mathbf{0}$. The imaginary part of eq. (2.1) amounts to $\mathrm{div}\,\mathbf{A} = 0$ and hence, $\mathbf{A}$ is a curl: $\mathbf{A} = \mathrm{curl}\,\omega$. The current (2.3) is then a curl as well: $\mathbf{J} = -2e\rho_0\,\mathrm{curl}\,\omega$. Taking the divergence of eq. (2.6) with $\mathbf{E}_t = 0$, we get

$$-\kappa\nabla\cdot\widetilde{\mathbf{E}} = -\kappa\nabla\times\mathbf{E} = 0, \tag{2.7}$$

which means that $\mathbf{E}$ is a pure gradient: $\mathbf{E} = -\nabla A_0$. Now evaluating the curl of both sides of eq. (2.6) and subtracting eq. (2.5), it is not difficult to obtain an inhomogeneous Helmholtz equation:

$$-\mu^2 \Delta B + (\kappa^2 + 2e^2\mu\rho_0)B = -e\kappa\rho_0. \tag{2.8}$$

The only bounded solution of eq. (2.8) is the constant magnetic field

$$B(\mathbf{x}) = B_0 = -\frac{e\kappa\rho_0}{\kappa^2 + 2e^2\mu\rho_0}. \tag{2.9}$$

For constant B, eq. (2.6) becomes $\mathbf{E} = \frac{2e^2\rho_0}{\kappa}\nabla\omega$, which allows to eliminate ω:

$$\Delta A_0 = \frac{2e^2\rho_0}{\kappa} B_0. \tag{2.10}$$

The real part of the nonlinear Schrödinger equation (2.1) is another equation for A_0:

$$eA_0 + e^2\left(\frac{\kappa}{2e^2\rho_0}\right)^2 (\nabla A_0)^2 + U'(\rho_0) = 0. \tag{2.11}$$

Now we show that the equations (2.10) and (2.11) are, in general, incompatible. In the Laplace coordinates, $z = (x+iy)/2$, $\overline{z} = (x-iy)/2$, these are rewritten as

$$\partial\overline{\partial} A_0 = \frac{2e^2\rho_0}{\kappa} B_0, \tag{2.12}$$

$$A_0 + e\left(\frac{\kappa}{2e^2\rho_0}\right)^2 \partial A_0 \overline{\partial} A_0 = \frac{1}{e} F(\rho_0), \tag{2.13}$$

where $\partial = \partial/\partial z$ and $\overline{\partial} = \partial/\partial\overline{z}$. Solution of the Poisson equation (2.12) is

$$A_0 = \frac{2e^2\rho_0}{\kappa} B_0 z\overline{z} + f(z) + \overline{f(z)}, \tag{2.14}$$

with $f(z)$ being an arbitrary analytic function. Substitution of eq. (2.14) into (2.13), produces

$$\frac{2e^2\rho_0}{\kappa} B_0 z\overline{z} + eB_0^2 z\overline{z} + e\left(\frac{\kappa}{2e^2\rho_0}\right)^2 f'(z)\overline{f'(z)}$$
$$+ \left[f(z) + \frac{\kappa B_0}{2e\rho_0} z f'(z) + \text{c. c.}\right] = -\frac{1}{e} U'(\rho_0). \tag{2.15}$$

Expanding $f(z)$ in a Taylor series in the neighbourhood of $z = 0$: $f(z) = f_0 + f_1 z + f_2 z^2 + \ldots$, and substituting this into eq. (2.15), we have

$$\left(\frac{2e^2\rho_0}{\kappa} B_0 + eB_0^2\right) z\overline{z} + e\left(\frac{\kappa}{2e^2\rho_0}\right)^2 \left[f_1\overline{f'(z)} + \overline{f}_1 f'(z) - |f_1|^2\right.$$
$$\left. + \sum_{i=1}^{\infty}\sum_{j=1}^{\infty} C_{ij} z^i \overline{z}^j\right] + \left[f(z) + \frac{\kappa B_0}{2e\rho_0} z f'(z) + \text{c. c.}\right] = -\frac{1}{e} U'(\rho_0), \tag{2.16}$$

with $C_{ij} = \text{const}$. In eq. (2.16) there are terms which are functions of z only; terms which are functions only of $\bar{z}$, and terms which are products of powers of z and $\bar{z}$. For the equation to be satisfied, each group of terms must be set equal to a constant. For instance, assembling all functions of z, we obtain

$$e\left(\frac{\kappa}{2e^2\rho_0}\right)^2 \bar{f}_1 f'(z) + \frac{\kappa B_0}{2e\rho_0} z f'(z) + f(z) = C_0. \tag{2.17}$$

Eq. (2.17) is separable and so its solution is straightforward: $f(z) = C_1(z - z_0)^n + C_0$, where $n = -(2e\rho_0)/(\kappa B_0)$, and $z_0 = -\kappa\bar{f}_1/(2e^2\rho_0 B_0)$. Coming back to the equation (2.15), it is elementary to see now that z_0 is necessarily zero, and the only admissible value of n is 2 which amounts simply to $\mu = 0$. This means that a solution to the system (2.10), (2.11) exists only in the pure Chern-Simons case, $\mu = 0$. However, this solution is quadratic in coordinates,

$$A_0 = -\frac{2e^3\rho_0^2}{\kappa^2} z\bar{z} + C_1 z^2 + \bar{C}_1 \bar{z}^2 + \text{const}, \tag{2.18}$$

and so the corresponding electric field grows linearly as x or $y \to \pm\infty$. Returning to the original system (2.1), (2.2), this implies that the condensate solution exists for no μ and κ.

2.2 Conserved quantities and nonvanishing backgrounds

The absence of the condensate in the standard nonrelativistic model (2.1)–(2.4) is encoded already in its Lagrangian. The inadequate Lagrangian, in its turn, is inherited from the nongauged version of the model. To trace the origin of the problem, consider the nonlinear Schrödinger equation in one dimension:

$$i\psi_t + \psi_{xx} - U'(\rho)\psi = 0. \tag{2.19}$$

Here $\rho = |\psi|^2$. It is easy to check that the Lagrangian

$$\mathcal{L} = \frac{i}{2}(\psi_t\bar{\psi} - \bar{\psi}_t\psi) - |\psi_x|^2 - U(\rho), \tag{2.20}$$

which is normally associated with eq. (2.19), does not *automatically* produce correct integrals of motion for solutions with $|\psi|^2 \to \rho_0$ at infinity.

Indeed, the number of particles integral, $N = \int (i\psi\partial\mathcal{L}/\partial\psi_t - i\bar{\psi}\partial\mathcal{L}/\partial\bar{\psi}_t)dx$, takes the form $N = \int \rho\, dx$ and obviously diverges. The regularized number of particles, $N = \int (\rho - \rho_0)\, dx$, has to be obtained by the *ad hoc* subtraction of the background contribution. The conventional definition of momentum, $P = \int (\psi_x\partial\mathcal{L}/\partial\psi_t + \bar{\psi}_x\partial\mathcal{L}/\partial\bar{\psi}_t)dx$, does not yield the correct expression either. For the Lagrangian (2.20) one obtains $P = \frac{i}{2}\int \left(\psi_x\bar{\psi} - \bar{\psi}_x\psi\right) dx$ which is not compatible with the Hamiltonian formulation of the model [7]. To obtain the definition compatible with the Hamiltonian structure we again have to make an *a posteriori* regularization [7] (see also [8]):

$$P = \frac{i}{2}\int\left(\overline{\psi}\psi_x - \psi\overline{\psi}_x\right)dx + \rho_0 \mathrm{Arg}\,\psi\big|_{-\infty}^{+\infty} = \frac{i}{2}\int(\psi_x\overline{\psi} - \overline{\psi}_x\psi)\left(1 - \frac{\rho_0}{\rho}\right)dx.$$

Proceeding to two dimensions, the standard definition $\mathbf{P} = \int \frac{i}{2}(\psi\nabla\overline{\psi} - \overline{\psi}\nabla\psi)d^2x$, is even less suitable, since for asymptotically nonvanishing configurations this functional is ill-defined. The same is true for the angular momentum, $L = \int \mathbf{x}\times\mathcal{P}d^2x$.

3 The regularized model and Hamiltonian structure

3.1 Coupling the condensate to the gauge field

Thus we have to conclude that the nonvanishing boundary conditions at infinity (or nonvanishing backgrounds) are not compatible with the internal structure of the standard model. On the one hand, this incompatibility manifests itself in the fact that there is no condensate solution. On the other hand, even if the condensate existed (as in the case of the non-gauged nonlinear Schrödinger), we would still have problems with integrals of motion. So let us turn to a nongauged model and see if the problem can be overcome at this more elementary level.

Adding the time derivative of some function to the standard one-dimensional Lagrangian (2.20) obviously does not change its Euler-Lagrange equation (2.19). This function can be chosen in such a way that the corresponding integrals of motion *automatically* arise in the regularized form. The appropriate time derivative is $\rho_0\partial/\partial t\,\mathrm{Arg}\,\psi$, and the regularized Lagrangian is

$$\widetilde{\mathcal{L}} = \frac{i}{2}(\psi_t\overline{\psi} - \overline{\psi}_t\psi)\left(1 - \frac{\rho_0}{\rho}\right) - |\psi_x|^2 - U(\rho). \tag{3.1}$$

The same prescription can be extended to two dimensions. The Lagrangian

$$\widetilde{\mathcal{L}} = \frac{i}{2}(\psi_t\overline{\psi} - \overline{\psi}_t\psi)\left(1 - \frac{\rho_0}{\rho}\right) - |\nabla\psi|^2 - U(\rho) \tag{3.2}$$

produces the two-dimensional nonlinear Schrödinger equation,

$$i\psi_t + \Delta\psi - U'(\rho)\psi = 0, \tag{3.3}$$

while the conserved quantities are: $N = \int(\rho - \rho_0)d^2x$, $\boldsymbol{P} = \int\mathcal{P}d^2x$, and $L = \int\mathbf{x}\times\mathcal{P}d^2x$, with

$$\mathcal{P} = \frac{i}{2}(\overline{\psi}\nabla\psi - \psi\nabla\overline{\psi})\left(1 - \frac{\rho_0}{\rho}\right). \tag{3.4}$$

It is easy to see that these integrals are convergent for fields with $\rho \to \rho_0$ as $r \to \infty$, and therefore the Lagrangian (3.2) is compatible with nonvanishing boundary conditions at infinity. (There are some problems with the differentiability of P and L at the vortex configurations but these problems are due to the *local* behaviour of the vortices and do not have to do with the asymptotic region.) It is natural to expect that if we couple ψ minimally to the gauge field, the resulting gauge theory will not be so hostile to asymptotically nonvanishing fields. Indeed, gauging the Lagrangian (3.2) yields

$$\begin{aligned}\widetilde{\mathcal{L}} &= \frac{i}{2}\left(\overline{\psi}D_0\psi - \psi\overline{D_0\psi}\right)\left(1 - \frac{\rho_0}{\rho}\right) - \overline{D_k\psi}D_k\psi \\ &\quad -\frac{\mu}{4}F_{\alpha\beta}F^{\alpha\beta} + \frac{\kappa}{4}\epsilon^{\gamma\alpha\beta}A_\gamma F_{\alpha\beta} - U(\rho). \qquad (3.5)\end{aligned}$$

Dropping the time derivative $\rho_0\partial_0 \operatorname{Arg}\psi$, this can be rewritten as

$$\begin{aligned}\widetilde{\mathcal{L}} &= \frac{i}{2}\left(\overline{\psi}D_0\psi - \psi\overline{D_0\psi}\right) - \overline{D_k\psi}D_k\psi + \rho_0 A_0 \\ &\quad -\frac{\mu}{4}F_{\alpha\beta}F^{\alpha\beta} + \frac{\kappa}{4}\epsilon^{\gamma\alpha\beta}A_\gamma F_{\alpha\beta} - U(\rho). \qquad (3.6)\end{aligned}$$

It is only one term, $\rho_0 A_0$, that makes eq. (3.6) different from the standard model (2.4). However, this term has a transparent physical interpretation (it describes the gauge field coupling to the uniformly charged static background), and produces a drastic change in the structure of solutions to the model. The equations of motion look the same, eqs. (2.1),(2.2), with the vector $\mathbf{J}$ being given by the standard expression (2.3). It is only the number density, J_0, that is changed by the addition of the term $\rho_0 A_0$ to the Lagrangian. This time J_0 is given not by eq. (2.3) but by $J_0 = \rho - \rho_0$. It is straightforward to check that the model regularized in this way does possess the condensate solution: $\psi = \sqrt{\rho_0}$, $A_0 = -(1/e)U'(\rho_0)$, $\mathbf{A} = 0$.

3.2 The constrained Hamiltonian formulation

The Hamiltonian formulation of the new model (as well as of the standard model of Jackiw and Pi) is crucial for the construction of self-dual solutions (see sec. 4). In the treatment of the singular Lagrangian (3.6) we follow Dirac-Bergmann's approach [9].

The momenta conjugate to the matter field ψ and its complex conjugate $\overline{\psi}$ are given by

$$\pi = \frac{\partial\mathcal{L}}{\partial(\partial_0\psi)} = \frac{i}{2}\overline{\psi}(1 - \frac{\rho_0}{\rho}), \quad \overline{\pi} = \frac{\partial\mathcal{L}}{\partial(\partial_0\overline{\psi})} = -\frac{i}{2}\psi(1 - \frac{\rho_0}{\rho}),$$

and give rise to two primary constraints:

$$\xi_1 = \pi - \frac{i}{2}\overline{\psi}(1 - \frac{\rho_0}{\rho}) = 0, \quad \xi_2 = \overline{\pi} + \frac{i}{2}\psi(1 - \frac{\rho_0}{\rho}) = 0. \qquad (3.7)$$

The momenta conjugate to the fields A^α are

$$\Pi_\alpha = \frac{\kappa}{2}\epsilon_{\alpha\beta}A^\beta - \mu F^{\alpha 0}, \tag{3.8}$$

and $\Pi_0 = 0$ is a third primary constraint. The canonical Hamiltonian is the Noetherian energy expressed in terms of canonical variables $\psi, \overline{\psi}, \pi, \overline{\pi}, A^\mu$ and Π_μ:

$$\begin{aligned} H_c &= \int \left\{ |\mathbf{D}\psi|^2 + U(\rho) + \frac{\mu}{2}(\mathbf{E}^2 + B^2) \right\} d^2x \\ &\quad - \int A_0 \{\{(\mu \operatorname{div} \mathbf{E} - \kappa B - e(\rho - \rho_0)\} d^2x \end{aligned} \tag{3.9}$$

Here $E^i = (\kappa\epsilon^{ij}A^j/2 - \Pi_i)/\mu$, and $B = \partial_1 A^2 - \partial_2 A^1$. The Poisson bracket between two functionals is

$$\{S,T\} = \int \left[\frac{\delta S}{\delta\psi}\frac{\delta T}{\delta\pi} - \frac{\delta S}{\delta\pi}\frac{\delta T}{\delta\psi} + \frac{\delta S}{\delta\overline{\psi}}\frac{\delta T}{\delta\overline{\pi}} - \frac{\delta S}{\delta\overline{\pi}}\frac{\delta T}{\delta\overline{\psi}} + \frac{\delta S}{\delta A^\alpha}\frac{\delta T}{\delta\Pi_\alpha} - \frac{\delta S}{\delta\Pi_\alpha}\frac{\delta T}{\delta A^\alpha} \right] d^2x. \tag{3.10}$$

The constraints (3.7) are second class, $\{\xi_1(\mathbf{x},t), \xi_2(\mathbf{y},t)\} = -i\delta^{(2)}(\mathbf{x}-\mathbf{y})$, and can be accommodated by the introduction of the Dirac bracket,

$$\{S,T\}_\mathrm{D} = \{S,T\} - \frac{1}{i}\int \Big[\{S,\xi_1(\mathbf{z})\}\{\xi_2(\mathbf{z}),T\} - \{S,\xi_2(\mathbf{z})\}\{\xi_1(\mathbf{z}),T\} \Big] d^2z.$$

In fact, one can solve (3.7) for π and $\overline{\pi}$ and treat S and T as functionals of ψ and $\overline{\psi}$ only:

$$S = S\left(\psi, \overline{\psi}, \pi(\psi,\overline{\psi}), \overline{\pi}(\psi,\overline{\psi})\right); \quad T = T\left(\psi, \overline{\psi}, \pi(\psi,\overline{\psi}), \overline{\pi}(\psi,\overline{\psi})\right).$$

(They also depend on A_i, Π_i, of course.) As a result, the Dirac bracket reads

$$\{S,T\}_\mathrm{D} = \int \left[\frac{1}{i}\left(\frac{\widetilde{\delta S}}{\delta\psi}\frac{\widetilde{\delta T}}{\delta\overline{\psi}} - \frac{\widetilde{\delta S}}{\delta\overline{\psi}}\frac{\widetilde{\delta T}}{\delta\psi} \right) + \frac{\delta S}{\delta A^\alpha}\frac{\delta T}{\delta\Pi_\alpha} - \frac{\delta S}{\delta\Pi_\alpha}\frac{\delta T}{\delta A^\alpha} \right] d^2x, \tag{3.11}$$

where

$$\frac{\widetilde{\delta}}{\delta\psi} = \frac{\delta}{\delta\psi} + \frac{\partial\pi}{\partial\psi}\frac{\delta}{\delta\pi} + \frac{\partial\overline{\pi}}{\partial\psi}\frac{\delta}{\delta\overline{\pi}}. \tag{3.12}$$

The constraint $\Pi_0 = 0$ is first class and the requirement of its conservation leads to a secondary constraint

$$\eta = \{\Pi_0, H_c\}_\mathrm{D} = \mu \operatorname{div} \mathbf{E} - \kappa B - e(\rho - \rho_0) = 0. \tag{3.13}$$

The secondary constraint $\eta = 0$ does not lead to further constraints since, as one can check, $\{\eta, H_c\}_\mathrm{D} = 0$.

The total Hamiltonian is now $H_\mathrm{T} = H_c + \int v\Pi_0 d^2x$, where $v = v(\mathbf{x},t)$ is an arbitrary multiplier. The variables A^0 and Π_0 have no physical significance as $\Pi_0 = 0$ all the time while A^0 can take arbitrary values. Accordingly, we may drop them out from the set of dynamical variables of the model. This can be accomplished by discarding the term $v\Pi_0$ in H_T (whose only role is to let A^0 vary arbitrarily) and by treating A^0 as an arbitrary multiplier. As a result, the Hamiltonian formulation of our model can be given in terms of three pairs of canonical fields: $\psi, \overline{\psi}; A^i, \Pi_i$. The dynamics is described by the Hamiltonian H_c with the canonical bracket (3.11). (Now α=1,2). The initial state should satisfy the constraint (3.13) while the Hamilton equations will contain an arbitrary function A^0. It can be determined only for a static problem, in which case the problem reduces to finding stationary points of energy

$$H = \int \left[|\mathbf{D}\psi|^2 + U(\rho) + \frac{\mu}{2}(\mathbf{E}^2 + B^2) \right] d^2x \tag{3.14}$$

on the constraint manifold (3.13). In this case A^0 plays the rôle of a Lagrange multiplier and H_c is the Lagrange function.

So far we were assuming $\mu \neq 0$. For $\mu = 0$ (pure Chern-Simons model) the situation is somewhat different. The theory has two more second class primary constraints,

$$\xi_3 = \Pi_1 - \frac{\kappa}{2}A^2 = 0, \quad \xi_4 = \Pi_2 + \frac{\kappa}{2}A^1 = 0, \tag{3.15}$$

$\{\xi_3(\mathbf{x},t), \xi_4(\mathbf{y},t)\} = -\kappa\delta^{(2)}(\mathbf{x}-\mathbf{y})$, resulting from the definition of momenta Π_i. These can be taken care of by modifying the bracket:

$$\begin{aligned} \{S,T\}_{\mathrm{D}} &= \{S,T\} + \int \Big[-\frac{1}{i}\{S,\xi_1(\mathbf{z})\}\{\xi_2(\mathbf{z}),T\} + \frac{1}{i}\{S,\xi_2(\mathbf{z})\}\{\xi_1(\mathbf{z}),T\} \\ &\quad - \frac{1}{\kappa}\{S,\xi_3(\mathbf{z})\}\{\xi_4(\mathbf{z}),T\} + \frac{1}{\kappa}\{S,\xi_4(\mathbf{z})\}\{\xi_3(\mathbf{z}),T\} \Big] d^2z. \end{aligned} \tag{3.16}$$

The reduced phase space in this case is spanned by $\psi, \overline{\psi}, A^1, A^2$, and the Dirac bracket can be worked out to be

$$\{S,T\}_{\mathrm{D}} = \int \left[\frac{1}{i}\left(\frac{\tilde{\delta} S}{\delta\psi}\frac{\tilde{\delta} T}{\delta\overline{\psi}} - \frac{\tilde{\delta} S}{\delta\overline{\psi}}\frac{\tilde{\delta} T}{\delta\psi} \right) + \frac{1}{\kappa}\left(\frac{\tilde{\delta} S}{\delta A^1}\frac{\tilde{\delta} T}{\delta A^2} - \frac{\tilde{\delta} S}{\delta A^2}\frac{\tilde{\delta} T}{\delta A^1} \right) \right] d^2x,$$

where

$$\frac{\tilde{\delta}}{\delta A^i} = \frac{\delta}{\delta A^i} + \frac{\partial \Pi_j}{\partial A^i}\frac{\delta}{\delta \Pi_j}. \tag{3.17}$$

In fact, in the pure Chern-Simons case a further reduction is possible. In this case the constraint (3.13) can be explicitly resolved which makes it possible to consider $\psi, \overline{\psi}$ as the only canonical variables. This is the approach utilised by Jackiw and Pi [4], [5]. In the general case, however, the fact that the constraint involves both Π_i and A^i makes such a reduction impossible.

4 Self-dual limit

Static solutions correspond to stationary points of the Hamiltonian (3.14) on the constraint manifold (3.13). In the mixed CS-Maxwell case, using the flux-vorticity relation, $\Phi = 2\pi n/e$ and the Bogomol'nyi decomposition,

$$|\mathbf{D}\psi|^2 = |(D_1 \pm iD_2)\psi|^2 \pm \frac{1}{2}\nabla \times \mathbf{J} \pm eB\rho, \tag{4.1}$$

the energy (3.14) takes the form

$$H = \int \Big\{ |(D_1 \pm iD_2)\psi|^2 + \frac{\mu}{2}\left[B \pm \frac{e}{\mu}(\rho - \rho_0)\right]^2$$
$$- \frac{e^2}{2\mu}(\rho - \rho_0)^2 + U(\rho) + \frac{\mu}{2}(\nabla A_0)^2 \pm \frac{1}{2}\nabla \times \mathbf{J}\Big\} d^2x \pm 2\pi\rho_0 n. \tag{4.2}$$

For $U(\rho) = \frac{e^2}{2\mu}(\rho - \rho_0)^2$ (which corresponds to the Bose gas with δ-function pairwise repulsion) and fields approaching the condensate background $\psi = \sqrt{\rho_0}e^{in\theta}$, $A_0 = 0$, $\mathbf{A} = n\hat{\theta}/(er)$, the energy can be rewritten as

$$H = \int \left\{ |(D_1 \pm iD_2)\psi|^2 + \frac{\mu}{2}\left[B \pm \frac{e}{\mu}(\rho - \rho_0)\right]^2 + \frac{\mu}{2}(\nabla A_0)^2 \right\} d^2x \pm 2\pi\rho_0 n.$$

The lower bound of the energy, $H = \pm 2\pi\rho_0 n$, is saturated when the following self-duality equations are satisfied:

$$(D_1 \pm iD_2)\psi = 0, \tag{4.3}$$
$$B \pm \frac{e}{\mu}(\rho - \rho_0) = 0, \tag{4.4}$$
$$A_0 = 0. \tag{4.5}$$

The upper (lower) sign should be associated with the positive (negative) vorticity n. Comparing (4.4) to (3.13), we see that solutions of eqs. (4.3), (4.4) lie on the constraint manifold only for $\kappa = \mu$ and only in the case of the upper sign. Thus, for $\kappa > 0$, only vortices with positive vorticities may exist. Eq. (4.3) yields

$$A^i = \pm\frac{1}{2e}\epsilon_{ij}\partial_j \ln\rho + \frac{1}{e}\partial_i \mathrm{Arg}\psi, \tag{4.6}$$

and we should retain only the upper sign here. Substituting this into (4.4), we arrive at

$$\nabla^2 \ln\rho = 2\frac{e^2}{\mu}(\rho - \rho_0). \tag{4.7}$$

Eq. (4.7) appeared previously in the self-dual limit of the relativistic Higgs model with Maxwell term [10] and is known to possess solutions with the "topological vortex" asymptotic behaviour: $\rho(\infty) = \rho_0$, $\rho(0) \sim r^{2n}$, $n \geq 1$. No explicit solutions have been found, however. For $\rho_0 \neq 0$, the equation does not pass the Painlevé test [11], and therefore is nonintegrable. Nevertheless, eq. (4.7) exhibits N-vortex solutions [12] which can be found, for instance, numerically.

The self-duality reduction is also possible in the pure Chern-Simons case ($\mu = 0$). Making use of the identity (4.1) and the constraint (3.13), the energy (3.14) can be represented as

$$H = \int \left[|(D_1 \pm iD_2)\psi|^2 \mp \frac{e^2}{\kappa}(\rho - \rho_0)^2 + U(\rho) \right] d^2x \pm 2\pi\rho_0 n. \tag{4.8}$$

With the choice of $U(\rho) = \pm e^2(\rho - \rho_0)^2/\kappa$, one observes that the energy is minimal, provided $(D_1 \pm iD_2)\psi = 0$, whence we have, as before, eq. (4.6). Substituting this into the constraint equation (3.13), we arrive at

$$\nabla^2 \ln \rho = \pm 2\frac{e^2}{\kappa}(\rho - \rho_0). \tag{4.9}$$

Note that no equation for A_0 arises here. This is not surprising as A_0 is not a dynamical variable. (In our Hamiltonian formulation, it is just a Lagrange multiplier.) For any combination of the Maxwell and Chern-Simons terms it can be determined from the spatial part of eq. (2.2):

$$\mu \nabla \times B + \kappa \nabla \times A_0 = e\mathbf{J}. \tag{4.10}$$

The matter current is calculated from eq. (4.6): $\mathbf{J} = \mp \nabla \times \rho$. Substituting this into eq. (4.10) for $\mu = 0$, we can readily solve for A_0: $A_0 = \mp e(\rho - \rho_0)/\kappa$.

References

[1] The vortex configurations of a scalar field minimally coupled to the Maxwell field, were discovered by Abrikosov within the (2+0)-dimensional Ginzburg-Landau model of superconductivity [A. A. Abrikosov (1957), *Sov. Phys. JETP*, **5** 1174]. This model can be considered the static limit of the (2+1)-dimensional Higgs theory. Therefore, Abrikosov's vortices are in fact *relativistic* in the sense that they are static solutions of the (2+1)-dimensional relativistic Higgs-Maxwell theory [H. B. Nielsen and P. Olesen (1973), *Nucl. Phys.*, **B61** 45]. The relativity of Abrikosov's vortices, and of the Ginzburg-Landau model itself, is obvious from the fact that this model does not include the Gauss law, $\Delta A_0 = J_0$. This omission is compatible with the assumption of the electrical neutrality ($A_0 = 0$) only when J_0 vanishes for static fields, i.e. when it is a *relativistic* charge: $J_0 = ie(\overline{\phi}D_0\phi - \phi\overline{D_0\phi}\}$.

[2] The possibility of existence of the *Chern-Simons*-Maxwell vortices was first discussed in S. K. Paul and A. Khare (1986), *Phys. Lett.*, **B174** 420, within the relativistic Higgs model. [For the actual numerical solutions of this model see D. Boyanovsky (1991), *Nucl. Phys.*, **B350** 906]. Multivortex solutions and selfduality limit were discovered (in the pure Chern-Simons case) in J. Hong, Y. Kim, and P. Y. Pac (1990), *Phys. Rev. Lett.*, **64** 2230, and R. Jackiw and E. J. Weinberg (1990), *Phys. Rev. Lett.*, **64** 2234.

[3] R. Jackiw, K. Lee, and E. J. Weinberg (1990), *Phys. Rev. D*, **42** 3488.

[4] The nonrelativistic gauge theory in (2+1) dimensions was formulated in R. Jackiw and S.-Y. Pi (1990), *Phys. Rev. Lett.*, **64** 2969. The Jackiw-Pi model comprises both the Maxwell and Chern-Simons interactions.

[5] R. Jackiw and S.-Y. Pi (1990), *Phys. Rev. D*, **42** 3500;
R. Jackiw and S.-Y. Pi (1992), *Prog. Theor. Phys. Suppl.*, **107** 1.

[6] Some of these results have been briefly announced in:
I. V. Barashenkov and A. O. Harin (1993), *JINR Rapid Communications*, **4**[61], 70; (1994), *Phys. Rev. Lett.*, **72** 1575.

[7] I. V. Barashenkov and E. Yu. Panova (1993), *Physica (Amsterdam)*, **69D** 114.

[8] M. M. Bogdan, A. S. Kovalev, and A. M. Kosevich (1989), *Fiz. Nizk. Temp.*, **15** 511 [Sov. J. Low Temp. Fiz. **15** 288].

[9] P. A. M. Dirac (1964), *Lectures on quantum mechanics*, No 2 of *Belfer graduate School of Science monograph series*, Yeshiva University, New York;
K. Sundermeyer (1982), *Constrained Dynamics*, Vol. 169 of *Lecture Notes in Physics*, Springer-Verlag, Berlin.

[10] E. B. Bogomol'nyi (1976), *Sov. J. Nucl. Phys.*, **24** 449.

[11] P. Winternitz, private communication.

[12] C. H. Taubes (1980), *Commun. Math. Phys.*, **72** 277.

Classical solutions of gravitating Chern-Simons electrodynamics

Gérard Clément

Abstract

We discuss the reduction of gravitating Chern-Simons electrodynamics with two commuting Killing vectors to a dynamical problem with four degrees of freedom, and construct regular particle-like solutions.

In three space-dimensions, the abelian Higgs model coupled to Einstein gravity is known to admit static multi-vortex solutions [1]. Stationary vortex solutions arise [2] when a Chern-Simons term for the gauge field [3, 4] is added. We shall show here that, in this last case, the Higgs field is not really necessary, as sourceless gravitating Chern-Simons electrodynamics admits regular particle-like solutions.

In the present talk, we first discuss the reduction of gravitating Chern-Simons electrodynamics with two commuting Killing vectors to a dynamical problem. We then present exact particle-like solutions to this problem in the case where the cosmological constant is negative. Finally, we briefly discuss the generalization to the case where Einstein gravity is replaced by topologically massive gravity [4].

The action for gravitating Chern-Simons electrodynamics is the sum

$$I = I_G + I_E \,, \tag{1}$$

of the action for Einstein gravity

$$I_G = -m \int \mathrm{d}^3x \, \sqrt{|g|} \, (g^{\mu\nu} \, R_{\mu\nu} + 2\,\Lambda) \tag{2}$$

(m is related to the Einstein gravitational constant κ by $m = 1/2\kappa$, while Λ is the cosmological constant), and of the action for Chern-Simons electrodynamics

$$I_E = -\frac{1}{4} \int \mathrm{d}^3x \, (\sqrt{|g|} \, g^{\mu\nu} \, g^{\rho\sigma} \, F_{\mu\rho} \, F_{\nu\sigma} - \mu \, \varepsilon^{\mu\nu\rho} \, F_{\mu\nu} \, A_\rho) \,, \tag{3}$$

where $F_{\mu\nu} \equiv A_{\nu;\mu} - A_{\mu;\nu}$, and μ is the topological mass constant.

Let us assume that our three-dimensional space-time has two commuting Killing vectors K_1, K_2. The metric ds^2 is then invariant under the SL(2,R) group of transformations in the plane (K_1, K_2), which is locally isomorphic to SO(2,1). A parametrization of the metric making this invariance manifest is

$$ds^2 = \lambda_{ab}(\rho)\, dx^a\, dx^b + \zeta^{-2}(\rho)\, R^{-2}(\rho)\, d\rho^2 . \tag{4}$$

In (4), λ is the 2×2 matrix

$$\lambda \equiv \begin{pmatrix} T+X & Y \\ Y & T-X \end{pmatrix} \tag{5}$$

of determinant $R^2 \equiv \boldsymbol{X}^2$, where

$$\boldsymbol{X}^2 \equiv T^2 - X^2 - Y^2 \tag{6}$$

is the Minkowski pseudo-norm in the minisuperspace spanned by the vector $\boldsymbol{X}$. The light-cone $R^2 = 0$ divides this minisuperspace in three regions. The signature of the metric (4) is Lorentzian with ρ space-like if $\boldsymbol{X}$ is space-like, Lorentzian with ρ time-like (cosmology) if $\boldsymbol{X}$ is past time-like, and Riemannian if $\boldsymbol{X}$ is future time-like. The function $\zeta(\rho)$ in (4) plays the role of the usual lapse function by allowing for arbitrary reparametrizations of the coordinate ρ. We complete this parametrization by assuming the ansatz for the gauge potentials:

$$A_\mu\, dx^\mu = \psi_a(\rho)\, dx^a . \tag{7}$$

Consider first the case of gravity alone [5]. The action I_G reduces with the parametrization (4) to

$$I_G = -\int d^2x \int d\rho\, [\zeta\, \frac{m}{2}\, \dot{\boldsymbol{X}}^2 + 2\,\zeta^{-1}\, m\,\Lambda] \tag{8}$$

(with $\dot{} = d/d\rho$), where ζ acts as Lagrange multiplier. Varying the action (8) with respect to $\boldsymbol{X}$ and fixing $\zeta = 1$, we find that the metric follows a geodesic

$$\boldsymbol{X} = \boldsymbol{\alpha}\,\rho + \boldsymbol{\beta} \tag{9}$$

of minisuperspace. The variation of I_G with respect to ζ leads to the Hamiltonian constraint (written for $\zeta = 1$)

$$H_0 \equiv -\frac{m}{2}\,\dot{\boldsymbol{X}}^2 + 2\,m\,\Lambda = 0 , \tag{10}$$

which fixes the squared slope $\boldsymbol{\alpha}^2$. A particularly interesting class of solutions are the BTZ black-hole solutions for a negative cosmological constant $\Lambda = -l^{-2}$ [6],

$$\begin{aligned} ds^2 \;=\; & (2\,l^{-2}\,\rho - \frac{M}{2})\, dt^2 + J\, dt\, d\theta - (2\,\rho + \frac{M\,l^2}{2})\, d\theta^2 \\ & - [4\,l^{-2}\,\rho^2 + \frac{J^2 - M^2\,l^2}{4}]^{-1}\, d\rho^2 , \end{aligned} \tag{11}$$

where θ is an angle. These are regular, with two horizons ($\boldsymbol{X}$ past light-like) if $|J| \leq Ml$. The parameters M and J may be interpreted [6] as the mass and spin of these 'particle-like' metrics.

We now consider the full action (1), which reduces with the parametrization (4), (7) to $I = \int \mathrm{d}^2x \int \mathrm{d}\rho L$, with

$$L = \frac{1}{2}[\zeta(-m\dot{\boldsymbol{X}}^2 + \dot{\overline{\psi}}\boldsymbol{\Sigma}\cdot\boldsymbol{X}\,\dot{\psi}) - \mu\overline{\psi}\dot{\psi} - 4\zeta^{-1}m\Lambda]. \tag{12}$$

The real Dirac-like matrices in (12) are defined by

$$\Sigma^0 = \begin{pmatrix} 0 & 1 \\ -1 & 0 \end{pmatrix}, \quad \Sigma^1 = \begin{pmatrix} 0 & -1 \\ -1 & 0 \end{pmatrix}, \quad \Sigma^2 = \begin{pmatrix} 1 & 0 \\ 0 & -1 \end{pmatrix}, \tag{13}$$

and

$$\overline{\psi} \equiv \psi^T \Sigma^0 \tag{14}$$

is the real adjoint of the 'spinor' ψ. Our reduced dynamical system (12) has five degrees of freedom, parametrized by the coordinates $\boldsymbol{X}$, ψ and the conjugate momenta $\boldsymbol{P} \equiv \partial L/\partial\dot{\boldsymbol{X}}$, $\Pi^T \equiv \partial L/\partial\dot{\psi}$. The invariance of the original action under diffeomorphisms implies the invariance of (12) under SO(2,1) transformations, which leads to the conservation of the angular momentum vector

$$\boldsymbol{J} = \boldsymbol{L} + \boldsymbol{S}, \tag{15}$$

sum of 'orbital' and 'spin' contributions given by

$$\boldsymbol{L} \equiv \boldsymbol{X} \wedge \boldsymbol{P}, \quad \boldsymbol{S} \equiv \frac{1}{2}\Pi^T\boldsymbol{\Sigma}\,\psi. \tag{16}$$

It follows that, as implied by the notation, the components g_{ab} of the metric tensor transform vectorially under the action of SO(2,1), while the gauge potentials transform spinorially.

In the case $\mu = 0$, the coordinates ψ_a are cyclic, so that the two corresponding degrees of freedom may be eliminated altogether [7]. If $\mu \neq 0$, variation of the Lagrangian (12) with respect to ψ leads to the first integrals

$$\Pi^T - \frac{\mu}{2}\overline{\psi} = \varpi^T. \tag{17}$$

Treating (17) as a pair of second-class constraints, we eliminate the Π^a in terms of the ψ_a. Actually it is more convenient to use, instead of ψ, the translated spinor $\hat{\psi} \equiv \psi - \mu^{-1}\overline{\varpi}^T$, in terms of which the spin vector $\boldsymbol{S}$ takes the form

$$\boldsymbol{S} = \hat{\boldsymbol{S}} + \frac{1}{4\mu}\varpi^T\boldsymbol{\Sigma}\,\overline{\varpi}^T, \tag{18}$$

with

$$\hat{\boldsymbol{S}} = \frac{\mu}{4}\hat{\overline{\psi}}\boldsymbol{\Sigma}\,\hat{\psi}. \tag{19}$$

It then follows from the Dirac brackets of the $\hat{\psi}$ (written in matrix form)

$$[\hat{\psi}, \hat{\psi}^T]_D = -\mu^{-1}\,\Sigma^0\,, \tag{20}$$

that the Dirac algebra of the $\hat{S}^i$ is the angular-momentum algebra

$$[\hat{S}^i, \hat{S}^j]_D = \varepsilon^{ijk}\,\hat{S}_k\,. \tag{21}$$

Note however that $\hat{\boldsymbol{S}}$ is a very special spin vector, as it is null,

$$\hat{\boldsymbol{S}}^2 = 0\,, \tag{22}$$

owing to the reality of the ψ_a. From now on, we shall choose the gauge $\varpi^T = 0$, which amounts to dropping the hats in eqs. (19)-(22).

The canonical Hamiltonian derived from (12) may be expanded in terms of the remaining variables $\boldsymbol{X}$, $\boldsymbol{P}$ and ψ as

$$H \equiv -\frac{\boldsymbol{P}^2}{2m} - 2\,\mu\,\boldsymbol{S}\cdot\frac{\boldsymbol{X}}{R^2} + 2\,m\,\Lambda\,, \tag{23}$$

corresponding to a dynamical system with four degrees of freedom (point particle coupled to a null spin vector). The energy of this system is fixed by the Hamiltonian constraint

$$H = 0\,. \tag{24}$$

The Hamiltonian (23) generates the equations of motion

$$\begin{aligned}
\dot{\boldsymbol{X}} &= -\frac{\boldsymbol{P}}{m}\,,\\
\dot{\boldsymbol{P}} &= \frac{2\,\mu}{R^2}\,\boldsymbol{S} - \frac{4\,\mu}{R^4}\,\boldsymbol{X}\,(\boldsymbol{S}\cdot\boldsymbol{X})\,,\\
\dot{\boldsymbol{S}} &= \frac{2\,\mu}{R^2}\,\boldsymbol{X}\wedge\boldsymbol{S} \quad\Longleftrightarrow\quad \dot{\psi} = \frac{\mu}{R^2}\,\boldsymbol{\Sigma}\cdot\boldsymbol{X}\,\psi\,.
\end{aligned} \tag{25}$$

Using these equations, we can easily prove the following
Theorem: The only static solution ($Y = 0$) to gravitating Chern-Simons electrodynamics is empty space ($\boldsymbol{S} = 0$, implying $\psi = 0$).

It does not seem possible to obtain the general solution of eqs. (24), (25) in closed form. However we can construct a class of planar stationary rotationally symmetric solutions such that $\boldsymbol{J}\cdot\boldsymbol{X} = 0$ for Λ negative or zero (we can show that the constant vector $\boldsymbol{J}$ must be null in this case). These solutions are, for $\Lambda = -l^{-2}$,

$$\begin{aligned}
\mathrm{d}s^2 &= (2l^{-2}\rho - \frac{M(\rho)}{2})\mathrm{d}t^2 - lM(\rho)\mathrm{d}t\mathrm{d}\theta - (2\rho + l^2\frac{M(\rho)}{2})\mathrm{d}\theta^2 - \frac{l^2}{4}\frac{\mathrm{d}\rho^2}{\rho^2}\,,\\
A_\mu\,\mathrm{d}x^\mu &= c\,\rho^{-\mu l/2}\,(\mathrm{d}t + l\,\mathrm{d}\theta)\,,
\end{aligned} \tag{26}$$

with

$$M(\rho) = M + \frac{c^2}{2m}\,\frac{\mu\, l}{\mu\, l + 1}\,\rho^{-\mu l}\,. \tag{27}$$

For $\mu > 0$, $m < 0$, $M > 0$ the metric (26) is regular ($\rho = 0$ is at infinite geodesic distance), horizonless, and asymptotic to the BTZ extreme black-hole metric (eq. (11) with $|J| = Ml$). Because of these properties, we suggest that solution (26) qualifies as the particle-like solution of gravitating Chern-Simons electrodynamics. In the limit $\mu \to 0$, this solution reduces to a dyon extreme black-hole solution of Einstein-Maxwell gravity [7]. The planar 'electrostatic' solutions for $\Lambda = 0$,

$$\begin{aligned} \mathrm{d}s^2 &= (a + b\,r - \frac{c^2}{4\,m}\,\mathrm{e}^{-2\mu r})\,\mathrm{d}t^2 - 2\,\sigma_0\,\mathrm{d}t\,\mathrm{d}\theta - \mathrm{d}r^2\,, \\ A_\mu\,\mathrm{d}x^\mu &= c\,\mathrm{e}^{-\mu r}\,\mathrm{d}t\,, \end{aligned} \tag{28}$$

are again regular for $\mu > 0$, $m < 0$.

We briefly discuss the generalization to the case where Einstein gravity is replaced by topologically massive gravity (TMG). The action (8) must be replaced by the action for TMG

$$\begin{aligned} I_G &= -m \int \mathrm{d}^3x[\sqrt{|g|}\,(g^{\mu\nu}\,R_{\mu\nu} + 2\,\Lambda) \\ &\quad - \frac{1}{2\,\mu'}\,\varepsilon^{\lambda\mu\nu}\,\mathrm{Tr}\,(\Gamma_\lambda\,\partial_\mu\Gamma_\nu + \frac{2}{3}\,\Gamma_\lambda\,\Gamma_\mu\,\Gamma_\nu)]\,, \end{aligned} \tag{29}$$

where the Γ_λ are the connections written in matrix form, and μ' is the topological mass constant for gravity. This gives rise, with the parametrization (4), to the generalized Lagrangian

$$L_G = -\frac{m}{2}\,[-\frac{1}{\mu'}\,\zeta^2\,(\boldsymbol{X}\wedge\dot{\boldsymbol{X}})\cdot\ddot{\boldsymbol{X}} + \zeta\,\dot{\boldsymbol{X}}^2 + 4\,\zeta^{-1}\,\Lambda]\,. \tag{30}$$

Solutions to the equations of motion deriving from (30) are discussed in [8]. The equations of motion for Chern-Simons electrodynamics coupled to TMG are currently under investigation [9]. We have been able to show that these equations again admit particle-like solutions of the form (26), with the 'mass function' (27) replaced by

$$M(\rho) = M - a\,\rho^{(1-\mu' l)/2} - b\,\rho^{-\mu l}\,. \tag{31}$$

To conclude, let us mention that extreme BTZ black holes (just as extreme Reissner-Nordström black holes) do not interact, so that stationary multi-extreme-black-hole systems (with auxiliary conical singularities) are possible [10]. It is very likely that stationary multi-particle solutions to gravitating Chern-Simons electrodynamics may likewise be constructed from the particle-like solution (26).

References

[1] Linet, B. (1988), *Gen. Rel. Grav.*, **20** 451.
Comtet, A. and Gibbons, G.W. (1988), *Nucl. Phys.*, B **299** 719.

[2] Linet, B. (1990), *Gen. Rel. Grav.*, **22** 469.
Valtancoli, P. (1992), *Int. J. Mod. Phys.*, A **7** 4335.
Cangemi, D. and Lee, C. (1992), *Phys. Rev.*, D **46** 4768.

[3] Schonfeld, J. (1981), *Nucl. Phys.*, B **185** 157.

[4] Deser, S., Jackiw, R. and Templeton, S. (1982), *Phys. Rev. Lett.*, **48** 975; (1982), *Ann. Phys., NY*, **140** 372.

[5] Clément, G. (1994), *Phys. Rev.*, D **49** 5131.

[6] Bañados, M., Teitelboim, C. and Zanelli, J. (1992), *Phys. Rev. Lett.*, **69** 1849.
Bañados, M., Henneaux, M., Teitelboim, C. and Zanelli, J. (1993), *Phys. Rev.*, D **48** 1506.

[7] Clément, G. (1993), *Class. Quantum Grav.*, **10** L49.

[8] Clément, G. (1994), *Class. Quantum Grav.*, in press.

[9] in collaboration with Ait Moussa, K.

[10] Clément, G. (1994), preprint GCR-94/02/01.

Exponentially localised instantons in a hierarchy of Higgs models

D.H. Tchrakian and G.M. O'Brien

1 Introduction

The systematic construction of hierarchies of Abelian (in d=2) and non-Abelian (in d>2) Higgs models in d dimensions which support finite action and topologically stable lump solutions, was reviewed in Ref.[1]. Very briefly, the method involves the construction of a hierarcy of Yang-Mills (YM) models in all even dimensions supporting such lump solutions, and then subjecting these to dimensional reduction where the residual systems are the hierarchies of Higgs models in question.

In this article we shall investigate in more detail, the asymptotic properties of the lumps of the Higgs models. These are very different from the asymptotic properties of the even (higher) dimensional YM hierarchies whose connection-fields have a pure-gauge type of behaviour at infinity, their lumps are localised with a power behaviour, and in those cases where these systems are scale invariant this localisation exhibits a further scale arbitrariness. By contrast, the lumps of the hierarchies of Higgs models are in general exponentially localised to an absolute scale. This property is potentially very important from the viewpoint of physical applications and hence is highlighted in the title of this article.

The material is presented in two sections below. In the first, we present the hierarchy of Higgs models in d dimensions in a formal way and examine particular asymptotic properties of the lumps by examining the fields in the (Dirac) string-gauge. We shall restrict to spherically symmetric solutions throughout. In the second section, we shall consider the two physically interesting cases of d=2 and d=4 in detail, where we shall explicitly demonstrate the exponential localisation of the lumps, interpret these as instantons, and finally present a preliminary attempt at constructing a dilute gas of these instantons.

2 Higgs hierarchies and string gauge

The starting point in the construction of these hierarchies is the introduction of the YM hierarchies in all even dimensions. We denote the n-fold antisymmetrised product of the curvature 2-form F(2) by F($2n$) and refer to it as a curvature $2n$-form. In $2(p+q)$ dimensions, the Hodge dual $*(F(2p))(2q)$ of the curvature $2p$-form is a $2q$-form. Taking $q>p$ and employing the constant κ with the inverse dimensions of a length, we can state the following inequality

$$tr[\kappa^{2(q-p)}F(2p) - (*F(2q))(2p)]^2 > 0 \tag{1}$$

from which it follows that

$$L_{p,q} = tr[\kappa^{4(q-p)}F(2p)^2 + (2q)!(2p)!F(2q)^2)] \geq 2\kappa^{2(p+q)}C_{p+q} \tag{2}$$

where C_{p+q} is the $(p+q)$-th Chern-Pontryagin density. The left hand side of (2) defines the Lagrange densities of the YM hierarchy. When p=q, these systems support self-dual solutions[2] localised to an arbitrary scale, and when $p \neq q$ they have non-self-dual solutions[3] which are localised to the absolute scale κ according to a power behaviour. These are finite action and topologically stable lumps.

The hierarchies of Higgs models are obtained by subjecting the systems of YM hierarchies defined by (2) to dimensional reduction. The residual models in d dimensions involve a Higgs field interacting minimally with the residual gauge field, after the coordinates of the (p+q−d) dimensional compact coset space are integrated out. We do not describe this descent procedure here and refer instead to Refs.[1] and [4]. Here we note the main criterion used, namely that we restrict to those modes of descent for which the residual inequality of (2) presents a lower bound for the residual action capable of forcing the solutions to be topologically stable. This question was considered in detail in Ref.[5], and can be summarised by the criterion that the residual Lagrange density does not feature a 'cosmological' constant term. This enables the enforcement of the usual finite action conditions as boundary conditions, and together with the restriction to those modes of descent for which the residual C-P density is a non-vanishing total divergence[4], enables the enforcement of asymptotic conditions that render the solution in question topologically stable. We express this symbolically by

$$L^H_{p,q}[A_\mu, \Phi] \geq \partial_\mu \Omega_\mu [A_\mu, \Phi] \tag{3}$$

where $\Omega[A, \Phi]$ is the residual Chern-Simons(C-S) density. It turns out that the above citerion is satisfied by the mode of descent which gives rise to the following Higgs multiplet

$$\Omega = \begin{bmatrix} 0 & \varphi \\ -\varphi & 0 \end{bmatrix} \tag{4}$$

which under the residual gauge group G_+ X G_- X $U(1)$, transforms covariantly. Suitable boundary conditions are of the usual type

$$|\Phi| \xrightarrow[r\to\infty]{} \eta \tag{5}$$

where the constant η with the dimensions of an inverse length is the vacuum expectation value of the Higgs field and the only form in which it features in the residual Lagrangian on the left hand side of (3) is as an even power of $(\eta^2 + \Phi^2)$, with antihermitian Φ. This feature of the residual Lagrangian is a result of our choice of descent[5], and it enables the vanishing of $L^H_{p,q}$ at infinity by requiring (5). It is this function of Φ and η that gives rise to the Higgs self-interaction potential and it is the 'mass' term η that is responsible for the exponential localisation of the ensuing lump. The Higgs potential is present in all $L^H_{p,q}$ arising from descents with $(p+q-\mathrm{d})>1$ and absent otherwise, in which cases the lump is not exponentially localised, for example in the case of the monopole in the Bogomol'nyi-Prasad-Sommerfield limit. In these cases, a suitable Higgs self-interaction potential can be added to $L^H_{p,q}$ without invalidating the inequality (3) which is responsible for the topological stability of the lump, and which then renders the lump exponentially localised.

The volume integral of the right hand side of (3), which can be expressed as a surface integral over the large (d−1)-sphere, is the topological charge pertaining to the lump solution. To find field configurations satisfying (5) which yields a non-trivial topological charge, we must choose the residual gauge group under which $L^H_{p,q}$ is invariant suitably. It turns out that for a d dimensional Higgs model, with d>2, the residual gauge group must be chosen to be $SO(d)$ X $U(1)$ or $SO_\pm(d)$ X $U(1)$, where $SO_\pm(d)$ denotes chiral $SO(d)$. For d=2 the required residual gauge group is $SO(2) \approx U(1)$ In the d>2 cases, the spherically symmetric field configurations

$$A_\mu = \frac{1}{r}(1 + f(r))\gamma_{\mu\nu}\hat{x}_\nu \tag{6a}$$

$$\Phi = h(r)\gamma_{d+1}\gamma_\mu\hat{x}_\mu \tag{6b}$$

where $\gamma_{\mu\nu}$ are the representations of the $SO(d)$ algebra and γ_{d+1} is the chirality matrix in d dimensions. It is straightforward, if tedious in the generic case, to verify that for the field configuration (6), the condition for finiteness of the action imposed on the Lagrangian of the residual Higgs model is

$$0 \xleftarrow[0\leftarrow r]{} h(r) \xrightarrow[r\to\infty]{} \eta \tag{7a}$$

$$-1 \xleftarrow[0\leftarrow r]{} f(r) \xrightarrow[r\to\infty]{} 0 \tag{7b}$$

provided that we restrict ourselves to those Lagrangians that are independent of the dimensional parameter κ, namely the Lagrangians $L^H_{p,p}$ descending from the scale invariant Lagrangians $L_{p,q}$ of the $4p$ dimensional YM hierarchy. This asymptotic behaviour, which incorporates (4) as well as satisfying requirements of singlevaluedness and smoothness of the solution, is also consistent with non-trivial topological charge. In the d=2 cases, the radially symmetric field configurations are

$$A_\mu = \frac{1}{r}(n + f(r))\varepsilon_{\mu\nu}\hat{x}_\nu \tag{8a}$$

$$\Phi = h(r)e^{-in\theta} \tag{8b}$$

where n is the vortex number and θ the azimuthal angle, and the asymptotic conditions (7) still hold. It is possible to see at this point that the solutions to the Higgs models considered here differ in their asymptotic properties from the hierarchy of YM models in (2), in that the solutions of the latter exhibit a pure-gauge asymptotic form while the connection fields in (6) and (8) do not. Indeed, subsituting the rightside of (7) into (6) we see that the asymptotic fields in this case have the form

$$A_\mu = -\frac{1}{2}\varpi^{-1}\delta_\mu\varpi, \Phi = \mu\varpi \tag{9a, b}$$

with $\varpi = \gamma_{d+1}\gamma_\mu\hat{x}_\mu$, and where the connection field is not of the form of a pure-gauge but one half of a pure-gauge. This is reminicent of the asymptotic behaviour of the monopole in three dimensions where we know that the gauge field at infinity coincides with the Abelian curvature on S^2.

We can see from (6) and (8) that the field configurations responsible for finite action topologically stable lumps occur in $SO(d)$ Higgs models involving $SO(d)$-vector valued Higgs fields. We recognise the simplest examples of these hierarchies, first in d=2 as the hierarchy of Abelian Higgs models presented in Ref.[6], the best known member of which is the usual Abelian Higgs model with vortex lumps, and in d=3 in the hierarchy of $SO(3) \approx SU(2)$ Higgs models with iso-vector Higgs fields, the best known amongst which is the usual YM-Higgs model with the monopole lump. This illustrates the field multiplet structure of our d dimensional Higgs models.

Having noted the particular asymptotic behaviour (9) of the lump solutions of of our Higgs models, we proceed to illustrate another interesting common feature of these lump fields following from (9). This is done in the (Dirac) string gauge in which the asymptotic connection field is singular along the (negative) x_d-axis, and the gauge group $SO(d)$ breaks down to $SO(d-1)$

$$\tilde{A}_i = \frac{1}{r(1+\hat{x}_d)}\gamma_{ij}\hat{x}_j, \tilde{A}_d = 0 \tag{10}$$

with the indices $i, j = 1, 2, ..., d-1$. The details of this gauge transformation are given in Ref. [7], and we limit ourselves here to noting that the non-singular field strength in this gauge coincides with the $SO(d-1)$ curvature on S^{d-1}. For example in the well known d=3 case, this $SO(2) \approx U(1)$ curvature corresponding to (10) is

$$\tilde{F}_{\theta\varphi} = \frac{1}{2}\gamma_3 sin\theta \tag{11}$$

in terms of the spherical polar coordinates on S^2, and in the d=4 case the components of the $SO(3) \approx SO(2)$ curvature corresponding to (10) are

$$\tilde{F}_{\psi\theta} = sin\psi(-\gamma_{31}cos\varphi + \gamma_{23}sin\varphi) \tag{12a}$$

$$\tilde{F}_{\theta\varphi} = -sin^2\psi sin\theta\,[\gamma_{12}cos\theta + (\gamma_{31}sin\varphi + \gamma_{23}cos\varphi)\,sin\theta] \qquad (12b)$$

$$\tilde{F}_{\varphi\psi} = -sin\psi sin\theta\,[-\gamma_{12}sin\theta + (\gamma_{31}sin\varphi + \gamma_{23}cos\varphi)\,cos\theta] \qquad (12c)$$

We are now in a position to make a peculiar obsrvation. It is possible to find a Higgs model in all dimensions except d=2 and d=4, whose Lagrange densitiy in the (Dirac) string gauge does not vanish on the (d−1)-sphere at infinity. (This information will be useful in the next section.)

To explain this we first note that in this gauge where the Higgs field is gauged to a constant value with the magnitude of the VEV, both the Higgs self interaction potentials and the covariant derivatives vanish, and hence the only term in the Lagrange density which can be non-vanishing is the one that depends exclusively on the curvarure. Now the generic residual Higgs model Lagrangian $L^H_{p,p}$ under consideration features only one such term, and that is the square of the curvature $2p$-form $F(2p)$. The condition that this quantity be non-vanishing in this gauge is that $2p \geq d-1$, since there are only d−1 (angular) coordinates on the (d−1)-sphere on which the curvature in this gauge takes its values. These conditions cannot be met in the d=2 and d=4 cases.

That this is so in a d=2 (Abelian) Higgs model is obvious since on the 1-sphere we have only one (angular) coordinate and hence the curvature 2-form vanishes. In the d=4 case, the first member of the hierarchy of $SO(4)$ Higgs models is $L^H_{2,2}$ the one pertaining to p=2. an hence the relevant term is the square of the curvature 4-form. But there are only three angular coordinates on the 3-sphere and hence the curvature must vanish. In all other dimensions d>4 it is always possible to find a residual Higgs Lagrangian which satisfies the condition d<$2p$+1.

3 Instantons of Higgs models in d=2 and d=4

To interpret the above lumps as instantons in d dimensions, we cconsider the relevant Higgs model in the temporal gauge with $A_0 \equiv A_d = 0$, and express the other vacuum fields A_i and Φ according to the asymptotic forms (9a,b) where now the quantities $\varpi = \varpi(\vec{x})$ are taken to be x_0-independent. This leads to the time independence of the vacuum field as required. To arrive at this via the requirement that as usual the Higgs self-interaction potential vanish for the vacuum field configuration, it is necessary to have the square of the covariant derivative $D_\mu\Phi$ in the Lagrangian. This is not the case for any of the models $L^H_{p,p}$ except for p=1 in d=2 and d=3. It is easy however to overcome this problem by simply adding an appropriate positive definite term to the Lagrangian, and we have verified that in both the following examples this can be done without changing any of the qualitative properties of the lumps. We therefore proceed with the understanding that the lumps we discuss below can be interpreted as the instantons of 1+1 and

3+1 dimensional prototype theories. Our main concern here will be to note the exponentially localised nature of these lumps.

In dimensions d=2, it is possible to express[6] the Lagrange of the residual Higgs model compactly for the whole hierarchy labeled by the integer ρ

$$L^H_{p,p} = (\eta^2 - \phi^2)^{2(p-2)}\{4p(2p{-}1)(2p{-}2)!^2\left[(\eta^2 - \phi^2)F_{\mu\nu} - i(p-1)D_{[i}\varphi D_{j]}\varphi^*\right]^2$$

$$+2p(\eta^2 - \phi^2)^2|D_\mu\varphi|^2 + 4\lambda(2p-1)^2(\eta^2 - \phi^2)^4\} \tag{13}$$

where φ is a complex valued field, or a two component $SO(2)$ vector, and $\phi = |\varphi|$ is its magnitude. The dimensionless coupling strength λ must be positive and when it equals one, the action can be minimised by a set of first order Bogomol'nyi equations. The topological charge in this case is the usual vortex number n appearing in (8), which can be computed from the surface integral of the C-S density Ω_μ appearing on the right side of (3). The only term in this C-S density which, subject to the asymptotic conditions (7), contributes to this surface integral is

$$\Omega_\mu \approx \eta^{2(p-1)}\varepsilon_{\mu\nu}A_\nu \tag{14}$$

The radially symmetric solutions of the form (8) in the asymptotic domains were found in Ref. [6]. In the region $r<<1$ these are

$$\begin{aligned} f(r) &= -1 - \left[(p-1)C_1^2 + \tfrac{1}{2}(1-2p)\right]r^2 + o(r^2) && (15a)\\ &= -n - \tfrac{1}{2}(1-2p)r^2 + (p-1)nC_n^2 r^{2n} + o(r^{2n}) && (15b)\\ h(r) &= C_1 r + o(r) && (16a)\\ &= C_n r^n + o(r^n) && (16b) \end{aligned}$$

and in the region $r>>1$

$$f(r) = -C\sqrt{2(2p-1)/pr}\left[K_0\left(\sqrt{2p(2p-1)}r\right)\right]^{\frac{1-p}{p}} K_1\left(\sqrt{2p(2p-1)}r\right) \tag{17a}$$

$$h(r) = \eta - C\eta\left[K_0\left(\sqrt{2p(2p-1)}r\right)\right]^{\frac{1}{p}} \tag{17b}$$

where the functions K_0, K_1 are the modified Bessel functions which decay exponentially. Thus the intanton solutions of the 1+1 dimensional hierarchy of Abelian Higgs models are exponentially localised.

In dimension d=4 the generic, arbitrary p, $SO(4)$ Higgs model in the hierarchy is too cumbersome to express so we restrict ourselves to the first member of the hierarchy, namely the p=2 case given in Ref. [8]

$$L^H_{2,2} = tr\left(F^2_{\mu\nu\rho\sigma} + 4\lambda_1\{F_{[\mu\nu}, D_{\rho]}\Phi\}^2 - 18\lambda_2\left(\{(\eta^2+\Phi^2), F_{\mu\nu}\} - [D_\mu\Phi, D_\nu\Phi]\right)^2\right.$$

$$\left. -54\lambda_3\{(\eta^2+\Phi^2), D_\mu\Phi\}^2 + 54\lambda_4\left(\eta^2+\Phi^2\right)^4\right) \quad (18)$$

the dimensionless coupling constants $\lambda_1, ...\lambda_4$ must all be positive, and when they are all equal to one, then the action density can be minimised by a set of first order Bogomol'nyi equations. These however are in this case overdetermined and hence we expect to find non-trivial solutions only to the second order Euler-Lagrange equations.

Here the asymptotic solutions of the spherically symmetric form (6) are found to be, in the $r<<1$ region,

$$f(r) = Ar^2 + o\left(r^2\right) \quad (19a)$$

$$h(r) = B\eta r + o(r) \quad (19b)$$

and in the $r>>1$ region

$$f(r) = \lambda_1\eta^2 r^2 K_2\left(\sqrt{3\lambda_1}\eta r\right) \quad (20a)$$

$$h(r) = \eta + \sqrt{\lambda_2/\lambda_1}\eta^2 r K_1\left(2\sqrt{2\lambda_2/\lambda_1}\eta r\right) \quad (20b)$$

where we note the modified Bessel functions of one order higher than the ones in (17a,b) respectively, occurring. Thus we have verified that the instanton of the p=1 member of the hierarchy of $SO(4)$ Higgs models in 3+1 dimensions is exponentially localised, and we expect that the instantons of the other members of this hierarchy should also be localised exponentially.

We finish with a speculative suggestion for extending the model (18) so that a dilute instanton gas can be constructed. This is because the model (18) as it stands is not capable of sustaining a non-vanishing contribution to the action, from the instanton-instanton interactions. By a dilute gas here, following Polyakov [9] we understand a collection of one-instanton fields which overlap only asymptotically. Thus as far as this field configuration is concerned, we need only use the fields corresponding to (10) in that singular gauge and hence can neglect the Higgs fields entirely for this purpose. The singularity free curvature field strength in this gauge coincides with the $SO(3)$ curvature on S^3, given by (12). Assuming that the instantons are far enough that to a good approximation we can treat the total field as the linear sum of these $SO(3)$ curvatures each centred on its 3-sphere we can compute their contribution to the action. Since the Higgs independent term in (18) is quartic in the curvature, this would feature the mutual interactions of four instantons. But we have only three non vanishing field strengths in (12), so the curvature 4-form occurring in (18) will vanish identically.

It is no use to probe the higher, $p>2$, members of the hierarchy to remedy this situation since these all feature a curvature $2p$-form which will also vanish on the 3-sphere. Furthermore, we are restricted to use a theory that features only the square of the derivative of any field, so the only choice left is the cubic term

$$\tilde{L} = \kappa^2 tr F_{\mu\nu} F_{\nu\rho} F_{\rho\mu} \tag{21}$$

which though is not expressible as the square of any antisymmetric p-form, nevertheless satisfies the constraint of involving no higher powers of the derivative of the connection field than the quadratic. This is a low-dimensional accident and does not occur for higher order analogues of (21). The only problem with (21) is that it is not positive definite and its addition to (18) may predjudice the existence of the instanton solution. On the other hand it may be possible to safeguard the instanton solution by arranging the magnitude of the dimensional coupling constant κ in (21) to be sufficiently small. This is not an easy question to settle and to date we have no reliable answers, except that we have verified that the asymptotic solution (19) in the $r<<1$ region still holds when (21) is incorporated.

Notwithstanding the speculative nature of the incorporation of (21) with (18), we finish with one further comment. The contribution to the action from the instanton-instanton interaction terms will come from the cubic term (21) and will thus feature 3-instanton interactions. From (12) we know that the asymptotic curvature decays as the inverse square of the radial variable, and integrating the cube of such fields in four dimensions, the result would have an inverse square dependence on the radial variable. Thus, the expected instanton-instanton interaction term is the inverse of the Laplacian operator in four dimensions. This would be an added bonus in the subsequent semiclassical analysis, and is reminiscent of the logarithmic interactions of the two dimensional Coulomb gas, considered by Berezinsky[10] and, Kosterlitz[11] and Thouless.

Acknowledgements

This research has been covered in part by the European Union under the Human Capital and Mobility programme.

References

[1] D.H. Tchrakian (1994), 'Skyrme like models in gauge theory', in *Constraint Theory and Quantization Methods*, F. Colomo, L. Lusanna and G. Marmo (eds), World Scientific.

[2] D.H. Tchrakian (1985), *Phys. Lett.*, **B150** 360.

[3] J. Burzlaff and D.H. Tchrakian (1994), *J. Phys.*, **A26** L1053.

[4] Diarmuid O'Se', T.N. Sherry and D.H. Tchrakian (1986),*J. Math. Phys.*, **27** 325;
Zhong-Qi Ma, G.M. O'Brien and D.H. Tchrakian (1986), *Phys. Rev.*, **D33** 1177;
Zhong-Qi Ma and D.H. Tchrakian (1988), *ibid*, **D38** 3827.

[5] T.N. Sherry and D.H. Tchrakian (1992), *Phys. Lett.*, **B295** 237.

[6] J. Burzlaff, A. Chakrabarti and D.H. Tchrakian (1994), *J. Phys.*, **A27** 1617.

[7] Zhong-Qi Ma and D.H. Tchrakian (1992), *Lett. Math. Phys.*, **26** 111.

[8] G. M. O'Brien and D.H. Tchrakian (1989),*Mod. Phys. Lett.*, **A4** 1389.

[9] A.M. Polyakov (1977), *Nucl. Phys.*, **B120** 429.

[10] V.L. Berezinskii (1970), *JETP*, **32** 493.

[11] J.M. Kosterlitz and D.J. Thouless (1973), *J. Phys.*, **C6** L97.

Obstructions to gauging WZ terms: a symplectic curiosity

José M Figueroa-O'Farrill

Dedicated to the memory of Arnoldo Ferrer Andreu (1916-1994)

0 As children we are taught to expect that behind any number of continuous symmetries of a dynamical system, there always lurk an equal number of conserved quantities. However at some point in our lives we find out that this is not necessarily the case. The correspondence between symmetries and conservation laws—equivalently, the existence of a moment mapping associated to a symplectic group action—must overcome a homological obstruction. That is, this obstruction takes the form of a class in a suitably defined cohomology theory which must vanish for the correspondence to go through. The purpose of this talk is to point out a curious coincidence. In my joint work with Sonia Stanciu trying to understand the gauging of nonreductive Wess-Zumino-Witten models, I came across the fact that the obstructions to gauging the Wess-Zumino term of a (toy) one-dimensional σ-model are none other than the obstructions for the existence of the moment mapping. Of course, as a physical system this σ-model is not very interesting, but I hope that this symplectic curiousity serves to bring a little *divertimento* to fit the occasion.

1 Let (M,ω) be a symplectic manifold; that is, the two-form ω is closed and is nondegenerate when thought of as a section of $\operatorname{Hom}(TM, T^*M)$. We say that a vector field ξ on M is **symplectic** if its flow fixes ω:

$$\mathcal{L}_\xi \omega = 0 \ .$$

Since $d\omega = 0$, this means that the one-form $\imath(\xi)\omega$ is closed. If $\imath(\xi)\omega$ is actually exact—so that there is a function f such that $\imath(\xi)\omega = df$—then ξ is called **hamiltonian**. We see then that in a symplectic manifold one has the following interpretation of the first de Rham cohomology:

$$H^1(M) = \frac{\text{closed one-forms}}{\text{exact one-forms}} = \frac{\text{symplectic vector fields}}{\text{hamiltonian vector fields}} \ .$$

In other words, we have an exact sequence of vector spaces

$$0 \longrightarrow \operatorname{Ham}(M) \longrightarrow \operatorname{Sym}(M) \longrightarrow H^1(M) \longrightarrow 0 \ , \tag{1.1}$$

where Ham(M) and Sym(M) denote the hamiltonian and symplectic vector fields, respectively. It is clear from its definition as the stabilizer of ω, that Sym(M) is a Lie algebra. Moreover, Ham(M) is a Lie subalgebra. Indeed, if ξ_f and ξ_g are hamiltonian vector fields associated to the functions f and g, then

$$[\xi_f, \xi_g] = \xi_{\{f,g\}} \tag{1.2}$$

where $\{f, g\}$ is the Poisson bracket. More is true, however, and Ham(M) is actually an ideal of Sym(M); for if η is a symplectic vector field

$$[\eta, \xi_f] = \xi_{\eta \cdot f} \ .$$

In other words, the exact sequence (1.1) is actually an exact sequence of Lie algebras. The induced Lie bracket on $H^1(M)$ is zero, however, because of the fact that Ham(M) contains the first derived ideal $\mathrm{Sym}(M)' \equiv [\mathrm{Sym}(M), \mathrm{Sym}(M)]$.

The assignment of a hamiltonian vector field to a function defines a map

$$\begin{aligned} C^\infty(M) &\to \mathrm{Ham}(M) \\ f &\mapsto \omega^{-1}(df) \end{aligned}$$

which by (1.2) is a Lie algebra morphism. Its kernel consists of the locally constant functions $df = 0$; that is, $H^0(M)$. This gives rise to another exact sequence of Lie algebras

$$0 \longrightarrow H^0(M) \longrightarrow C^\infty(M) \longrightarrow \mathrm{Ham}(M) \longrightarrow 0 \ , \tag{1.3}$$

where $H^0(M)$ is the center of $C^\infty(M)$ and hence abelian. Putting this sequence together with (1.1) we find the following 4-term exact sequence of Lie algebras interpolating between $H^0(M)$ and $H^1(M)$:

$$0 \longrightarrow H^0(M) \longrightarrow C^\infty(M) \longrightarrow \mathrm{Sym}(M) \longrightarrow H^1(M) \longrightarrow 0 \ . \tag{1.4}$$

2 Now let G be a connected Lie group acting on M in such a way that ω is G-invariant. Let $\mathbf{g}$ denote the Lie algebra of G. Every $X \in \mathbf{g}$ gives rise to a Killing vector field on M which we denote ξ_X. The map $X \mapsto \xi_X$ is a Lie algebra morphism. Since ω is G-invariant, ξ_X is symplectic. In other words, a symplectic G-action on M gives rise to a Lie algebra morphism $\mathbf{g} \longrightarrow \mathrm{Sym}(M)$. There will be conserved charges associated to these continuous symmetries if and only if this map lifts to a Lie algebra morphism $\mathbf{g} \longrightarrow C^\infty(M)$ in such a way that the resulting diagram

$$\begin{array}{ccccccccc} 0 & \to & H^0(M) & \to & C^\infty(M) & \to & \mathrm{Sym}(M) & \to & H^1(M) & \to & 0 \\ & & & & \nwarrow & & \nearrow & & & & \\ & & & & & \mathbf{g} & & & & & \end{array}$$

commutes. The obstruction to the existence of such a lift follow easily from the exactness of (1.4). First of all, the image of $\mathbf{g}$ in $\mathrm{Sym}(M)$ will come from $C^\infty(M)$ if it is sent to zero in $H^1(M)$. That is, if there exists functions ϕ_X such that $\imath(\xi_X)\omega = d\phi_X$. This is not enough because we want the map $X \mapsto \phi_X$ to be a Lie algebra morphism. Because the map $X \mapsto \xi_X$ is a Lie algebra morphism, the map $X \mapsto \phi_X$ is at most a projective representation characterized by the $H^0(M)$-valued cocycle $c(X,Y) \equiv \{\phi_X, \phi_Y\} - \phi_{[X,Y]}$. If and only if this cocycle is a coboundary is the representation an honest representation. Indeed, if there exists some map $X \mapsto b_X \in H^0(M)$ such that $c(X,Y) = -b_{[X,Y]}$, then one straightens the map $X \mapsto \phi'_X = \phi_X - b_X$ and the resulting map $\mathbf{g} \to C^\infty(M)$ is a morphism.

If this is case then one can define the **moment(um) mapping**

$$\Phi : M \longrightarrow \mathbf{g}^*$$

by $\langle \Phi(m), X\rangle = \phi'_X(m)$ for all $m \in M$. This map is equivariant in that it intertwines between the G-action on M and the coadjoint action on $\mathbf{g}^*$.

3 We can understand the conditions

$$\imath(\xi_X)\omega = d\phi_X \tag{3.1a}$$

$$\{\phi_X, \phi_Y\} = \phi_{[X,Y]} \tag{3.1b}$$

purely in terms of cohomology as follows. First of all notice that the map $\mathbf{g} \to H^1(M)$ defined by $X \mapsto [\imath(\xi_X)\omega]$ annihilates the first derived ideal $\mathbf{g}'$, since $[\mathrm{Sym}(M), \mathrm{Sym}(M)] \subset \mathrm{Ham}(M)$. Therefore it induces a map $\mathbf{g}/\mathbf{g}' \to H^1(M)$; or, in other words, it defines an element in

$$(\mathbf{g}/\mathbf{g}')^* \otimes H^1(M) \cong H^1(\mathbf{g}) \otimes H^1(M)\ .$$

Then (3.1a) simply says that this element is zero. Similarly the cocycle $c : \bigwedge^2 \mathbf{g} \to H^0(M)$ defined above defines a class in $H^2(\mathbf{g}) \otimes H^0(M)$. Then (3.1b) says that this class should be zero. In other words, the obstruction to the existence of a moment mapping defines a class

$$[O] \in \left(H^1(\mathbf{g}) \otimes H^1(M)\right) \oplus \left(H^2(\mathbf{g}) \otimes H^0(M)\right)\ . \tag{3.2}$$

In fact, we can understand this class as a single class in a different cohomology theory. Let us start by considering the G-action on M as a map

$$\alpha : G \times M \longrightarrow M$$

and let us define a G-action on $G \times M$ to make α equivariant. One convenient way to do so is

$$\beta : G \times G \times M \longrightarrow G \times M$$

where $\beta(g,h,m) = (gh,m)$; that is, G acts via left translations on the first factor and ignores the second. Equivariance of α allows us to pull back G-invariant forms on M to G-invariant forms on $G \times M$. The G-invariant forms on $G \times M$ form a subcomplex $\Omega^{\cdot}(G \times M)^G$ of the de Rham complex. Therefore $\alpha^*\omega \in \Omega^2(G\times M)^G$. Similarly if we denote by $\pi : G\times M \to M$ the Cartesian projection onto the second factor, $\pi^*\omega$ is also a G-invariant form on $G\times M$. Define then $\omega_\alpha \equiv \alpha^*\omega - \pi^*\omega$. This is a closed form in $\Omega^2(G\times M)^G$ and hence defines a class in $H^2(G \times M)^G$. The complex $\Omega^{\cdot}(G \times M)^G$ is isomorphic to the double complex $\Omega^{\cdot}(G)^G \otimes \Omega^{\cdot}(M)$. Applying the Künneth theorem to this complex, one finds that

$$H^n(G \times M)^G \cong \bigoplus_{p+q=n} H^p(\mathbf{g}) \otimes H^q(M) \ . \tag{3.3}$$

It is then an easy computational matter to prove that under this isomorphism the class of ω_α goes over to the class $[O]$ in (3.2). (The $H^0(\mathbf{g})\otimes H^2(M)$ component is zero precisely because in ω_α we subtract $\pi^*\omega$ from $\alpha^*\omega$.)

As an example, if $(T^*N, d\theta)$ is the phase space of some configuration space N on which G acts, the action of G lifts naturally to a symplectic action on M. In fact, the tautological one-form θ is already invariant. In this case, $\omega_\alpha = d(\alpha^*\theta - \pi^*\theta)$ and since $(\alpha^*\theta - \pi^*\theta)$ is G-invariant, the class $[\omega_\alpha]$ in $H^2(G \times M)^G$ is trivial. Our "classical" intuition on the correspondence between continuous symmetries and conservation laws is borne out of this example.

4 What does this have to do with gauging σ-models? Let B be a two-manifold with boundary $\partial B = \Sigma$. Let (M,ω) be as before except that we drop the nondegeneracy condition on ω. The Wess-Zumino term of the σ-model in question is given by the function

$$S_{\mathrm{WZ}}[\varphi] = \int_B \varphi^*\omega \tag{4.1}$$

on the space of maps $\varphi : B \to M$; but because ω is closed, the resulting equations of motion only depend on the restriction of φ to the boundary Σ. Therefore it defines a variational problem in the space $\mathrm{Map}(\Sigma, M)$ of maps $\Sigma \to M$ (which extend to B). The σ-model also comes with a kinetic term defined on Σ, but since the gauging of this term is simply accomplished via minimal coupling we shall disregard it in what follows. It should also be mentioned that we are ignoring for the present purposes the topological obstructions concerning the well-definedness of the WZ term itself. Similarly we will consider only gauging the algebra: demanding invariance under "large" gauge transformations invariably brings about other topological obstructions.

Let G be a connected Lie group, acting on M in such a way that it fixes ω. The action of G on M induces an action of G on $\mathrm{Map}(B, M)$ under which the action (4.1) is invariant. For our purposes, gauging the WZ term will consist in promoting (4.1) to an action which is invariant under $\mathrm{Map}(\Sigma, \mathbf{g})$ via the addition of further terms involving a gauge field. We do this in steps following the Noether procedure.

5 Let $\lambda \in \mathrm{Map}(\Sigma, \mathbf{g})$. More explicitly, if we fix a basis $\{X_a\}$ for $\mathbf{g}$, then $\lambda = \lambda^a X_a$ with λ_a functions on Σ. The action of λ on the pull-back of any form Ω on M, is given by

$$\delta_\lambda \varphi^* \Omega = d\lambda^a \wedge \varphi^* \imath_a \Omega + \lambda^a \varphi^* \mathcal{L}_a \Omega$$

where $\imath_a$ and $\mathcal{L}_a$ denote respectively the contraction and Lie derivative relative to the Killing vector corresponding to X_a. In particular since ω is closed and $\mathbf{g}$-invariant, we find that $\delta_\lambda \varphi^* \omega = d(\lambda^a \varphi^* \imath_a \omega)$, whence the variation of (3.1) becomes

$$\delta_\lambda S_{\mathrm{WZ}}[\varphi] = \int_\Sigma \lambda^a \, \varphi^* \imath_a \omega \ .$$

Let us now introduce a $\mathbf{g}$-valued gauge field $A = A^a X_a$ on Σ, which transforms under $\mathrm{Map}(\Sigma, \mathbf{g})$ as

$$\delta_\lambda A = d\lambda + [A, \lambda] \ .$$

The most general (polynomial) term we can add to (3.1) involving the gauge field is given by

$$S_{\mathrm{extra}}[\varphi, A] = \int_\Sigma A^a \, \varphi^* \phi_a$$

for some functions $\phi_a \in C^\infty(M)$. It is then a small computational matter to work out the conditions under which the total action

$$S_{\mathrm{GWZ}}[\varphi, A] = \int_B \varphi^* \omega + \int_\Sigma A^a \, \varphi^* \phi_a$$

is gauge-invariant; that is, $\delta_\lambda S_{\mathrm{GWZ}} = 0$. Doing so one finds that the conditions are

$$\imath_a \omega = d\phi_a$$
$$\mathcal{L}_a \phi_b = f_{ab}{}^c \phi_c$$

which are none other than (3.1a) and (3.1b) relative to the chosen basis for $\mathbf{g}$.

We therefore conclude that, for ω a symplectic form, the WZ term (4.1) can be gauged if and only if one can define an equivariant moment mapping for the G-action.

6 Bibliography

The homological obstructions to defining a moment mapping have been well-known since at least the mid nineteen-seventies. The treatment here follows in spirit the one in Weinstein's 1976 lectures [1]. The conditions for gauging reductive (two-dimensional) WZW models were obtained independently by Hull and Spence in [2] and by Jack, Jones, Mohammedi and Osborne in [3]. The conditions for gauging higher-dimensional σ-models with WZ term were later considered by Hull and Spence in [4]. The conditions for gauging nonreductive WZW models have been obtained in [5] as part of a general analysis of such models. The homological (re)interpretation of the obstructions to gauging a general WZ term will appear in [6].

[1] A Weinstein (1977), *Lectures on Symplectic Manifolds*, CBMS Regional Conference Series on Mathematics, **29**.
[2] CM Hull and B Spence (1989), *Phys. Lett.* **232B** (204) .
[3] I Jack, DR Jones, N Mohammedi and H Osborne (1990), *Nucl. Phys.* **B332** (359) .
[4] CM Hull and B Spence (1991), *Nucl. Phys.* **B353** (379-426) .
[5] JM Figueroa-O'Farrill and S Stanciu, *Nonreductive WZW models*, in preparation.
[6] JM Figueroa-O'Farrill and S Stanciu, *Homological obstructions to gauging Wess-Zumino terms in arbitrary dimensions*, to appear.

7 Postscript

After the talk, J. Cariñena pointed out to me another way to understand the obstructions in (3.1) in terms of Lie algebra cohomology with coefficients in the exact one-forms (equivalently, the hamiltonian vector fields). If we think of (1.1) and (1.3) as exact sequences of **g**-modules, we obtain two long exact sequence in Lie algebra cohomology. The map $X \mapsto \xi_X$ defines a class in $H^1(\mathbf{g}; \mathrm{Sym}(M))$. By exactness of the sequence induced by (1.1), we see that it comes from $H^1(\mathbf{g}; \mathrm{Ham}(M))$ if and only if its image in $H^1(\mathbf{g}; H^1(M))$ vanishes. Supposing it does and using now the exactness of the sequence induced by (1.3), we see that this class in $H^1(\mathbf{g}; \mathrm{Ham}(M))$ comes from $H^1(\mathbf{g}; C^\infty(M))$ precisely when its image in $H^2(\mathbf{g}; H^0(M))$ vanishes. These two obstructions precisely correspond to the classes in (3.2). Finally, I was informed by G. Papadopoulos, that the obstructions in (3.2) can also be understood as "anomalies" to global symmetries in the quantization of a particle interacting with a magnetic field. The details appear in *Comm. Math. Phys.* **144** (1992) 491-508. I am grateful to them both for letting me know of their results during the conference.

Acknowledgements It is a pleasure to thank Chris Hull, Takashi Kimura, and Sonia Stanciu for discussions on this topic; and especially Jim Stasheff for comments on a previous version of the TEXscript.

Global aspects of symmetries in sigma models with torsion

G. Papadopoulos

Abstract

It is shown that non-trivial topological sectors can prevent the quantum mechanical implementation of the symmetries of the classical field equations of sigma models with torsion. The associated anomaly is computed, and it is shown that it depends on the homotopy class of the topological sector of the theory and the group action on the sigma model manifold that generates the symmetries of the classical field equations.

1 Introduction

Two-dimensional sigma models have been extensively studied the last ten years because of their applications to string theory and the theory of integrable systems. More recently, attention has been focused on sigma models with symmetries. Such sigma models arise naturally in the investigation of sigma model duality (see for example refs. [1]) and the study of supersymmetric sigma models with potentials [2,3,4].

The fields of a sigma model are maps ϕ from the two-dimensional spacetime Σ into a Riemannian manifold $(\mathcal{M}, h)$ which is called sigma model manifold or target space, and the couplings are tensors on $\mathcal{M}$. In the following, we will focus on sigma models with couplings a metric h, a scalar function V, and an antisymmetric two-form b. The form b may be locally defined on $\mathcal{M}$ and the part of the sigma model action I that contains this coupling is called Wess-Zumino (WZ) term [5] or WZ action. The action I is either *locally* or *globally* defined depending on whether the WZ term b is a locally or globally defined form correspondingly. The classical field equations, however, are always *globally* defined on $\mathcal{M}$ because the coupling b enters in them through its exterior derivative $H = \frac{3}{2}db$ and H is a *globally* defined closed three-form on $\mathcal{M}$ (see section 3). From the classical field theory point of view, the latter is enough to guarantee that the above theory is covariant under reparameterisations of the sigma model manifold and so well defined. In the path integral quantisation of this theory, however, $\exp(2\pi i I)$ is required to be globally defined as well, *i.e.* the sigma model action I must be globally defined up to an integer. If the action I is not globally defined, this additional requirement leads to a certain quantisation condition for the WZ coupling b [6,7].

The symmetries of sigma models that we will examine are those induced by vector fields on the sigma model manifold. A sufficient condition for the transformations generated by vector fields on the sigma model manifold to be symmetries of the *field equations* is to leave the tensors h, V, H invariant. In the quantum theory though, we need to know the conditions for the invariance of the sigma model *action* under the above transformations. This is straightforward for the terms in the action that contain couplings h and V; the conditions are the same as those for the invariance of the associated terms in the field equations. *New* conditions are required though for the invariance of the WZ action in addition to those necessary for the invariance of the corresponding term in the field equations; I will call these new conditions anomalies. This is due to the fact that the WZ action may be locally defined on $\mathcal{M}$ and the symmetries of the field equations may leave this action invariant up to surface terms that cannot be integrated away. (The latter can happen even in the special case where b *is* globally defined on $\mathcal{M}$).

The main point of this talk is to present the conditions under which the action of a sigma model with a WZ term is invariant (up to an integer) under the transformations generated by a group G acting on the sigma model manifold $\mathcal{M}$. I will then introduce the (1,1)-supersymmetric two-dimensional massive sigma model and show that the above anomaly cancels for the symmetries that arise naturally in this model.

The conditions for the invariance of the WZ action have been presented before by the author in ref. [8]. However the original publication does not contain the proof of key statements regarding these conditions and there is no mention of the conditions that are necessary for the WZ action to be invariant under infinitesimal transformations. These will be included here. The action of massive (1,1)-supersymmetric sigma models was given in ref. [3] in collaboration with C.M. Hull and P.K. Townsend.

In section two, I will briefly review the conditions for the symmetries of the equations of motion of a charged particle coupled to a magnetic field to be implemented in the quantum theory of the system. In section three, I will give the action of a bosonic sigma model with a WZ term and a scalar potential, and the conditions for its field equations to be invariant under transformations generated by a group acting on the sigma model manifold. In section four, I will present a global definition of the WZ action and give the conditions for this action to be invariant under the symmetries of its field equations, and in section five, I will discuss the symmetries of supersymmetric massive sigma models.

2 A Quantum Mechanical Model

It is instructive to present the conditions under which the symmetries of the equations of motion of a charged particle coupled to a magnetic field can be implemented quantum mechanically [9]. This is because there is a close relation between these conditions and some of the conditions that I will derive in section 4 for the case of the two-dimensional sigma model with a WZ term. The action of a charged particle coupled to a magnetic field b is

$$I = \int dt \ \frac{1}{2} h_{ij} \partial_t \phi^i \partial_t \phi^j + b_i \partial_t \phi^i , \tag{2.1}$$

where t is a parameter of the world-line, ϕ are the co-ordinates of the particle and h is the metric of the manifold $\mathcal{M}$ in which the particle propagates. The coupling b is a locally defined one-form on $\mathcal{M}$ with patching conditions $b_1 = b_2 + da_{12}$ on the intersection $U_1 \cap U_2$ of any two open sets U_1, U_2 of $\mathcal{M}$, and a_{12} is a function on $U_1 \cap U_2$. Because of these patching conditions the last term of the above action is locally defined on $\mathcal{M}$. The equations of motion are

$$\nabla_t \partial_t \phi^i - h^{ij} \omega_{jk} \partial_t \phi^k = 0 , \tag{2.2}$$

where

$$\omega = db , \tag{2.3}$$

∇_i is the Levi-Civita covariant derivative of the metric h and $\nabla_t \equiv \partial_t \phi^i \nabla_i$. The equations of motion are globally defined on $\mathcal{M}$ because $\omega \equiv db$ is a globally defined closed two-form, *i.e.* $\omega_1 = \omega_2$ on the intersection of any two open sets U_1 and U_2 of $\mathcal{M}$.

Let G be a group and $f : G \times \mathcal{M} \to \mathcal{M}$ be a group action of G on $\mathcal{M}$; $f_{gg'} = f_g f_{g'}$ and $f_e = Id_{\mathcal{M}}$ where $g, g', e \in G$ (e is the identity element of G). Sufficient conditions for the invariance of the equations of motion (2.2) under the group action f are

$$(f_g^* h)_{ij} = h_{ij}, \qquad (f_g^* \omega)_{ij} = \omega_{ij}, \qquad g \in G , \tag{2.4}$$

i.e. the transformations f_g are isometries and leave the closed two-form ω invariant.

In the quantum theory, it is required that ω be the curvature of a line bundle over the manifold $\mathcal{M}$ and the wave functions be sections of this line bundle; this property of ω is a quantisation condition for the coupling b of the action (2.1) and it is called Dirac's quantisation condition. If in addition this theory has symmetries generated by a group action as above, the conditions (2.4) are not enough to guarantee that these symmetries can be implemented with unitary transformations on the Hilbert space of the theory. For this, *additional* conditions are necessary. To describe the additional conditions, let $[\omega]$ be the cohomology class of the curvature ω in $H^2(\mathcal{M}, \mathbb{Z})$ and G be compact and connected. We first pull-back $[\omega]$ using the group action f on the manifold $G \times \mathcal{M}$ and then decompose the pulled-back cohomology class $f^*[\omega]$ as

$$f^*[\omega] = [\omega] + [\sigma_1] + [\sigma_2] , \tag{2.5}$$

where $[\sigma_1] \in H^1(G, H^1(\mathcal{M}, \mathbb{Z}))$ and $[\sigma_2] \in H^2(G, \mathbb{Z})$. (We have used the Künneth formula to perform this decomposition). Apart from (2.4), the *additional* conditions to implement the classical symmetries with unitary transformations on the Hilbert space of the above theory are

$$[\sigma_1] = 0, \qquad [\sigma_2] = 0 . \tag{2.6}$$

After some computation, we can show that a consequence of the first condition of (2.6) is that the $\exp(2i\pi I)$ of the action I (eqn. (2.1)) and the charges of this theory associated with the above symmetries are globally defined on the manifold $\mathcal{M}$, and a consequence of the second condition of (2.6) is that the Poisson bracket algebra of these charges is isomorphic to the Lie algebra of G. For the proof of all the above statements as well as the study of the case where G is disconnected using the theory of universal classifying spaces see refs. [9].

In section 4, I will examine the analogue of the first condition of (2.6) for the case of two-dimensional sigma models with WZ term. It is worth pointing out though that a condition similar to the second of (2.6) appears in the case of two-dimensional sigma models as well but this will be presented elsewhere [10].

3 Two-dimensional Sigma Models with Torsion

The action of a two-dimensional sigma model with WZ term b and target space a Riemannian manifold $(\mathcal{M}, h)$ is

$$I = \int d^2x (h + b)_{ij} \partial_{\ddagger} \phi^i \partial_{=} \phi^j - V(\phi) , \tag{3.1}$$

where the sigma model fields ϕ are maps from the two-dimensional spacetime Σ with light-cone co-ordinates $\{x^{\ddagger} = t + x, x^{=} = t - x\}$ into $\mathcal{M}$ and V is a real function on $\mathcal{M}$. The two-form b is *locally* defined on $\mathcal{M}$ with patching conditions $b_1 = b_2 + dm_{12}$ at the intersection $U_1 \cap U_2$ of any two open sets U_1, U_2 of $\mathcal{M}$ where m_{12} is an one-form defined on $U_1 \cap U_2$. The WZ term in the action is then locally defined on $\mathcal{M}$. One can define a closed three-form H on $\mathcal{M}$ as

$$H = \frac{3}{2} db . \tag{3.2}$$

Observe that H is globally defined on $\mathcal{M}$ and it is called torsion for reasons that will become apparent below. The field equations are

$$\nabla^{(+)}_{=} \partial_{\ddagger} \phi^i + \frac{1}{2} h^{ij} \partial_j V = 0 , \tag{3.3}$$

where the connections of the covariant derivatives $\nabla^{(\pm)}$ are

$$\Gamma^{(\pm)i}{}_{jk} = \{{}^{i}_{jk}\} \pm H^i_{jk} . \tag{3.4}$$

The tensor H is the torsion of the connection $\Gamma^{(+)}$ and it is globally defined on $\mathcal{M}$.

Let G be a group and $f : G\times\mathcal{M} \to \mathcal{M}$ be a group action of G on the target manifold $\mathcal{M}$ as in the previous section; $f_{gg'} = f_g f_{g'}$ and $f_e = Id_{\mathcal{M}}$ where $g, g', e \in G$. Sufficient conditions for the invariance of the field equations (3.3) under the group action f are

$$(f_g^* h)_{ij} = h_{ij}, \qquad (f_g^* H)_{ijk} = H_{ijk}, \qquad f_g^* V = V, \qquad g \in G \,, \tag{3.5}$$

i.e. the group action leaves invariant the metric h (so the group action f generates isometries of the Riemannian manifold $(\mathcal{M}, h)$), the closed three-form H and the scalar potential V. Let in addition G be a Lie group with Lie algebra $\mathcal{L}(G)$. The infinitesimal form of the above conditions is then

$$(L_a h)_{ij} = 0, \qquad (L_a H)_{ijk} = 0, \qquad L_a V = 0 \,, \tag{3.6}$$

where $\{L_a; a = 1, \ldots, \dim\mathcal{L}(\mathrm{G})\}$ is the Lie derivative with respect to the vector field $\{X_a; a = 1, \ldots, \dim \mathcal{L}(\mathrm{G})\}$ generated by the group action f on $\mathcal{M}$. The vector fields X_a are Killing $(\nabla_{(i} X_{j)a} = 0)$.

In the path-integral quantisation of this theory, one needs to know the conditions for the action (3.1) to be invariant (up to an integer) under the transformations generated by a group action. The invariance of the terms in the action (3.1) involving the metric h and the scalar potential V follows directly from the conditions on h, V given in eqn. (3.5) for the invariance of the field equations. However for the invariance of the Wess-Zumino action in (3.1), we need additional conditions besides those of eqn. (3.5) due partly to the fact that this term is not globally defined on $\mathcal{M}$. There are two main approaches to define globally the Wess-Zumino term. The first is the homotopy approach due to Wess and Zumino [5], and Rohm and Witten [6], and, the second is the Čech Cohomology approach due to O. Alvarez [7]. In the next section, I will use the former to examine the symmetries of the Wess-Zumino term. A study of the symmetries of the WZ action in the Čech Cohomology approach will be presented elsewhere [10].

4 Symmetries and the Wess-Zumino Action

Let Σ, the two-dimensional space-time, be a closed manifold, *i.e.* compact and without boundary, and $[\Sigma, \mathcal{M}]$ be the homotopy classes of maps from Σ into the sigma model target manifold $\mathcal{M}$. To define the Wess-Zumino term in the homotopy approach, we choose a 'background' map ϕ_0 from Σ into $\mathcal{M}$ such that ϕ_0 is homotopic to ϕ, *i.e.* there is a map F $(F : [0,1]\times\Sigma \to \mathcal{M})$ and

$$F(s, x) = \begin{cases} \phi_0(x) & \mathrm{s{=}0}\,, \\ \phi(x) & \mathrm{s{=}1}\,. \end{cases} \tag{4.1}$$

The homotopy F interpolates between ϕ_0 and ϕ. In the following, we will use the notation $\phi_0 \simeq_F \phi$ to denote that the map ϕ_0 is homotopic to ϕ with respect to F. The action of the Wess-Zumino term is defined [6] as

$$S_{WZ}[\phi, \phi_0; F] = \int_{[0,1]\times\Sigma} F^* H \; . \tag{4.2}$$

As indicated, the action of the Wess-Zumino term depends on the choice of the homotopy F that interpolates between ϕ and ϕ_0. To determine the dependence of S_{WZ} on F, we take two different homotopies F_1 and F_2 that interpolate between ϕ and ϕ_0 and compute the difference

$$\Delta S_{WZ} = S_{WZ}[\phi, \phi_0; F_1] - S_{WZ}[\phi, \phi_0; F_2] \; . \tag{4.3}$$

This difference can be rewritten as

$$\Delta S_{WZ} = S_{WZ}[\phi_0, \phi_0; F_3] = \int_{S^1\times\Sigma} F_3^* H \; , \tag{4.4}$$

where

$$F_3(s_3, x) = \begin{cases} F_1(2s_3, x) & 0 \le s_3 \le \frac{1}{2} \\ F_2(-2s_3 + 2, x) & \frac{1}{2} \le s_3 \le 1 \; . \end{cases} \tag{4.5}$$

Note that $F_3(0, x) = F_3(1, x) = \phi_0(x)$. The difference ΔS_{WZ} of (4.4) is the integral of a closed three-form over a compact three-manifold without boundary and in general its value is a real number. However if $[H] \in H^3(M, \mathbb{Z})$, the difference is an integer and thus the functional

$$A[\phi, \phi_0] = e^{2i\pi S_{WZ}[\phi, \phi_0; F]} \tag{4.6}$$

becomes independent of the choice of homotopy F. This property of A is *sufficient* for the consistency of the path-integral quantisation of this theory. In the following, we will take $[H] \in H^3(M, \mathbb{Z})$ and so the WZ action will be independent from the choice of homotopy F mod 1.

The invariance of the field equations of a sigma model with a Wess-Zumino term under the group action f of a (connected) group G on $\mathcal{M}$ does not necessarily imply the invariance of the action $S_{WZ}[\phi, \phi_0; F]$. The transformation $\phi^g \equiv f_g(\phi)$ of the field ϕ induces the transformation $S_{WZ}[\phi^g, \phi_0; F_4]$ on the Wess-Zumino action which can be rewritten as

$$\begin{aligned} S_{WZ}[\phi^g, \phi_0; F_4] &= S_{WZ}[\phi^g, \phi_0; F_3] + \text{integer} \\ &= S_{WZ}[\phi^g, \phi_0^g; F_2] + S_{WZ}[\phi_0^g, \phi_0; F_1] + \text{integer} \; , \end{aligned} \tag{4.7}$$

where

$$F_3(s, x) = \begin{cases} F_1(2s, x) & 0 \le s \le \frac{1}{2} \; , \\ F_2(2s - 1, x) & \frac{1}{2} \le s \le 1 \; . \end{cases} \tag{4.8}$$

The first equality in (4.7) follows from the observation that $\phi_0 \simeq_{F_3} \phi^g$ and the property of the WZ action to be independent of the choice of homotopy F up to an integer, and the second from the definition of the WZ term. Using again the property of the WZ action to be independent of the choice of homotopy mod 1, we can choose $F_2 = F^g$ and then use the property of H to be invariant under the group action f (this comes from the invariance of the field equations eqn. (3.5)) to reexpress (4.7) as

$$S_{WZ}[\phi^g, \phi_0; F_4] = S_{WZ}[\phi, \phi_0; F] + S_{WZ}[\phi_0^g, \phi_0; F_1] + \text{integer}' \,. \qquad (4.9)$$

From (4.9) it is clear that in *addition* to the conditions (3.5) for the invariance of the field equations, the vanishing of

$$c[\phi_0, g] = S_{WZ}[\phi_0^g, \phi_0; F_1] \mod 1 \,. \qquad (4.10)$$

is a necessary condition [8] in order for the group action f of G to be a symmetry of the action $S_{WZ}[\phi, \phi_0; F]$. Note that $c[\phi_0, g]$ is independent of the choice of homotopy F_1. I will refer to $c[\phi_0, g]$ as anomaly.

The anomaly $c[\phi_0, g]$ has some novel properties. In particular, we can show that $c[\phi_0, g]$ depends on the homotopy class $[\phi_0]$ of ϕ_0 rather than ϕ_0 itself. To prove this, let $\phi_1 \simeq_{F_1} \phi_0$. We write

$$\begin{aligned} S_{WZ}[\phi_0^g,\phi_0; F] &= S_{WZ}[\phi_0^g, \phi_0; F_5] + \text{integer} \\ &= S_{WZ}[\phi_0^g, \phi_1^g; F_2] + S_{WZ}[\phi_1^g, \phi_1; F_3] + S_{WZ}[\phi_1, \phi_0; F_4] + \text{integer} \,, \end{aligned} \qquad (4.11)$$

where

$$F_5(s_5, x) = \begin{cases} F_4(3s_5, x) & 0 \le s_5 \le \frac{1}{3} \,, \\ F_3(3s_5 - 1, x) & \frac{1}{3} \le s_5 \le \frac{2}{3} \,, \\ F_2(3s_5 - 2, x) & \frac{2}{3} \le s_5 \le 1 \,. \end{cases} \qquad (4.12)$$

Observe from the expression for F_5 that $\phi_0 \simeq_{F_5} \phi_0^g$. Using the property of the Wess-Zumino action to be independent of the choice of homotopy mod 1 and the invariance of H, we write $S_{WZ}[\phi_0^g, \phi^g; F_2] = S_{WZ}[\phi_0, \phi; F_1] + \text{integer}$. The equation (4.11) then becomes

$$\begin{aligned} S_{WZ}[\phi_0^g, \phi_0; F] &= S_{WZ}[\phi_0, \phi_1; F_1] + S_{WZ}[\phi_1^g, \phi_1; F_3] \\ &\quad + S_{WZ}[\phi_1, \phi_0; F_4] + \text{integer}' \\ &= S_{WZ}[\phi_1^g, \phi_1; F_1] + \text{integer}'' \,. \end{aligned} \qquad (4.13)$$

So if ϕ_0 is homotopic to ϕ_1, we have shown that $c[\phi_1, g] = c[\phi_0, g]$ and thus the anomaly c is a map from $[\Sigma, \mathcal{M}] \times G$ into $\mathbb{R}/\mathbb{Z}$.

We can also show that

$$c[[\phi], g_1 g_2] = (c[[\phi], g_1] + c[[\phi], g_2])\text{mod } 1, \qquad g_1, g_2 \in G \,. \qquad (4.14)$$

This immediately follows from

$$\begin{aligned} Swz[\phi_0^{g_1g_2},\phi_0;F] &= Swz[\phi_0^{g_1g_2},\phi_0;F_3] + \text{integer} \\ &= Swz[(\phi_0^{g_1})^{g_2},\phi_0^{g_2};F_2] + Swz[\phi_0^{g_2},\phi_0;F_1] + \text{integer} \\ &= Swz[\phi_0^{g_1},\phi_0;F_4] + Swz[\phi_0^{g_2},\phi_0;F_1] + \text{integer}' \ , \end{aligned} \quad (4.15)$$

where F_3 is related to F_1 and F_2 as in (4.8) and $F_2 = F_4^{g_2}$. In addition, $c\big[[\phi],e\big] = 0$, so the anomaly is represented by a group homomorphism from G into $\mathbb{R}/\mathbb{Z}$ for every homotopy class of maps from Σ into $\mathcal{M}$.

An important consequence of the above properties of the anomaly c is that c vanishes whenever it is evaluated on the trivial topological sector of the theory, *i.e.* on the trivial class of $[\Sigma,\mathcal{M}]$. To prove this, we observe that the trivial topological sector can be represented by the constant maps from Σ into $\mathcal{M}$. Since the anomaly c depends on the homotopy classes of the maps rather than the maps themselves, we can use any representative of the trivial homotopy class, hence any constant map from Σ into $\mathcal{M}$, to do the computation. Let ϕ_0 be such a map, the expression (4.10) for the anomaly involves the pull-back F^*H of the three-form H with respect to the homotopy F that interpolates between ϕ_0^g and ϕ_0. Since ϕ_0^g and ϕ_0 are constants maps from Σ into $\mathcal{M}$, they can be thought of as points in $\mathcal{M}$ and so we can take F to be any path in $\mathcal{M}$ that interpolates between them. In which case, F is a map of one variable and therefore the pull-back form F^*H of the three-form H vanishes ($F^*H = 0$). In fact using the same arguments, we can show that the anomaly c vanishes for all the topological sectors $[\phi]$ of the theory that admit a homotopy F which interpolates between ϕ and ϕ^g with differential map $dF : T([0,1]\times\Sigma) \to T(\mathcal{M})$ that has rank less than three. Thus only the *non-trivial* topological sectors can break the symmetries of the field equations of a sigma model with a WZ term.

The anomaly c vanishes for all sigma models with torsion for which $[\Sigma,\mathcal{M}]$ has only one element, the trivial class. For example, this is the case for the sigma models with $\Sigma = S^2$ and $\mathcal{M} = SU(N)$ because $\pi_2(SU(N)) = 0$.

It is well known that the homotopy classes of maps $[\Sigma,\mathcal{M}]$ under certain conditions can be given a group structure. So it is natural to ask whether or not the anomaly c is a group homomorphism from $[\Sigma,\mathcal{M}]\times G$ into $\mathbb{R}/\mathbb{Z}$. We have already shown above that $c\big[[\phi],g\big] = 0$ if $[\phi]$ is the trivial class of $[\Sigma,\mathcal{M}]$. So it remains to show whether $c\big[[\phi_1\cdot\phi_2],g\big] = \big(c\big[[\phi_1],g\big] + c\big[[\phi_2],g\big]\big)$ mod 1. Examination of the particle model in section 2 indicates that indeed the anomaly c is a group homomorphism but this will not be pursued further here.

Next we will calculate the form of the anomaly for infinitesimal transformations. For this, let us consider an one-parameter subgroup of G generated by the vector $v \in \mathcal{L}(G)$. The elements of this subgroup can be written as $g(r) = \exp(rv)$ where r is a real number and g can be thought as a map from the real line $\mathbb{R}$ into G. Using this subgroup, we define another map that we will call again g from $[0,1]\times(-\epsilon,\epsilon)$ into G by setting $r=sz$, i.e. $g(s,z)=\exp(szv)$, where ϵ is a positive real number. We observe that $g(s,z)$ has the following properties:

$$g(s,z)=\begin{cases} e & \text{if either s=0 or z=0} \\ g(z)=\exp zv & \text{s=1} . \end{cases} \tag{4.16}$$

The 'infinitesimal' anomaly is given by

$$\Delta[\phi,v] := \frac{d}{dz} S_{WZ}[\phi^{g(z)},\phi;F]|_{z=0} , \tag{4.17}$$

where

$$F(s,x) := \phi^{g(s,z)}(x) . \tag{4.18}$$

Because of (4.16), the homotopy F interpolates between ϕ and $\phi^{g(z)}$. Using the definition (4.17) of the infinitesimal anomaly and the Wess-Zumino action, we can show after some further computation that

$$\Delta[\phi,v] = \int_\Sigma v^a \phi^*(\eta_a) , \tag{4.19}$$

where

$$\eta_a \equiv i_a H , \tag{4.20}$$

and i_a is the inner derivation with respect to the Killing vector field X_a. Using $dH=0$ from (3.2), $L_aH=0$ from (3.6) and $L_a=di_a+i_ad$, we can show that $\eta_a\equiv i_aH$ is a closed two-form

$$d\eta_a = 0 . \tag{4.21}$$

Therefore the infinitesimal anomaly $\Delta[\phi,v]$ is the integral of the pull-back with the map ϕ of a closed two-form of $\mathcal{M}$ with domain the space-time Σ and so it depends on the homotopy class $[\phi]$ of ϕ, i.e. $\Delta=\Delta\big[[\phi],v\big]$.

The infinitesimal anomaly $\Delta\big[[\phi],v\big]$ vanishes whenever $[\phi]\in[\Sigma,\mathcal{M}]$ is the trivial class (in agreement with the behaviour of $c\big[[\phi],g\big]$ discussed above). The anomaly $\Delta\big[[\phi],v\big]$ also vanishes provided that the two-forms $\{\eta_a\}$ are exact. Note though that $\Delta\big[[\phi],v\big]$ does *not* vanish if b is merely globally defined on $\mathcal{M}$.

The Noether charges associated with the symmetries of the action (3.1) generated by the group action f are

$$Q_a = \int dx\ (X_{ai}\partial_t\phi^i + u_{ai}\partial_x\phi^i) , \tag{4.22}$$

where the one-forms $\{u_a\}$ are locally defined and they are related to the closed two-forms $\{\eta_a\}$ as follows:

$$\eta_a = du_a . \tag{4.23}$$

If the two-forms $\{\eta_a\}$ are exact, then $\{u_a\}$ are globally defined and consequently the Noether charges $\{Q_a\}$ are globally defined as well. However even if $\{u_a\}$ are globally defined, it is not expected that the Poisson bracket algebra of these charges to be necessarily isomorphic to $\mathcal{L}(G)$. Recall that the Poisson bracket algebra of the charges of the particle model in section 2 has similar behaviour.

The methods developed above to examine the symmetries of the WZ action for two-dimensional sigma models can be extended to the case of n-dimensional ones [8].

5 Concluding Remarks

Symmetries generated by vector fields on the sigma model manifold arise naturally in the study of (p,q)-supersymmetric two-dimensional massive sigma models because the charges of such symmetries appear as central charges in the Poisson bracket algebra of supersymmetry charges of these models. The above vector fields also enter in the expression for the scalar potential.

The simplest massive supersymmetric sigma model with a central charge is the one with (1,1)-supersymmetry. Let X be a Killing vector field that leaves invariant both the torsion H of eqn. (3.2) and u of eqn. (4.23). It can be shown that the most general action of a massive sigma model with (1,1) supersymmetry [3] is

$$I = \int d^2x \{ \partial_{\ddagger} \phi^i \partial_{=} \phi^j (h_{ij} + b_{ij}) + i h_{ij} \lambda^i_+ \nabla^{(+)}_{=} \lambda^j_+ - i h_{ij} \psi^i_- \nabla^{(-)}_{\ddagger} \psi^j_- - \frac{1}{2} \psi^k_- \psi^l_- \lambda^i_+ \lambda^j_+ R^{(-)}_{ijkl} + m \nabla^{(-)}_i (u - X)_j \lambda^i_+ \psi^j_- - V(\phi) \} \,, \tag{5.1}$$

where λ_+ and ψ_- are real chiral fermions and

$$V(\phi) = \frac{m^2}{4} h^{ij} (u - X)_i (u - X)_j \,, \tag{5.2}$$

is the scalar potential. The rest of the notation follows from sections 3 and 4. Observe that the requirement for the action (5.1) to be (1,1)-supersymmetric imposes strong restrictions on the form of the scalar potential $V(\phi)$.

To define globally the field equations derived from this action on the sigma model target space $\mathcal{M}$, we have to assume that u is globally defined one-form on $\mathcal{M}$. As we have established in the previous section, this is a sufficient condition for the WZ action to be invariant under the infinitesimal transformations generated by X.

Computation of the Poisson bracket algebra of charges of the above model reveals that

$$\{S_+, S_+\} = E + P, \qquad \{S_-, S_-\} = E - P, \qquad \{S_+, S_-\} = Q_X \,. \tag{5.3}$$

where S_+, S_- are the supersymmetry charges, E is the energy, P is the momentum and Q_X is the Noether charge associated with the symmetries generated by X (eqn. (4.22)). To derive (5.3), we have assumed $X^i u_i = 0$. Observe that Q_X appears in the Poisson bracket of left- with right- supersymmetry charges. The form of the charges as well as details of this computation can be found in refs. [4].

In conclusion, non-trivial topological sectors can break quantum mechanically the symmetries of the field equations of a sigma model with torsion. This is partly due to the WZ action which may not be globally defined and/or may be invariant up to surface terms that cannot be integrated away. An associated anomaly is computed and it is found to be a WZ-like action that depends on the homotopy class of the topological sector of the theory and the group action on the sigma model manifold that generates the symmetries of the field equations. Sufficient conditions for the vanishing of this anomaly are also given. In particular, it is shown that the anomaly vanishes whenever the theory has only one topological sector, *i.e.* the trivial one. Finally it is worth pointing out that, as in the case for the one-dimensional sigma model, other conditions are necessary in addition to the vanishing of the above anomaly in order for the symmetries of the field equations of n-dimensional ($n \geq 2$) sigma models with torsion to be implemented quantum mechanically.

Acknowledgements

Discussions with C.M. Hull, H. Nicolai, P.K. Townsend and A. van de Ven are gratefully acknowledged. I was funded by a grant from the European Union.

References

[1] T.H. Buscher (1985), *Phys. Lett.*, **159B** 127; (1987), *ibid.*, **194B** 51.
M. Roček and E. Verlinde (1992), *Nucl. Phys.*, **B373** 630.
A. Giveon and M. Roček (1992), *Nucl. Phys.*, **B380** 128.
E. Alvarez, L. Alvarez-Gaumé and Y. Lozano (1994), CERN-TH-7204/94, hep-th/9403155.

[2] L. Alvarez-Gaumé and D.Z. Freedman (1983), *Commun. Math. Phys.*, **91** 87.

[3] C.M. Hull, G. Papadopoulos and P.K. Townsend (1993), *Phys. Lett.*, **316B** 291.

[4] G. Papadopoulos and P.K. Townsend (1994), *Class. Quantum Grav.* **11** 515; 'Massive (p,q)-supersymmetric sigma models revisited', R/94/9, DESY 94-092, hep-th/9406015.

[5] J. Wess and B. Zumino (1971), *Phys. Lett.*, **37B** 96.

[6] R. Rohm and E. Witten (1986), *Ann. Phys.*, 170 454.

[7] O. Alvarez (1985), *Commun. Math. Phys.*, 100 279.

[8] G. Papadopoulos (1990), *Class. Quantum Grav.*, 7 L41.

[9] G. Papadopoulos (1992), *Commun. Math. Phys.*, 144 491; (1990), *Phys. Lett.*, 248B 113.

[10] C.M. Hull and G. Papadopoulos, in preparation.

Canonical structure of the non-linear σ-model in a polynomial formulation

C. D. Fosco and T. Matsuyama

Abstract

We study the canonical formulation of the $SU(N)$ non-linear σ-model in a polynomial, first-order representation. The fundamental variables in this description are a non-Abelian vector field L_μ and a non-Abelian antisymmetric tensor field $\theta_{\mu\nu}$, which constrains L_μ to be a 'pure gauge' ($F_{\mu\nu}(L) = 0$) field. The second-class constraints that appear as a consequence of the first-order nature of the Lagrangian are solved, and the corresponding reduced phase-space variables explicitly found. We also treat the first-class constraints due to the gauge-invariance under transformations of the antisymmetric tensor field, constructing the corresponding most general gauge-invariant functionals, which are used to describe the classical dynamics of the physical degrees of freedom. We present these results in detail in $1+1$, $2+1$ and $3+1$ dimensions, mentioning some properties of the $d+1$-dimensional case. We show that there is a kind of duality between this description of the non-linear σ-model and the massless Yang-Mills theory. The duality is further extended to more general first-class systems.

1 Introduction

One of the distinctive properties of the non-linear σ-model [1], is that its dynamical variables belong to a non-linear manifold [2], thus realising the corresponding symmetry group in a non-linear fashion [3]. Whence either the Lagrangian becomes non-polynomial in terms of unconstrained variables, or it becomes polynomial but in variables which satisfy a non-linear constraint. It is often convenient to work in a polynomial or 'linearized' representation of the model, where the symmetry is linearly realised. For example, in the $O(N)$ models, where the field is a N-component vector constrained to have constant modulus, a polynomial representation is constructed simply by introducing a Lagrange multiplier for that quadratic constraint. However, this simplicity is not present in general because the constraints required to define the manifold can be much more complex.

In references [4, 5] a polynomial representation of the non-linear σ-model was introduced; let us briefly explain it for the $SU(N)$ model in d+1 dimensions.

The usual presentation [8] of this model is in terms of a $SU(N)$ field $U(x)$, with Lagrangian

$$\mathcal{L} = \frac{1}{2} g^{d-1} tr(\partial_\mu U^\dagger \partial^\mu U) \ , \tag{1.1}$$

where g is a coupling constant with dimensions of mass (the constant f_π in its application to Chiral Perturbation Theory in 3+1 dimensions). The polynomial description [5] of this model is constructed in terms of a non-Abelian ($SU(N)$) vector field L_μ plus a non-Abelian antisymmetric tensor field $\theta_{\mu\nu}$[1] with the Lagrangian

$$\mathcal{L} = \frac{1}{2} g^2 L_\mu \cdot L^\mu + g\, \theta_{\mu\nu} \cdot F^{\mu\nu}(L) \tag{1.2}$$

where the fields L_μ and $\theta_{\mu\nu}$ are defined by their components in the basis of generators of the adjoint representation of the Lie algebra of $SU(N)$; i.e., $L_\mu(x)$ is a vector with components $L^a_\mu, a = 1, \cdots, N^2-1$, and analogously for $\theta_{\mu\nu}$. The components of $F_{\mu\nu}$ in the same basis are: $F^a_{\mu\nu}(L) = \partial_\mu L^a_\nu - \partial_\nu L^a_\mu + g^{\frac{3-d}{2}} f^{abc} L^b_\mu L^c_\nu$, where the dots mean $SU(N)$ scalar product, for example: $L_\mu \cdot L^\mu = \sum_{a=1}^{N^2-1} L^a_\mu L^\mu_a$. The d-dependent exponents in the factors of g are chosen in order to make the fields have the appropriate canonical dimension for each d.

The Lagrange multiplier $\theta_{\mu\nu}$ imposes the constraint $F_{\mu\nu}(L) = 0$, which is equivalent [9] to $L_\mu = g^{\frac{d-3}{2}} U \partial_\mu U^\dagger$, where U is an element of $SU(N)$. When this is substituted back in (1.2), (1.1) is obtained [2]. This polynomial formulation could be thought of as a concrete Lagrangian realization of the Sugawara theory of currents [7], where all the dynamics is defined by the currents, the energy-momentum tensor, and their algebra. Indeed, L_μ corresponds to one of the conserved currents of the non-polynomial formulation, due to the invariance of $\mathcal{L}$ under global (left) $SU(N)$ transformations of $U(x)$. The energy-momentum tensor corresponding to (1.2) is indeed a function of L_μ only:

$$T^{\mu\nu} = \frac{1}{2} g^2 (L^\mu \cdot L^\nu - \frac{1}{2} g^{\mu\nu} L^2) \ . \tag{1.3}$$

The action corresponding to Lagrangian (1.2) is invariant under the gauge transformations

$$\begin{aligned} \theta_{\mu\nu}(x) &\rightarrow \theta_{\mu\nu}(x) + \delta_\omega \theta_{\mu\nu}(x) \\ \delta_\omega \theta_{\mu\nu}(x) &= D^\rho \omega_{\rho\mu\nu}(x) \ , \end{aligned} \tag{1.4}$$

[1] To avoid the proliferation of indices, we frequently work in terms of the *dual* of $\theta_{\mu\nu}$, which in $1+1$ is a pseudo-scalar, in $2+1$ a pseudovector, etcetera.

[2] For a complete derivation of the equivalence between the theories defined by (1.1) and (1.2) within the path integral framework, see ref. [5].

where $\omega_{\rho\mu\nu}(x)$ is any completely antisymmetric tensor field, because $D_\mu \delta_\omega \theta^{\mu\nu}$ vanishes as a consequence of the Bianchi identity for L_μ[3]. Obviously d must be larger than one in order to this transformation be well defined, since at least three different indices are needed to have a Bianchi identity. The Hamiltonian formulation of the model has then a rich structure, since there are second-class constraints ($\mathcal{L}$ is first-order), first-class constraints (for $d > 1$), and moreover they are reducible for $d > 2$.

In section 2 we discuss the Hamiltonian formulation of the $1+1$, $2+1$ and $3+1$ models, following the Dirac algorithm [6]. In section 3 we construct the general gauge invariant functionals for the transformations generated by the first-class constraints found in section 2, and in section 4 we apply the Dirac's brackets method to the second-class system formed by the first-class constraints plus some canonical gauge-fixing conditions. In Appendix A we discuss a duality between first-class systems.

2 Hamiltonian formalism and constraints

2.1 $1+1$ dimensions

From Section 1, the polynomial Lagrangian in $1+1$ dimensions becomes

$$\mathcal{L} = \frac{1}{2} g^2 L_\mu L^\mu + \frac{1}{2} g\, \theta\, \epsilon_{\mu\nu} F^{\mu\nu}(L) \tag{2.1}$$

where θ is a pseudoscalar field. It is clear from (2.1) that there is no gauge symmetry in this case. Thus there will not be first-class constraints in the corresponding Hamiltonian formulation. However, there are second-class constraints, because $\mathcal{L}$ is of first-order in the derivatives. This property will also appear in higher dimensions, so we will only discuss it in some detail for this case. Defining the canonical momenta the primary constraints appear:

$$\begin{aligned} \pi_0^a(x) &\equiv \frac{\partial \mathcal{L}}{\partial(\partial_0 L_0^a)} \approx 0 \\ \pi_1^a &\equiv \frac{\partial \mathcal{L}}{\partial(\partial_0 L_1^a)} \approx g\theta^a(x) \\ \pi_\theta^a(x) &\equiv \frac{\partial \mathcal{L}}{\partial_0 \theta^a} \approx 0 \end{aligned} \tag{2.2}$$

and the canonical Hamiltonian becomes

$$H = \int dx[-\frac{1}{2} g^2 L_0^a L_0^a + \frac{1}{2} g^2 L_1^a L_1^a - g L_0^a (D_1 \theta)^a] \,. \tag{2.3}$$

[3]This kind of symmetry also appears when considering the dynamics of a two-form gauge field, see for example references [10]

Following the usual Dirac algorithm one obtains one more constraint:

$$gL_0^a(x) \approx -(D_1\theta)^a(x) \tag{2.4}$$

and the Lagrange multipliers become fully determined. The full set of (primary plus secondary) constraints is second-class, and its particular form allows us to eliminate the canonical pairs corresponding to L_0^a and θ^a, thus effectively eliminating the associated degrees of freedom. The Dirac bracket becomes equal to the Poisson bracket for the remaining degrees of freedom. The resulting Hamiltonian is

$$H = \int dx[\frac{1}{2g^2}D_1\pi_1 \cdot D_1\pi_1 + \frac{1}{2}g^2 L_1 \cdot L_1] , \tag{2.5}$$

with canonical brackets between the L_1^a's and their momenta π_1^a. Thus these two variables become symplectic coordinates on the reduced phase-space, or constraint surface.

2.2 $2+1$ dimensions

The polynomial Lagrangian in this case becomes

$$\mathcal{L} = \frac{1}{2}g^2 L_\mu \cdot L^\mu + \frac{1}{2}g\,\theta_\mu \cdot \epsilon^{\mu\nu\lambda}F_{\nu\lambda}(L). \tag{2.6}$$

The constraint algorithm produces here the 2+1 analogous of the second-class constraints we showed in Section 1, allowing us to eliminate the 0-component of L_μ and all the components of θ_μ. However, there will remain a set of first-class constraints

$$G^a(x) = \frac{1}{2}\epsilon_{jk}F_{jk}^a \approx 0 , \tag{2.7}$$

with the first-class Hamiltonian

$$H = \int d^2x[\frac{1}{2g^2}D_j\pi_j \cdot D_k\pi_k + \frac{1}{2}g^2 L_j \cdot L_j] . \tag{2.8}$$

They satisfy the algebra

$$\{G^a(x), G^b(y)\} = 0 \quad \{H, G^a(x)\} = V^{ab}G^b(x) , \tag{2.9}$$

where $V^{ab} \equiv g^{-\frac{3}{2}}f^{acb}(D_j\pi_j)^c(x)$.

Now we show in what sense we can relate the massless Yang-Mills theory to the non-linear σ-model in this formulation. The $SU(N)$ Yang-Mills theory is defined by the Lagrangian

$$\mathcal{L}_{YM} = -\frac{1}{4}F_{\mu\nu}(L) \cdot F^{\mu\nu}(L) , \tag{2.10}$$

which in the temporal gauge gives rise to the canonical Hamiltonian

$$H = \int d^2x [\frac{1}{2}\pi_j \cdot \pi_j + \frac{1}{4}F_{jk} \cdot F_{jk}] , \tag{2.11}$$

and the first-class constraints ('Gauss' laws'):

$$H_a(x) = (D_j \pi_j)_a(x) \approx 0 , \tag{2.12}$$

which satisfy the $SU(N)$ algebra

$$\{H_a(x), H_b(y)\} = \delta(x-y) f_{abc} H_c(x) . \tag{2.13}$$

Note that (2.11) can be rewritten as

$$H = \int d^2x [\frac{1}{2}\pi_j \cdot \pi_j + \frac{1}{2} G \cdot G] , \tag{2.14}$$

where the G_a's are the ones defined in (2.7). We then see that the first-class systems corresponding to the Yang-Mills model and the non-linear σ-model can be related by: 1) Interchanging the constraints:

$$H_a(x) \leftrightarrow G_a(x) , \tag{2.15}$$

and 2) Interchanging L_j by $\frac{1}{g}\pi_j$ in the non-derivative terms in the Hamiltonian. The generalization of this mapping is constructed in Appendix A.

2.3 3 + 1 dimensions

After eliminating the second-class constraints, one obtains a first-class Hamiltonian which looks exactly like the one of the 2 + 1-dimensional case:

$$H = \int d^3x [\frac{1}{2g^2} D_j \pi_j \cdot D_k \pi_k + \frac{1}{2} g^2 L_j \cdot L_j] , \tag{2.16}$$

and the set of first-class constraints

$$G_j^a(x) = \frac{1}{2}\epsilon_{jkl} F_{kl}^a(x) \approx 0 \quad 1 \leq i < j \leq 3 . \tag{2.17}$$

Although the system seems to be the obvious generalization to the 2 + 1-dimensional one, there is an essential difference: The constraints (2.17) are not all independent, but verify the Bianchi identity:

$$(D_j G_j)^a(x) = 0 , \ \forall a . \tag{2.18}$$

This implies that the set of constraints is reducible, containing only two independent functions. The counting of degrees of freedom then gives 1 for the number of physical dynamical variables ($1 = 3 - 2$). We mention that the elimination of the second-class constraints applies in a similar way to the general d-dimensional case, and that the corresponding Hamiltonian and constraints are the obvious generalizations of (2.16) and (2.17), respectively. Due to the existence of the Bianchi identity in general, the number of independent constraints in an arbitrary dimension is just enough to kill $d-1$ out of the d degrees of freedom in H, leading to only one physical variable, as it should be for a model which describes the dynamics of a scalar field.

3 Gauge invariant functionals

To construct the gauge invariant functionals, we make use of the concept of gauge invariant projection, defined as follows: Let $I = I[\pi, L]$ be an arbitrary functional of the phase-space fields. Then its gauge-invariant projection $\mathcal{P}(I)[\pi, L]$ is defined by:

$$\mathcal{P}(I)[\pi, L] = \frac{1}{\mathcal{N}} \int \mathcal{D}\omega \, I[\pi^{\omega}, L] \,, \tag{3.1}$$

where π^{ω} is the gauge-trasnformed of π by the gauge group element $\omega(x)$ (for example, in $2+1$ dimensions, $\pi_j^{\omega} = \pi_j + \epsilon_{jk} D_k \omega$) and the functional integration is over all the possible configurations for ω. The normalisation factor $\mathcal{N}$ is just the volume of the gauge group: $\mathcal{N} = \int \mathcal{D}\omega$. It is then easy to see that the gauge invariant projection of an arbitrary functional is indeed gauge invariant:

$$\mathcal{P}(I)[\pi^{\omega}, L] = \mathcal{P}(I)[\pi, L] \,, \tag{3.2}$$

and that $\mathcal{P}$ is a linear projection operator:

$$\mathcal{P}(\lambda_1 I_1 + \lambda_2 I_2) = \lambda_1 \mathcal{P}(I_1) + \lambda_2 \mathcal{P}(I_2) \quad , \quad \mathcal{P}^2 = \mathcal{P} \ , \ \forall I \,. \tag{3.3}$$

A functional F is gauge invariant if and only if $\mathcal{P}(F) = F$. This can be shown to be equivalent to saying that F belongs to the image of $\mathcal{P}$. We then construct the most general gauge-invariant functional by applying $\mathcal{P}$ to an arbitrary functional.

In $2+1$ dimensions, we further decompose the momentum as

$$\pi_j(x) = D_j \alpha(x) + \epsilon_{jk} D_k \beta(x) \,, \tag{3.4}$$

(where α and β are scalar and pseudoscalar, respectively) to show that

$$\begin{aligned} \mathcal{P}(I)[\pi, L] &= \mathcal{P}(I)[\alpha, \beta, L] = \frac{1}{\mathcal{N}} \int \mathcal{D}\omega \; I[D_j \alpha + \epsilon_{jk} D_k(\beta + \omega), L] \\ &= I[\alpha, 0, L] \end{aligned} \tag{3.5}$$

where the last line was obtained by performing the shift $\omega \to \omega - \beta$. (3.5) shows that any gauge invariant functional is independent of β; the reciprocal is immediate. The conclusion can be put as follows: The general gauge invariant functional depends arbitrarily on L, an on π only through the combination $D_j \pi_j$.

This result is generalizable to $3+1$ dimensions. F is shown to depend only on $D_j \pi_j$ and L_j, by using the same argument as in the $2+1$ case. The decomposition of π_j is now

$$\pi_j(x) = D_j \alpha(x) + \epsilon_{jkl} D_k \beta_l(x) \,, \tag{3.6}$$

and the β dependence is removed as before by a shift in ω. The only difference appears in the actual construction of the projection operator, which appears to be ill-defined at first sight. This is so because the gauge transformations in $d = 3$:

$$\pi_j^\omega = \pi_j + \epsilon_{jkl} D_k \omega_l , \tag{3.7}$$

are invariant under $\omega_j(x) \to \omega_j(x) + D_j\lambda(x)$, for any λ. This produces an infinite factor when one integrates over ω in the definition (3.1) of $\mathcal{P}(I)$. Of course, this factor is also present in $\mathcal{N}$, but to explicitly cancel them on needs to 'fix the gauge' for the integration over ω. A convenient way to do that is by using the Faddeev-Popov trick, which gives the 'gauge fixed' projector

$$\begin{aligned} \mathcal{P}(I)[\pi, L] &= \frac{1}{\int \mathcal{D}\omega \det M_f[\omega]\delta[f(\omega)]} \\ &\times \int \mathcal{D}\omega \det M_f[\omega]\delta[f(\omega)] I[D_j\alpha + \epsilon_{jkl}D_k\omega_l] , \end{aligned} \tag{3.8}$$

where $M_f[\omega] = \frac{\delta}{\delta\lambda} f(\omega^\lambda)$. We have seen that the gauge invariant functionals depend on $D_j\pi_j$ and L_j (the result is indeed true in any number of dimensions). However, there is still a degree of redundancy in this description because one is interested only in gauge invariant functions *on-shell*, i.e., on the surface $F_{jk}(L) = 0$. Thus we do not need the full L_j, but only its restriction to the constraint surface. As it was shown in ref. [8], it is possible to solve that kind of equation using a perturbative approach. The main result we need to recall is that that perturbative expansion allows one to express L_j as a function of the scalar $\partial_j L_j$ only. Then we obtain a more symmetrical description in terms of the gauge-invariant, scalar variables:

$$(D_j\pi_j)^a(x) \ , \ (D_jL_j)^a(x) = \partial_j L_j^a(x) \, . \tag{3.9}$$

(3.9) is true in any number of dimensions. Their equations of motion link each other:

$$\begin{aligned} \frac{\partial}{\partial t}(D_j\pi_j) &= -g^2\partial_j L_j \quad , \quad \frac{\partial}{\partial t}(\partial_j L_j) = -g^{-2}\partial_j D_j(D_k\pi_k) \\ F_{jk}(L) &\approx 0 , \end{aligned} \tag{3.10}$$

(where we have included the constraints). They imply the second order equations

$$(\partial_t^2 - \partial_j D_j)D_k\pi_k = 0 \ , \ (\partial_t^2 - \partial_j D_j)\partial_k L_k = 0, \tag{3.11}$$

which shows the scalar nature of the (only) physical degree of freedom.

4 Dirac brackets method

As an alternative to the previous approach, we apply here the 'Dirac brackets method' to the treatment of the first-class constraints in the 2 + 1 model (it can however be straightforwardly generalized to the $d+1$ model). It consists in constructing the Dirac brackets corresponding to the set of *second-class* constraints containing all the original first-class constraints plus a suitable set of gauge fixing conditions. We chose the canonical gauge fixing functions:

$$\chi^a(x) = \pi_2^a(x) = 0 . \tag{4.1}$$

The basic ingredient to calculate the Dirac brackets is the Poisson bracket between χ^a and $G^b(x)$: $\{\chi^a(x), G^b(y)\} = (D_1)^{ab}\delta(x-y)$. From this it follows that the only non-trivial Dirac's brackets between canonical variables are

$$\begin{aligned} \{L_1^a(x), \pi_1^b(y)\}_D &= \delta_{ab}\delta(x-y) , \\ \{L_2^a(x), \pi_1^b(y)\}_D &= \langle x, a \mid D_1^{-1} D_2 \mid y, b\rangle . \end{aligned} \tag{4.2}$$

The second one is a complicated non-local function. It is more convenient to take advantage of the results of the previous section to work with L_j and $D_1\pi_1$. Then the Dirac brackets become local

$$\begin{aligned} \{L_1^a(x), (D_1\pi_1)^b(y)\}_D &= -D_1^{ab}\delta(x-y) \\ \{L_2^a(x), (D_1\pi_1)^b(y)\}_D &= -D_2^{ab}\delta(x-y) . \end{aligned} \tag{4.3}$$

A Duality transformation for first-class systems

The kind of 'duality' that exists between the massless Yang-Mills theory and the polynomial version of the non-linear σ-model is a particular case of a more general concept, which we define in this Appendix. Let us consider a constrained dynamical system defined on a phase-space of coordinates $q_j, p_j,\ j = 1 \cdots N$, with first-class Hamiltonian H and a complete set of irreducible first-class constraints $G_a \approx 0,\ a = 1, \cdots, N$. We assume that the first-class constraints satisfy the closed algebra

$$\{G_a, G_b\} = g_{abc}(q,p)\, G_c \tag{A.1}$$

and regarding the Hamiltonian, we impose on it the requirement of having the structure

$$H = \frac{1}{2}\, F_a(q,p)\, F_a(q,p) \tag{A.2}$$

where the functions $F_a(q,p),\ a = 1, \cdots, N$ verify the relations

$$\begin{aligned} \{F_a, F_b\} &= f_{abc}(q,p)\ F_c \\ \{G_a, F_b\} &= \lambda_{abc}(q,p)\ F_c \end{aligned} \tag{A.3}$$

and λ is completely antisymmetric with respect to the last two indices. This implies that the Poisson bracket of H and each of the G_a's will be *strongly* equal to zero, what is stronger than what we need in a general first-class system. Indeed, Equations (A.2) and (A.3) select among all the possible first-class systems the class which admit a duality of the kind we are going to define.

The associated dual first-class system is defined on the same phase-space, and its Hamiltonian and constraints (denoted with a tilde) are defined by

$$\tilde{H} = \frac{1}{2} G_a G_a \quad , \quad \tilde{G}_a = F_a \approx 0 \, , \tag{A.4}$$

where F_a and G_a are the ones introduced in (A.1), (A.2) and (A.3). We then verify that the new system is also first-class, since

$$\begin{aligned} \{\tilde{G}_a, \tilde{G}_b\} &= \tilde{g}_{abc}(q,p)\, \tilde{G}_c \\ \{\tilde{G}_a, \tilde{F}_b\} &= V_{ab}(q,p)\, \tilde{G}_b \, , \end{aligned} \tag{A.5}$$

where:

$$\begin{aligned} \tilde{g}_{abc}(q,p) &= f_{abc}(q,p) \\ V_{ab}(q,p) &= \lambda_{abc}(q,p)\, G_c(q,p) \, . \end{aligned} \tag{A.6}$$

Thus evidently this mapping leaves the first-class nature of the system invariant. Note however, that the irreducibility of the new constraints is not guaranteed. That will depend upon the particular form of the F_a's. An interesting property of the new system is that, because of (A.3),

$$\{G_a, \tilde{G}_b\} \approx 0 \;\; \forall a, b \, , \tag{A.7}$$

which proves that the G_a's constitute a set of M independent gauge invariant functions.

Note that the transformation we defined is not necessarily involutive; to guarantee that we would need a completely antisymmetric λ_{abc} in Equation (A.3).

Acknowledgements

C. D. F. was supported by an European Community Postdoctoral Fellowship. T. M. was supported in part by the Daiwa Anglo-Japanese Foundation. We also would like to express our acknowledgement to Dr. I. J. R. Aitchison for his kind hospitality.

References

[1] For early references, see for example:
M. Gell-Mann and M. Lévy (1960), *Nuovo Cimento*, **16** 705;
S. Weinberg (1968), *Phys. Rev.*, **166** 1568;
J. Schwinger (1968), *Phys. Rev.*, **167** 1432.

[2] J. Zinn-Justin (1993), *Quantum Field Theory and Critical Phenomena*, Oxford University Press, Second Edition.

[3] S. Gasiorowicz and D. Geffen (1969), *Rev. Mod. Phys.*, **41** 531;
S. Coleman, J. Wess and B. Zumino (1969), *Phys. Rev*, **177** 2239.

[4] D. Z. Freedman and P. K. Townsed (1981), *Nucl. Phys.*, B**177** 282.

[5] G. L. Demarco, C. D. Fosco and R. C. Trinchero (1992), *Phys. Rev. D*, **45** 3701.

[6] P. A. M. Dirac (1950), *Can. J. Math.*, **2** 129; (1951), *ibid.*, **3** 1; (1958), *Proc. Roy. Soc. (London)*, A**246** 326; (1967), *Lectures on Quantum Mechanics*, Yeshiva University, New York, Academic Press.

[7] H. Sugawara (1968), *Phys. Rev.*, **120** 1659.

[8] A. A. Slavnov (1971), *Nucl. Phys.*, B**31** 301.

[9] C. Itzykson and J. B. Zuber (1986), *Quantum Field Theory*, McGraw-Hill.

[10] W. Siegel (1980), *Phys. Lett.*, B**93** 170;
J. Thierry-Mieg (1990), *Nucl. Phys.*, B**335** 334.

A manifestly gauge-invariant approach to quantum theories of gauge fields

Abhay Ashtekar, Jerzy Lewandowski, Donald Marolf, José Mourão and Thomas Thiemann

Abstract

In gauge theories, physical histories are represented by space-time connections modulo gauge transformations. The space of histories is thus intrinsically non-linear. The standard framework of constructive quantum field theory has to be extended to face these *kinematical* non-linearities squarely. We first present a pedagogical account of this problem and then suggest an avenue for its resolution.

1 Introduction

As is well-known, for over 40 years, quantum field theory has remained in a somewhat peculiar situation. On the one hand, perturbative treatments of realistic field theories in four space-time dimensions have been available for a long time and their predictions are in excellent agreement with experiments. It is clear therefore that there is something "essentially right" about these theories. On the other hand, their mathematical status continues to be dubious in all cases (with interactions), including QED. In particular, it is generally believed that the perturbation series one encounters here can be at best asymptotic. However, it is not clear what exactly they are asymptotic to.

This overall situation is in striking contrast with, for example, non-relativistic quantum mechanics. There, we know well at the outset what the Hilbert space of states is and what the observables are. In physically interesting models, we can generally construct the Hamiltonian operator and show that it is self-adjoint. We take recourse to perturbation theory mainly to calculate its eigenvalues and eigenvectors. Therefore, if the perturbation series turns out not to be convergent but only asymptotic, we know what it is asymptotic to. The theories exist in their own right and perturbative methods serve as approximation techniques to extract answers to physically interesting questions. In realistic quantum field theories, we do not yet know if there is an underlying, mathematically meaningful framework whose predictions are mirrored in the perturbative answers. When one comes to QCD, the problem is even more severe. Now, it is obvious from observations that the physically relevant phase of the theory is the one in which quarks and gluons are confined. And this phase lies beyond the grasp of standard perturbative treatments.

To improve this situation, the program of constructive quantum field theory was initiated in the early seventies. This approach has had remarkable success in certain 2 and 3 dimensional models. From a theoretical physics perspective, the underlying ideas may be summarized roughly as follows. Consider, for definiteness, a scalar field theory. The key step then is that of giving meaning to the Euclidean functional integrals by defining a rigorous version $d\mu$ of the heuristic measure "$[\exp(-S(\phi))] \prod_x d\phi(x)$" on the space of histories of the scalar field, where $S(\phi)$ denotes the action governing the dynamics of the model. The appropriate space of histories turns out to be the space $\mathcal{S}'$ of (tempered) distributions on the Euclidean space-time and regular measures $d\mu$ on this space are in one to one correspondence with the so-called generating functionals χ, which are functionals on the Schwarz space $\mathcal{S}$ of test functions satisfying certain rather simple conditions. (Recall that the tempered distributions are continuous linear maps from the Schwarz space to complex numbers.)

Thus, the problem of defining a quantum scalar field theory can be reduced to that of finding suitable measures $d\mu$ on $\mathcal{S}'$, or equivalently, "appropriate" functionals χ on $\mathcal{S}$. Furthermore, there is a succinct set of axioms –due to Osterwalder and Schrader– which spells out the conditions that χ must satisfy for it to be "appropriate," i.e., for it to lead to a consistent quantum field theory which has an associated Hilbert space of states, a Hamiltonian, a vacuum and an algebra of observables. This strategy has led to the rigorous construction of a number of interesting theories in 2 and 3 dimensions such as the $\lambda\phi^4$ and the Yukawa models and to an understanding of their relation to perturbative treatments. A more striking success of these methods is that they have led to a rigorous construction of the Gross-Neveau model in 3-dimensions which, being non-renormalizable, fails to exist perturbatively in the conventional sense.

These successes are remarkable. However, we believe that the framework has an important limitation: As in the heuristic, theoretical physics treatments [1, 2], it is based on the assumption that the theories under consideration can be considered to be *kinematically linear* [3, 4, 5]. That is, even though dynamical non-linearities are properly incorporated, it is assumed that the *space $\mathcal{S}'$ of histories is a vector space.* This assumption permeates the whole framework. In particular, in their standard form, all of the key Osterwalder-Schrader axioms use this property. Now, for unconstrained systems –such as the $\lambda\phi^4$-model– this assumption is not restrictive. However, for constrained systems it is generally violated. An outstanding example is provided by the Yang-Mills theory. Now, because of the presence of constraints, the system has gauge freedom and the space of physically distinct histories is provided by $\mathcal{A}/\mathcal{G}$, the space of connections on the Euclidean space-time modulo gauge transformations. In dimensions $d+1>2$, this is a highly non-linear space with complicated topology. Rather than facing this non-linearity squarely, one often resorts to gauge fixing, ignores global problems such as those associated with Gribov ambiguities, obtains a linear space and proceeds to apply the standard techniques of constructive quantum field theory. To an outsider the disparity between the "roughness" with which $\mathcal{A}/\mathcal{G}$ is steam-rollered into a linear space and the sophistication with which functional analysis is then used to construct measures seems rather striking. It is natural to ask if one can not modify the general framework itself and tailor it to the kinematical non-linearities of $\mathcal{A}/\mathcal{G}$.

The purpose of this contribution is to suggest an avenue towards this goal. (For earlier work with the same motivation, see [6].) We should emphasize, however, that ours is only an approach: we will not be able to present a definitive generalization of the Osterwalder-Schrader axioms. Furthermore, the key steps of our framework are rather loose. However, it *is* tailored to facing the kinematical non-linearities of gauge theories squarely from the very beginning.

None of the authors is an expert in constructive quantum field theory. The main ideas behind this approach came rather from an attempt to construct quantum general relativity non-perturbatively. Consequently, certain notions from quantum gravity –such as diffeomorphism invariance– play a non-trivial role in the initial stages of our constructions. This is a strength of the framework in the context of diffeomorphism invariant gauge theories. Examples are: the Yang-Mills theory in 2 dimensions which is invariant under area-preserving diffeomorphisms and the Chern-Simons theories in 3 dimensions and the Husain-Kuchař model [7] in 4 dimensions which are invariant under all diffeomorphisms. In higher dimensional Yang-Mills theories, on the other hand, the action does depend on the background Euclidean or Minkowskian geometry; it is not diffeomorphism invariant. In the general program sketched here, this geometry *is* used in subsequent steps of our constructions. However, we expect that, in a more complete and polished version, it would play an important role from the beginning. It is clear that considerable work is still needed to make the framework tight and refined, tailored more closely to quantum Yang-Mills fields. As the title of the contribution indicates, our primary aim is only to suggest a new approach to the problem, thereby initiating a re-examination of the appropriateness of the "standard" methods beyond the kinematically linear theories.

This contribution is addressed to working theoretical –rather than mathematical– physicists. Therefore, the presentation will be somewhat pedagogical. In particular, *we will not assume prior knowledge of the methods of constructive quantum field theory.* We begin in section 2 with a summary of the idea and techniques used in this area and indicate in particular how the appropriate measure is constructed for the $\lambda\phi^4$ model. As indicated above, this discussion will make a strong use of the kinematical linearities of models considered. In section 3, we turn to gauge theories and show how certain recent advances in the development of calculus on the space of connections modulo gauge transformations can be used to deal with the intrinsic kinematical non-linearities. In section 4, we indicate how these techniques can be used for Yang-Mills theories. In particular, we will construct the 2-dimensional Euclidean Yang-Mills theory and indicate its relation to the Hamiltonian framework. In section 5, we summarize the main results, point out some of the strengths and the limitations of this approach and discuss directions for further work.

2 Kinematically linear theories

Several of the ideas we will use in the construction of measures on infinite dimensional non-linear spaces are similar (but not identical!) to those used in the linear case. Therefore, in this section we will recall from [3, 4, 8, 9] some well known results about measures on infinite dimensional linear spaces in the context of constructive quantum field theory.

Let us consider an Euclidean scalar field theory on flat Euclidean space-time $M = I\!R^{d+1}$. It is natural to choose as the space $\mathcal{S}H$ of *classical* histories of the theory the *linear* space of suitably regular (say C^2 and rapidly decreasing at infinity) scalar field configurations; $\mathcal{S}H = \{\phi(x)\}$. The dynamics of the theory is determined by the action functional S on $\mathcal{S}H$

$$S(\phi) = \int_{I\!R^{d+1}} \left(\frac{1}{2}\partial_\mu\phi(x)\partial^\mu\phi(x) + V(\phi(x))\right) d^{d+1}x \ , \tag{2.1}$$

where $V(\phi(x))$ denotes the self-interaction potential (which is assumed to be bounded from below). In the "classical" theory [1] we are interested in *points* in $\mathcal{S}H$ (i.e. particular histories) that correspond to the extrema of S, i.e. in solutions of the (Euclidean) equation of motion

$$\Delta\phi - \frac{\partial V}{\partial\phi} = 0 \ . \tag{2.2}$$

If V is cubic or higher order in ϕ the equation (2.2) is non-linear and we have an example of a dynamically non-linear theory with a linear space of

[1]We used quotation marks in the word "classical" because, strictly speaking, the solutions of the Euclidean equations of motion do not play a direct role in classical physics; they have physical interpretation only in the semi-classical approximation.

histories. In our terminology, this is an example of a *kinematically linear* but *dynamically non-linear* theory. Theories of non-abelian gauge fields, on the other hand, are examples of theories in which non-linearities are present already at the kinematical level.

In quantum field theory, the interest lies not in particular histories satisfying (2.2) but in summing over all histories, i.e. in defining measures on SH that correspond to the heuristic expression [3, 4, 5]

$$d\mu(\phi) = \text{``}\frac{1}{Z} e^{-S(\phi)} \prod_{x \in I\!R^{d+1}} d\phi(x)\text{''} \ . \tag{2.3}$$

We will begin in section 2.1 with a brief review of how measures are constructed on infinite dimensional linear spaces. In section 2.2, these techniques will be applied to two illustrative examples in constructive quantum field theory; the massive free scalar field in $d+1$ dimensions and the $\lambda\phi^4$-model in 2 dimensions.

2.1 Integration on SH

We will first present an "algebraic" approach to integration and then summarize the situation from a measure-theoretic viewpoint. In the algebraic approach, the main idea is to reduce the problem of integration over infinite dimensional spaces to a series of integrations over finite dimensional spaces by judiciously choosing the functions one wants to integrate.

In a linear space like SH the simplest functions that one can introduce are the linear ones. Let S denote the space of all test (or smearing) functions on $I\!R^{d+1}$, i.e. the space of all infinitely differentiable functions which fall off sufficiently rapidly at infinity. S is called Schwarz space. Its elements $e \in S$ can be used to *probe* the structure of scalar fields $\phi \in SH$ through linear functions F_e on SH defined by

$$F_e(\phi) = \int_{I\!R^{d+1}} \phi(x) e(x) \, d^{d+1}x \ ; \tag{2.4}$$

the test field e probes the structure of ϕ because it captures part of the information contained in ϕ, namely, the "component of ϕ along e". We can probe the behaviour of ϕ in the neighbourhood of a point in M by choosing test fields e_n supported in that neighbourhood.

To begin with, we want to define integrals of such simple functions. Let $e_1, ..., e_n$ denote arbitrarily chosen but fixed linearly independent probes. Consider the projection $p_{e_1,...,e_n}$ they define

$$\begin{aligned} p_{e_1,...,e_n} : SH &\rightarrow I\!R^n \\ \phi &\mapsto (F_{e_1}(\phi), ..., F_{e_n}(\phi)) \ . \end{aligned} \tag{2.5}$$

Next, consider functions f on $\mathcal{S}H$ that depends on ϕ only through their "n-components" $F_{e_1}(\phi), ..., F_{e_n}(\phi)$, i.e. functions of the type

$$f(\phi) = \tilde{f}\,(F_{e_1}(\phi), ..., F_{e_n}(\phi)) \tag{2.6}$$

or equivalently

$$f = p^*_{e_1,...,e_n} \tilde{f} \,, \tag{2.7}$$

where $\tilde{f}$ is a (well-behaved) function on $I\!R^n$. The function f is said to be *cylindrical* with respect to the finite dimensional subspace $V_{e_1,...,e_n}$ (of $\mathcal{S}$) spanned by the probes $\{e_1, ..., e_n\}$. These are the functions we first want to integrate.

This task is easy because the cylindrical functions are "fake" infinite dimensional functions: although defined on $\mathcal{S}H$, their "true" dependence is only on a finite number of variables. Fix a (normalized, Borel) measure $d\mu_{e_1,...,e_n}$ on $I\!R^n$ and simply define the integral of f over $\mathcal{S}H$ to be the integral of $\tilde{f}$ over $I\!R^n$ with respect to $d\mu_{e_1,...,e_n}$:

$$\int_{\mathcal{S}H} f(\phi) d\mu(\phi) := \int_{I\!R^n} \tilde{f}(\eta_1, ..., \eta_n)\; d\mu_{e_1,...,e_n}(\eta_1, ..., \eta_n) \,. \tag{2.8}$$

Next, in order to be able to integrate functions cylindrical with respect to *any* finite dimensional subspace of the probe space, we select, for every collection $\{e_1, ..., e_n\}$ of linearly independent probes, and every $n \in I\!N$, a measure $d\mu_{e_1,...,e_n}$ on $I\!R^n$ and define the integral of that cylindrical function over $\mathcal{S}H$ by (2.8). This is the key technique in the algebraic approach.

The procedure seems trivial at first sight. There is, however, a catch. Since a function which is cylindrical with respect to a linear subspace V of $\mathcal{S}H$ is necessarily cylindrical with respect to a linear subspace V' if $V \subseteq V'$, the representation (2.6) is *not* unique. (Indeed, even for fixed V, the explicit form of (2.6) depends on the basis we use). For the left side of (2.8) to be well-defined, therefore, the finite dimensional measures $(d\mu_{e_1,...,e_n})$ must satisfy non-trivial consistency conditions; these serve precisely to ensure that the integral of $f(\phi)$ over $\mathcal{S}H$ is independent of the choice of a particular representation of f as a cylindrical function. When these consistency conditions are satisfied, $\mathcal{S}H$ is said to be equipped with a *cylindrical measure.*

Let us consider a simple but representative example of consistency conditions. Consider $e, \hat{e} \in \mathcal{S}$ and let V_1, V_2 be the one and two-dimensional spaces spanned by e and e, $\hat{e}$ respectively and f be the function on $\mathcal{S}H$, cylindrical with respect to V_1, given by

$$f(\phi) = \tilde{f}_1\,(F_e(\phi)) := \exp\,[i\lambda \int_{I\!R^{d+1}} e(x)\phi(x)\; d^{d+1}x] \,. \tag{2.9}$$

This function is clearly cylindrical also with respect to V_2, i.e. it is a function of F_e and $F_{\hat{e}}$ that just happens not to depend on $F_{\hat{e}}$:

$$f(\phi) = \tilde{f}_2\,(F_e(\phi), F_{\hat{e}}(\phi)) := \exp\,[i\lambda \int_{I\!R^{d+1}} e(x)\phi(x)\; d^{d+1}x] \,. \tag{2.10}$$

To obtain a cylindrical measure, therefore, f, $d\mu_e$,and $d\mu_{e,\hat{e}}$ must satisfy:

$$\int_{SH} f(\phi)\, d\mu(\phi) = \int_{I\!R} e^{i\lambda\eta} d\mu_e(\eta) = \int_{I\!R^2} e^{i\lambda\eta_1} d\mu_{e,\hat{e}}(\eta_1, \eta_2)\,. \quad (2.11)$$

It is easy to see that this equality holds for our choice of f *and* for any other integrable function, cylindrical with respect to V_1, if and only if the measures satisfy the following consistency condition

$$d\mu_e(\eta) = \int_{I\!R} d\mu_{e,\hat{e}}(\eta, \hat{\eta})\,. \quad (2.12)$$

A natural solution to the consistency conditions is obtained by choosing all the $d\mu_{e_1,...,e_n}$ to be normalized Gaussian measures. Then, the resulting $d\mu$ on SH is called a *Gaussian cylindrical measure.*

Associated with every cylindrical measure (not necessarily Gaussian) on SH, there is a function χ on the Schwarz space $\mathcal{S}$ of probes, called the Fourier transform of the measure by analysts and the generating function (with imaginary current) by physicists:

$$\chi(e) := \int_{SH} \exp\,[i \int_{I\!R^{d+1}} e(x)\phi(x) d^{d+1}x]\; d\mu(\phi) = \int_{I\!R} e^{i\eta} d\mu_e(\eta)\,. \quad (2.13)$$

To see that χ is the generating functional used in the physics literature, let us substitute the heuristic expression (2.3) of $d\mu$ in (2.13) to obtain:

$$\begin{aligned} \chi(e) \;=\; & \text{``}\frac{1}{Z}\int_{SH} \exp[i \int_{I\!R^{d+1}} e(x)\phi(x) d^{d+1}x] \;\times \quad (2.14) \\ & \exp[-\Big(\int_{I\!R^{d+1}} \frac{1}{2}\partial_\mu\phi(x)\partial^\mu\phi(x) + V(\phi(x))\Big)\, d^{d+1}x] \prod_{x\in I\!R^{d+1}} d\phi(x)\text{''}\,. \end{aligned}$$

From the properties of these heuristic generating functionals discussed in the physics literature, one would expect χ to contain the complete information about $d\mu$. This is indeed the case.

In fact, such generating functionals can serve as a powerful tool to *define* non-Gaussian measures $d\mu$. This is ensured by a key result in the subject, the *Bochner theorem* [8, 9], which has the following consequence :
Let $\chi(e)$ be a function on the Schwarz space $\mathcal{S}$ satisfying the following three conditions :

(*i*) $\chi(0) = 1$

(*ii*) χ is continuous in every finite dimensional subspace of$\mathcal{S}$

(*iii*) For every $e_1, ..., e_N \in \mathcal{S}$ and $c_1, ..., c_N \in \mathbb{C}$ we have

$$\sum_{i,j=1}^{N} \overline{c_i} c_j \chi(-e_i + e_j) \geq 0\,. \quad (2.15)$$

Then, there exists a unique cylindrical measure $d\mu$ on SH such that χ is its generating functional.

Thus, functions χ satisfying (2.15) are generating functionals of cylindrical measures on $\mathcal{S}H$ and every generating functional is of this form. This concludes the "algebraic" part of our discussion.

We now turn to the measure-theoretic part and ask if the above procedure for integrating cylindrical functions actually defines a genuine measure –i.e., a σ-additive set function (see below)– on $\mathcal{S}H$ or a related space. This issue is important because a proper measure theoretic understanding would enable us to integrate functions on $\mathcal{S}H$ that are genuinely infinite dimensional, i.e., depend on infinitely many probes. Indeed, the classical action is invariably such a function. It turns out that one *can* define genuine measures, but to do so, we have to extend the space $\mathcal{S}H$. For later convenience, we will proceed in two steps. First, we will present a "maximal" extension and arrive at a space $\overline{\mathcal{S}H}$ which serves as "the universal home" for all cylindrical measures on $\mathcal{S}H$. In practice, however, the space $\overline{\mathcal{S}H}$ is too large in that the measures of interest to constructive quantum field theory are generally supported on a significantly smaller subspace $\mathcal{S}'$ of $\overline{\mathcal{S}H}$ (which are still larger than $\mathcal{S}H$). Then, in the second step, we will discuss this "actual home," $\mathcal{S}'$, for physically interesting measures.

The universal home, $\overline{\mathcal{S}H}$, is simply the algebraic dual of $\mathcal{S}H$: the space of *all* linear functionals on the probe space $\mathcal{S}$. This space is "very large" because we have not required the maps to be continuous in any topology on $\mathcal{S}$. It is easy to check that $\mathcal{S}$ also serves as the space of probes for $\overline{\mathcal{S}H}$ and that, given a consistent family of measures $d\mu_{e_1,\dots,e_n}$ that defines a cylindrical measure $d\mu$ on $\mathcal{S}H$, it also defines a cylindrical measure, say $d\bar{\mu}$, on $\overline{\mathcal{S}H}$. ($\overline{\mathcal{S}H}$ is the "largest" space for which this result holds.) Now, for a cylindrical measure to define a genuine, infinite dimensional measure, it has to be extendible in the following sense. For a cylindrical measure, the measurable sets are all cylindrical, i.e., inverse images, under projections $p_{e_1,\dots,e_n}$ of (2.5), of measurable sets in $(IR^n, d\mu_{e_1,\dots,e_m})$. The measure of a cylindrical set is just the measure of its image under the projection. Now, for $d\bar{\mu}$ to be a genuine measure, it has to be σ-additive, i.e., the measure of a *countable* union of non-intersecting measurable sets has to equal the sum of their measures. Unfortunately, although the union of a finite number of cylindrical sets is again cylindrical, in general the same is *not* true for a countable number. Thus, the question is whether we can add countable unions to our list of measurable sets and *extend* $d\bar{\mu}$ consistently. It turns out that *every* cylindrical measure on $\overline{\mathcal{S}H}$ can be so extended. (This is in general *not* true for $\mathcal{S}H$; even the Gaussian cylindrical measures on $\mathcal{S}H$ may not be extendable to σ-additive measures thereon. In particular this happens for physically interesting measures.) This is why $\overline{\mathcal{S}H}$ can be regarded as the "universal home" for cylindrical measures.

Unfortunately, the space $\overline{\mathcal{S}H}$ is typically too big for quantum scalar field theories: The *actual* home of a given measure $d\bar{\mu}$ is generally a smaller space, say $\widehat{\mathcal{S}H}$, of better behaved generalized histories, in the sense that $\bar{\mu}(\overline{\mathcal{S}H} -$

$\widehat{\mathcal{S}H}) = 0$. (More precisely, every measurable set U such that $U \subset \overline{\mathcal{S}H} - \widehat{\mathcal{S}H}$ has zero measure $\bar{\mu}(U) = 0$.) Since $\widehat{\mathcal{S}H}$ has a richer structure, it is most natural (and, in practice, essential) to ignore $\overline{\mathcal{S}H}$ and work directly with $\widehat{\mathcal{S}H}$.

The key result which helps one determine the actual home of a measure is the Bochner-Minlos theorem [9]. The version of this theorem which is most useful in scalar field theories can be stated as follows :
σ-additive measures on the space $\mathcal{S}'$ of continuous linear functionals on the probe space $\mathcal{S}$ (equipped with its natural nuclear topology $\tau_{(n)}$) are in one to one correspondence with generating functions χ on $\mathcal{S}$, satisfying the conditions (i), (iii) of (2.15) and

$$(ii')\chi \text{ is continuous with respect to } \tau_{(n)}. \tag{2.16}$$

(The space $\mathcal{S}'$ is of course the space of tempered distributions.) Since the topology on $\mathcal{S}$ used in (ii') is weaker than that in (ii), the Bochner-Minlos theorem expresses the general trend that the *weaker* the topology with respect to which the generating function χ is continuous, the *smaller* is the support of the resulting measure $d\bar{\mu}$.

This concludes our discussion of mathematical preliminaries. Let us summarize. One can define cylindrical measures on $\mathcal{S}H$ which enable one to integrate cylindrical functions. However, to integrate "genuinely infinite dimensional" functions, one needs a genuine measure. The algebraic dual $\overline{\mathcal{S}H}$ of the space $\mathcal{S}$ of probes is the univeral home for such measures in the sense that every cylindrical measure on $\mathcal{S}H$ can be extended to a genuine measure on $\overline{\mathcal{S}H}$. In practice, however, physically interesting measures have a much smaller support $\widehat{\mathcal{S}H}$ which, however, is larger than $\mathcal{S}H$. (Indeed, typically $\mathcal{S}H$ has *zero* measure.) The Bochner-Milnos theorem provides a natural avenue to constructing such measures.

2.2 Scalar field theories

For a measure $d\bar{\mu}$ to correspond to a physically interesting quantum scalar field theory, it has to satisfy (some version) of the Osterwalder-Schrader axioms [3, 4]. These axioms guarantee that from the measure it is possible to construct the physical Hilbert space with a well defined Hamiltonian and Green functions with the appropriate properties. They are based on the assumption that the actual home for measures that correspond to quantum scalar field theories is the space $\mathcal{S}'$ of tempered distributions. Thus, the appropriate histories for quantum field theories are distributional. In fact, typically, the measure is *concentrated* on genuine distributions in the sense that $\bar{\mu}(\mathcal{S}' - \mathcal{S}H) = 1$. This is the origin of ultra-violet divergences: while the measure is concentrated on distributions, the action (2.1) is ill-defined if the histories are distributional.

We will denote the measures on $\mathcal{S}'$ by $d\hat{\mu}$. The Osterwalder-Schrader axioms restrict the class of possible measures. The most important of these axioms are: Euclidean invariance and reflection positivity. The first requires that the measure $d\hat{\mu}$ be invariant under the action of the Euclidean group on $I\!R^{d+1}$. Reflection positivity is the axiom that allows to construct a physical Hilbert space with a non-negative self-adjoint Hamiltonian acting on it. Let θ denote the time reflection, i.e. reflection with respect to the hyperplane

$$(x^0 = 0, x^1, ..., x^d) \ . \tag{2.17}$$

Consider the subspace $\mathcal{R}^+$ of $L^2(\mathcal{S}', d\hat{\mu})$ of (cylindrical) functions of the form

$$f(\hat{\phi}) = \sum_{j=1}^{N} c_j e^{i\hat{\phi}(e_j^+)} \ , \tag{2.18}$$

where $c_j \in \mathbb{C}$ and e_j^+ are arbitrary probes with support in the $x^0 > 0$ half-space. Then, the reflection positivity of the measure is the condition that

$$< \theta f, f >_{L^2} = \int_{\mathcal{S}'} [(\theta f)(\hat{\phi})]^* \, f(\hat{\phi}) \, d\hat{\mu}(\hat{\phi}) \ \geq 0 \ . \tag{2.19}$$

(If reflection invariance is satisfied for one choice of Euclidean coordinates, by Euclidean invariance of the measure, it is satisfied for any other choice.)

The Hilbert space and the Hamiltonian can then be constructed as follows. (2.19) provides a degenerate inner product $(.,.)$ on $\mathcal{R}^+$, given by

$$(f_1, f_2) = < \theta f_1, f_2 >_{L^2} \ . \tag{2.20}$$

Denoting by $\mathcal{N}$ the subspace of $(.,.)$-null vectors on $\mathcal{R}^+$ we obtain a Hilbert space $\mathcal{H}$ by taking the quotient $\mathcal{R}^+/\mathcal{N}$ and completing it with respect to $(.,.)$:

$$\mathcal{H} = \overline{\mathcal{R}^+/\mathcal{N}} \ . \tag{2.21}$$

The Euclidean invariance provides an unitary operator $\widehat{T}_t$, $t > 0$, of time translations on $L^2(\mathcal{S}', d\hat{\mu})$. It in turn gives rise to the self-adjoint contraction operator on $\mathcal{H}$:

$$e^{-Ht} \ , \ t > 0 \ , \tag{2.22}$$

where H is the Hamiltonian.

Finally, we can formulate the two axioms in terms of the generating functional. Let $E \ : \ I\!R^{d+1} \ \rightarrow \ I\!R^{d+1}$ denote an Euclidean transformation. Then the condition that the measure be Euclidean invariant is equivalent to demanding:

$$\hat{\chi}(e \circ E) = \hat{\chi}(e) \ , \ \text{for every } E \ . \tag{2.23}$$

Next, let us consider reflection positivity. From the definition of the generating functional $\hat{\chi}$ (2.13) (here with $\mathcal{SH} \equiv \mathcal{S}'$) it follows that the condition of reflection positivity is equivalent to

$$\sum_{i,j=1}^{N} \bar{c}_i c_j \hat{\chi}(e_j^+ - \theta e_i^+) \ \geq \ 0 \tag{2.24}$$

for all $N \in I\!N$, $c_1, ..., c_N \in \mathbb{C}$ and for all $e_1^+, ..., e_N^+ \in \mathcal{R}^+$.

To summarize, constructive quantum field theory provides an elegant and compact characterization of quantum field theory as a measure $d\hat{\mu}$ on $\mathcal{S}'$ or as a (generating) function $\hat{\chi}$ on the space $\mathcal{S}$ of test functions, satisfying certain conditions. So all the work can be focussed on finding (or at least proving the existence of) appropriate $d\hat{\mu}$ or $\hat{\chi}$.

To conclude this section, we will provide two examples of such measures.

The first example is that of a free, massive scalar field on $I\!R^{d+1}$. Note first, that, it follows from the discussion in section 2.1 that a measure $d\hat{\mu}$ is Gaussian if and only if its generating functional is Gaussian, i.e. if and only if

$$\hat{\chi}(e) = \exp[-\frac{1}{2}(e, \widehat{C}e)] , \tag{2.25}$$

where $\widehat{C}$ is a positive definite, linear operator defined everywhere on $\mathcal{S}$. $\widehat{C}$ is called the *covariance* of the resulting Gaussian measure $d\hat{\mu}$. The free massive quantum scalar field theory corresponds to the Gaussian measure $d\hat{\mu}_m$ with covariance

$$\left(\widehat{C}_m e\right)(x) = \int_{I\!R^{d+1}} (-\Delta + m^2)^{-1}(x,y)\; e(y)\; d^{d+1}y , \tag{2.26}$$

where the integral kernel is defined by:

$$(-\Delta + m^2)^{-1}(x,y) = \frac{1}{(2\pi)^{d+1}} \int_{I\!R^{d+1}} d^{d+1}p\; \frac{e^{ip(x-y)}}{p^2 + m^2}.$$

Our next example is more interesting in that it includes interactions: the $\lambda\phi^4$ model in 2-dimensions. The classical action is now given by:

$$S(\phi) = \int_{I\!R^2} \left(\frac{1}{2}\partial_\mu\phi\partial^\mu\phi + \frac{m^2}{2}\phi^2 + \lambda\phi^4\right) d^2x \tag{2.27}$$

Therefore, the heuristic expression for the quantum measure is

$$\begin{aligned} d\mu_{\lambda,m} &= \frac{1}{Z_{\lambda,m}}\; e^{-\lambda\int_{I\!R^2}\hat{\phi}^4(x)d^2x}\; e^{-S_m(\hat{\phi})} \prod_{x\in I\!R^2} d\hat{\phi}(x) \\ &= \frac{1}{\widetilde{Z}_{\lambda,m}} \left(e^{-\lambda\int_{I\!R^2}\hat{\phi}^4(x)d^2x}\right) d\hat{\mu}_m(\hat{\phi}) , \end{aligned} \tag{2.28}$$

where S_m is the action of the free field with mass m and $d\hat{\mu}_m$ denotes the (rigorous) Gaussian measure discussed above. The problem with this expression is that while $d\hat{\mu}$ is concentrated on distributional connections, the factor in the exponent is ill-defined for distributional $\hat{\phi}$.

One way of trying to make sense of (2.28) is by substituting the Gaussian measure $d\hat{\mu}_m$ by a regulated Gaussian measure with a smaller support on which the integrand is well-defined. This can be achieved, e.g., by replacing the covariance C_m in (2.26) with the cutoff covariance C_m^k, given in the momentum space by [4]

$$\tilde{C}_m^k(p) = \frac{1}{(2\pi)^2} \frac{e^{-k(p^2+m^2)}}{p^2+m^2}, \; k > 0 \; . \tag{2.29}$$

Then the resulting measure $d\mu_m^k$ "lives" on the space $\mathcal{L}$ of *infinitely differentiable* functions that grow at most logarithmically at infinity (more precisely as $\sqrt{\ln(|x|)}$). (This is an extremely useful property. Nonetheless, we cannot just use this measure as the physical one because, among other reasons, it does not satisfy reflection positivity.) However, the function

$$\lambda \int_{I\!R^2} \phi^4(x) d^2x \tag{2.30}$$

is still infrared divergent (almost everywhere with respect to $d\mu_m^k$). Therefore, we have to put an infrared cutoff by restricting space-time to a box of volume V. Of course if we now take the regulators away, $V \to \infty$ and $k \to 0$, then we return to the ill-defined expression (2.28). Therefore, the question is whether it is possible to change $\lambda \int_{I\!R^2} \phi^4(x) d^2x$ by an interacting term with the same leading order dependence on ϕ but such that the limit, when the regulators $V \to \infty$, $k \to 0$, exists, is non trivial and satisfies the Osterwalder-Schrader axioms.

In 2-dimensions the answer is in the affirmative if we substitute (2.30) by [4]

$$\lambda \int_{I\!R^2} : \phi^4(x) : d^2x \; , \tag{2.31}$$

where $: \phi^4(x) :$ denotes normal ordering with respect to the Gaussian measure $d\mu_m^k$

$$: \phi^4(x) : = \; \phi^4(x) - 6C_m^k(0)\phi^2(x) + 3(C_m^k(0))^2 \; . \tag{2.32}$$

The expression for the regulated generating functional then reads

$$\chi_{m,\lambda}^{k,V} := \frac{1}{\tilde{Z}_{\lambda,m}^{k,V}} \int_{\mathcal{L}} \exp[i \int_V e\phi d^2x] \; \exp[-\lambda \int_V : \phi^4(x) : d^2x] \; d\mu_m^k(\phi) \; , \tag{2.33}$$

where $\mathcal{L}$ denotes the support of the measure (i.e., the space of C^∞ functions on $I\!R^2$ which grow at worst as $\sqrt{ln(|x|)}$). Finally, it can be now shown that the limit

$$\chi_{m,\lambda}(e) = \lim_{V\to\infty} \lim_{k\to 0} \chi_{m,\lambda}^{k,V}(e) \tag{2.34}$$

exists, is non-Gaussian and has the appropriate properties. Thus, while the extraordinarily difficult problem of rigorously constructing a quantum field theory is formulated succinctly in this approach as that of finding a suitable measure or generating functional, the actual task of finding physically interesting measures is correspondingly difficult.

Finally, note that the entire framework is "soaked in" kinematic linearity. The fact that the space $\mathcal{S}$ of probes and and home $\mathcal{S}'$ of measures are linear was exploited repeatedly in various steps.

3 Calculus on $\mathcal{A}/\mathcal{G}$

We now turn to gauge theories which are kinematically non-linear. Here, the classical space of histories is the infinite dimensional space $\mathcal{A}/\mathcal{G}$ of smooth connections modulo gauge transformations on the $d+1$ dimensional space-time M; $\mathcal{S}H = \mathcal{A}/\mathcal{G}$. We will assume the gauge group G to be a compact Lie group and, in specific calculations, take it to be $SU(2)$. In this section, we will summarize how one develops calculus on $\mathcal{A}/\mathcal{G}$ by suitably extending the arguments of section 2.1 [12, 15]. (Analogous methods for $d+1=2$ were first used in the second reference of [26]. For an alternative approach, see [6]). In the section 4, we will apply these ideas to a specific model along the lines of section 2.2

3.1 Cylindrical measures

The main idea ([11, 12]) is to substitute the linear duality between scalar field histories and test functions by the "non-linear duality" between connections and loops in M. This duality is provided by the parallel transport or holonomy map H, evaluated at a base point $p \in M$:

$$H : \ (\alpha, A) \to H_\alpha(A) = \mathcal{P}\exp(\oint_\alpha A.dl) \ \in \ G \ , \tag{3.1}$$

where α is a piecewise analytic loop in M and $A \in \mathcal{A}$. (For definiteness we will assume that A is explicitly expressed using (one of) the lowest dimensional representation(s) of the Lie algebra of G.) In view of this duality, loops will now be used to probe the structure of the space of connections. In the kinematically linear theories, the probes and the histories were objects of the same type; in the scalar field theories, for example, they were both functions on M. In gauge theories, the roles are played by quite different objects.

We will begin by specifying the precise structure of the space of probes. Fix a point $p \in M$ and consider only those loops α in M which are based at p. It is natural to define the following equivalence relation on the space of these loops:

$$\alpha' \sim \alpha \text{ iff } H_{\alpha'}(A) = H_\alpha(A), \ \forall A \in \mathcal{A} \ , \tag{3.2}$$

where $\mathcal{A}$ is the space of smooth connections on M. Denote the equivalence class by $[\alpha]_p$ and call it a (based) holonomic loop or a *hoop*. The set of all hoops, $\mathcal{H}G_p$, in M forms a group with respect to the product

$$[\alpha_1]_p \cdot [\alpha_2]_p = [\alpha_1 \circ_p \alpha_2]_p \ , \tag{3.3}$$

where $\alpha_1 \circ_p \alpha_2$ denotes the usual composition of loops at the base point p. For gauge theories, the hoop group $\mathcal{H}G_p$ will serve as the space of probes.

Let us now turn to connections. For simplicity, we will first consider the space $\mathcal{A}/\mathcal{G}_p$ where $\mathcal{G}_p$ is the subgroup of gauge transformations which are equal to the identity $1_G \in G$ at p. For each $\alpha \in \mathcal{H}G_p$ the holonomy H_α defines a G-valued function on $\mathcal{A}/\mathcal{G}_p$, i.e. a G-valued function on $\mathcal{A}$ which is invariant under $\mathcal{G}_p$-gauge transformations. These functions are sufficient to separate the points of $\mathcal{A}/\mathcal{G}_p$. That is, for every $[A_1]_p \neq [A_2]_p$ there exists a $\alpha \in \mathcal{H}G_p$ such that $H_\alpha(A_1) \neq H_\alpha(A_2)$, where $[A]_p$ denotes the $\mathcal{G}_p$ equivalence class of the connection A. In this sense, hoops play the role of non-linear probes for histories $[A]_p \in \mathcal{A}/\mathcal{G}_p$. Finally, note that each smooth connection $A \in \mathcal{A}$ defines a homomorphism H of groups,

$$H_.(A): \mathcal{H}G_p \rightarrow G; \quad \alpha \rightarrow H_\alpha(A), \tag{3.4}$$

which is smooth in an appropriate sense ([16]).

As in section 2.2, we can now define cylindrical functions and cylindrical measures on $\mathcal{A}/\mathcal{G}_p$ using the "probe functions" F_α defined by $F_\alpha([A]_p) := H_\alpha(A)$. Our first task is to introduce the analogs of the projections (2.5). For this, we need the notion of strongly independent hoops. Following [12], we will say that hoops $[\beta_i]_p$ are strongly independent if they have representative loops β_i such that each contains an open segment that is traced exactly once and which intersects other representatives at most at a finite number of points. The notion of strong independence turns out to be the appropriate non-linear analog of the linear independence of probes used in section 2.1. In particular, we have the following result. Given n strongly independent hoops, $[\beta_i]_p, i = 1, ..., n$, there exists a set of projections, $p_{\beta_1,...,\beta_n} : \mathcal{A}/\mathcal{G}_p \rightarrow G^n$, given by:

$$p_{\beta_1,..,\beta_n}([A]_p) = (H_{\beta_1}(A), .., H_{\beta_n}(A)) \tag{3.5}$$

which are surjective [12].

Now, a function f on $\mathcal{A}/\mathcal{G}_p$ is called *cylindrical* with respect to the subgroup of $\mathcal{H}G_p$ generated by $\beta_1, .., \beta_n$ if it is the pull-back by $p_{\beta_1,..,\beta_n}$ of a (well-behaved) complex valued function $\tilde{f}$ on G^n

$$f([A]_p) = \tilde{f}(H_{\beta_1}(A), .., H_{\beta_n}(A)). \tag{3.6}$$

Consider now a family $\{d\mu_{\beta_1,..,\beta_n}\}$ of (positive, normalized) measures on G^n, one for each set $\{\beta_1, .., \beta_n\}$, $n \in I\!N$ of strongly independent hoops. As in section 2.1, this family of measures on finite dimensional spaces allows us to define in a unique way a *cylindrical measure* $d\mu$ on the infinite dimensional space $\mathcal{A}/\mathcal{G}_p$, provided that appropriate consistency conditions are satisfied. These conditions are again a consequence of the fact that the representation (3.6) of f as a cylindrical function is not unique. If the consistency conditions are satisfied, then the family $(d\mu_{\beta_1,..,\beta_n})$ defines a unique cylindrical measure

$d\mu$ on $\mathcal{A}/\mathcal{G}_p$ through

$$\int_{\mathcal{A}/\mathcal{G}_p} d\mu([A]_p) f([A]_p) := \int_{G^n} d\mu_{\beta_1,..,\beta_n}(g_1,..,g_n)\tilde{f}(g_1,..,g_n) \qquad (3.7)$$

where f is the pull-back of $\tilde{f}$ as in (3.6).

In the linear case, Gaussian measures on $I\!R^n$ provided a natural way to meet the consistency conditions. In the present case, one can use the Haar measure on G^n. More precisely, the consistency conditions are satisfied if one chooses [12]

$$d\mu^0_{\beta_1,..,\beta_n}(g_1,..,g_n) = d\mu_H(g_1)..d\mu_H(g_n)\ , \qquad (3.8)$$

where $d\mu_H$ is the normalized Haar measure on G. This family $(d\mu^0_{\beta_1,..,\beta_n})$ leads to a measure cylindrical $d\mu^0$ on $\mathcal{A}/\mathcal{G}_p$ which is in fact invariant under the action of diffeomorphisms [2] of M.

3.2 "Universal home" and "actual home" for measures

As in the linear case, not every cylindrical measure on $\mathcal{A}/\mathcal{G}_p$ is a genuine, infinite dimensional measure. It turns out, however, that they can be extended to genuine measures on a certain completion, $\overline{\mathcal{A}/\mathcal{G}_p}$. This comes about as follows. Since $\mathcal{A}/\mathcal{G}_p$ is in one to one correspondence with the set of smooth homomorphisms from $\mathcal{H}G_p$ to G [16], a natural analog of the algebraic dual $\overline{\mathcal{S}H}$ of the space $\mathcal{S}$ of linear probes is now

$$\overline{\mathcal{A}/\mathcal{G}_p} := Hom(\mathcal{H}G_p, G), \qquad (3.9)$$

i.e., the set of *all* homomorphisms (without any continuity condition) from the hoop group to the gauge group [12]. Elements of $\overline{\mathcal{A}/\mathcal{G}_p}$ will be denoted by $\bar{A}$ and called *generalized connections.* Just as $\mathcal{S}$ continues to serve as the space of probes for $\overline{\mathcal{S}H}$ in the linear case, the group $\mathcal{H}G_p$ continues to provide non-linear probes for the space $\overline{\mathcal{A}/\mathcal{G}_p}$ of generalized connections. Furthermore, as in the linear case, every consistent family $(d\mu_{\beta_1,..,\beta_n})$ of measures on G^n that defines a cylindrical measure $d\mu$ on $\mathcal{A}/\mathcal{G}_p$ through (3.7) also defines a measure $d\bar{\mu}$ on $\overline{\mathcal{A}/\mathcal{G}_p}$ by

$$\int_{\overline{\mathcal{A}/\mathcal{G}_p}} \bar{f}(\bar{A})d\bar{\mu}(\bar{A}) = \int_{G^n} \tilde{f}(g_1,..,g_n)d\mu_{\beta_1,..,\beta_n}(g_1,..,g_n) \qquad (3.10)$$

where $\bar{f}$ is an arbitrary cylindrical function on $\overline{\mathcal{A}/\mathcal{G}_p}$ with respect to the subgroup of $\mathcal{H}G_p$ generated by $\beta_1,..,\beta_n$; $\bar{f}(\bar{A}) = \tilde{f}(\bar{A}(\beta_1),..,\bar{A}(\beta_n))$. Finally,

[2] Additional solutions of the consistency conditions leading to diffeomorphism invariant measures on $\mathcal{A}/\mathcal{G}_p$ were found by Baez in [13]. The resulting measures are sensitive to certain kinds of self-intersections of loops. Such measures are relevant for quantum gravity, formulated as a dynamical theory of connections [18],[19].

every measure $d\bar{\mu}$ on $\overline{\mathcal{A}/\mathcal{G}_p}$ defined as in (3.10) can be extended to a σ-additive measure on $\overline{\mathcal{A}/\mathcal{G}_p}$ and is thus a genuine, infinite dimensional measure [9, 12, 14].

From a physical point of view, however, we should still factor out by the gauge freedom *at* the base point p. A generic gauge transformation $g(.) \in \mathcal{G}$ acts on $\overline{\mathcal{A}/\mathcal{G}_p}$ simply by conjugation

$$(g \circ \bar{A})(\beta) = g(p) \cdot \bar{A}(\beta) \cdot g^{-1}(p) \tag{3.11}$$

The physically relevant space $\overline{\mathcal{A}/\mathcal{G}}$ is therefore the quotient of $\overline{\mathcal{A}/\mathcal{G}_p}$ by this action. The classical space of histories $\mathcal{A}/\mathcal{G}$ is naturally embedded in $\overline{\mathcal{A}/\mathcal{G}}$ through

$$[A] \rightarrow \{gH_.(A)g^{-1},\ g \in G\} \tag{3.12}$$

where $[A]$ denotes the gauge-equivalence class of the connection A. By integrating gauge-invariant functions on $\overline{\mathcal{A}/\mathcal{G}_p}$ with the help of $d\bar{\mu}$ we obtain a measure on $\overline{\mathcal{A}/\mathcal{G}}$ that we will denote also by $d\bar{\mu}$. This measure is of course just the push-forward of the measure on $\overline{\mathcal{A}/\mathcal{G}_p}$ under the canonical projection $\pi\ :\ \overline{\mathcal{A}/\mathcal{G}_p} \rightarrow \overline{\mathcal{A}/\mathcal{G}}$. Again, these measures $d\bar{\mu}$ on $\overline{\mathcal{A}/\mathcal{G}}$, associated with consistent families $(d\mu_{\beta_1,..,\beta_n})$, are always extendible to $\sigma-$additive measures. Thus, for gauge theories, $\overline{\mathcal{A}/\mathcal{G}}$ serves as the universal home for measures (for which the traces of holonomies are measurable functions) in the same sense that $\overline{\mathcal{SH}}$ is the universal home for measures in the linear case.

It is therefore natural to ask if there is a "non-linear" analog of the Bochner theorem. The answer is in the affirmative. In fact the arguments are now simpler because both $\overline{\mathcal{A}/\mathcal{G}_p}$ and $\overline{\mathcal{A}/\mathcal{G}}$ are compact Hausdorff spaces (with respect to the natural Gel'fand topologies). To see this explicitly, let us now restrict ourselves to the gauge group $G = SU(2)$. In this case, the Mandelstam identities imply that the vector space generated by traces of holonomies T_α is closed under multiplication. This in turn implies that the entire information about the measure is contained in the image of the "loop transform:"

$$\chi(\beta) = \int_{\overline{\mathcal{A}/\mathcal{G}}} \bar{T}_\beta([\bar{A}])\ d\bar{\mu}([\bar{A}])\ , \tag{3.13}$$

where $\bar{T}_\alpha[\bar{A}] = \frac{1}{2} tr \bar{A}[\alpha]$ is the natural extension to $\overline{\mathcal{A}/\mathcal{G}}$ of the "trace of the holonomy (or Wilson loop) function" on $\mathcal{A}/\mathcal{G}$. Note that (3.13) is the natural non-linear analog of the Fourier transform (2.13). The generating function χ is again a function on the space of probes which now happens to be the hoop group $\mathcal{H}G$ rather than the Schwarz space $\mathcal{S}$. From the normalization and the positivity of $d\bar{\mu}$ it is easy to see that $\chi(\alpha)$ satisfies

$$(i) \qquad \chi(p) = 1$$

$$(ii) \qquad \sum_{i,j=1}^{N} \bar{c}_i c_j [\chi(\beta_i \circ \beta_j) + \chi(\beta_i \circ \beta_j^{-1})] \geq 0,\ \forall \beta_i \in \mathcal{H}\ , \tag{3.14}$$

where p denotes the trivial (i.e., identity) hoop, $c_i \in \mathbb{C}$, are arbitrary complex numbers and $N \in \mathbb{N}$ is an arbitrary integer. Finally, the Riesz-Markov theorem [17] implies that every generating functional χ satisfying $\sum_i c_i \chi(\beta_i) = 0$ whenever $\sum_i c_i T_{\beta_i} = 0$ is the loop transform of a measure $d\bar{\mu}$ so that there is a one to one correspondence between positive, normalized (regular, Borel) measures on $\overline{\mathcal{A}/\mathcal{G}}$ and generating functionals on $\mathcal{H}G_p$ [11]. This result is analogous to the Bochner theorem in the linear case.

As in section 2.2, a given measure $\bar{\mu}_0$ on $\overline{\mathcal{A}/\mathcal{G}}$ may be supported on a smaller space of better behaved generalized connections (support or actual home for $d\bar{\mu}_0$). Indeed, just as $\overline{\mathcal{S}H}$ is "too large" for scalar field theories, we expect that $\overline{\mathcal{A}/\mathcal{G}}$ is "too large" for Yang-Mills theories. To see this, note that in (3.14), no continuity condition was imposed on the generating functional $\chi_{\bar{\mu}_0}$ (or, equivalently, the generating functionals are assumed to be continuous only with the discrete topology on $\mathcal{H}G_p$). Now, in Yang-Mills theories (with $d{+}1 > 2$), a background space-time metric *is* available and it can be used to introduce suitable topologies on $\mathcal{H}G_p$ which would be much weaker than the discrete topology. It would be appropriate to require that the Yang-Mills generating functional χ be continuous with respect to one of these topologies. Now, from our experience in the linear case, it seems reasonable to assume that the following pattern will emerge: the weaker the topology with respect to which χ is continuous, the smaller will be the support of $d\bar{\mu}$. Thus, we expect that, in Yang-Mills theories, physically appropriate continuity conditions will have to be imposed and these will restrict the support of the measure considerably. What is missing is the non-linear analog of the Bochner-Minlos theorem which can naturally suggest what the domains of the physically interesting measures should be.

The situation is very different in diffeomorphism invariant theories of connections such as general relativity. For these theories, diffeomorphism invariant measures such as $d\bar{\mu}^0$ are expected to play an important role. There are indications that the generating functionals of such measures will be continuous only in topologies which are much stronger than the ones tied to background metrics on M. Finally, since $d\bar{\mu}^0$ is induced by the Haar measure on the gauge group, it is very closely related to the measure used in the lattice gauge theories. Therefore, it is possible that even in Yang-Mills theories, it may serve as a "fiducial" measure –analogous to μ^k_m of the scalar field theory– in the actual construction of the physical measure. We will see that this is indeed the case in $d+1=2$.

4 Example: quantum Yang Mills theory in two dimensions

In this section we will show how the mathematical techniques introduced in section 3 can be used to construct Yang-Mills theory in 2 dimensions in the cases when the underlying space-time M is topologically IR^2 and when it is topologically $S^1 \times IR$. In particular, we will be able to show equivalence rigorous [3] between the Euclidean and the Hamiltonian theories; to our knowledge this was not previously demonstrated. The analysis of this case will also suggest an extension of the Osterwalder-Schrader axioms for gauge theories. Due to space limitations, however, we will present here only the main ideas; for details see [22].

4.1 Derivation of the continuum measure

We wish to construct the Euclidean quantum field theory along the lines of section 2.2. As indicated in section 3.2, the analog of the generating functional $\chi(e)$ on $\mathcal{S}$ is the functional $\chi(\alpha)$ on the hoop group $\mathcal{H}G_p$. In the present case, the heuristic expression of $\chi(\alpha)$ is:

$$\text{“}\chi(\alpha) = \int_{\mathcal{A}/\mathcal{G}} T_\alpha([A])\, e^{-S_{YM}([A])} \prod_{x\in M} d[A](x)\text{”}\,, \tag{4.1}$$

where $[A]$ is the gauge equivalence class to which the connection A belongs, $T_\alpha([A])$ is the trace of the holonomy of A around the closed loop α and S_{YM} is the Yang-Mills action. To obtain the rigorous analog of (4.1), we proceed, as in section 2.2, in the following steps: i) regulate the action using suitable cut-offs; ii) replace $\mathcal{A}/\mathcal{G}$ by a suitable completion thereof; iii) introduce a measure on this completion with respect to which the regulated action is measurable; iv) carry out the integration; and, v) take the appropriate limits to remove the cut-offs.

To carry out the first step, we will use a lattice regularization. Introduce a finite, square lattice of total length L_x and L_τ in x and τ direction respectively, and lattice spacing a, using the Euclidean metric on M. Thus we have imposed both infrared and ultraviolet regulators. There are $(N_x+1)(N_\tau+1)$ vertices on that finite lattice in the plane (and $(N_\tau+1)N_x$ in the cylinder) where $N_x a := L_x$, $N_\tau a := L_\tau$. Denote the holonomy associated with a plaquette $\Box$ by $p_\Box$. There are no boundary conditions for the plane while on the cylinder we identify [24] the vertical link variables l associated with the open paths starting at the vertices $(0,\tau)$ and $(a\ N_x,\tau)$. The regularized action is then given by $S_{reg} = \beta S_W$ where $\beta = 1/(g_0 a^2)$, with g_0, the bare coupling

[3] Similar results were obtained via the heuristic Fadeev-Popov approach in [23].

constant, and S_W is the Wilson action given by:

$$S_W := \sum_{\square}[1 - \frac{1}{N}\mathrm{Re}\ \mathrm{tr}(p_{\square}))] , \tag{4.2}$$

where Re tr stands for "real part of the trace." Finally, we note certain consequences of this construction. Choose a vertex and call it the base point p. One can show that in two dimensions the based plaquettes form a complete set of loops which are independent in the sense that one can separately assign to each holonomy associated with a plaquette an arbitrary group element. This means that they can, in particular, be used as independent integration variables on the lattice. They can also be used to define projection maps of (3.5).

The next step is to find the appropriate extension of $\mathcal{A}/\mathcal{G}$ and a measure thereon. For this, we note that since $\overline{\mathcal{A}/\mathcal{G}}$ of section 3.2 is the space of all homomorphisms from $\mathcal{H}G_p$ to G, for any given lattice, the Wilson action (4.2) can be regarded as a bounded cylindrical function $\bar{S}_W$ on $\overline{\mathcal{A}/\mathcal{G}}$. Therefore, this function is integrable with respect to the measure $d\bar{\mu}^0$. Since the push-forward of $d\bar{\mu}^0$ under the projections $p_{\beta_1,...,\beta_n}$ of (3.5) is just the Haar measure on G^n, using for β_i the plaquettes, we have:

$$\begin{aligned}\chi^{a,L_x,L_\tau}(\alpha) &:= \frac{1}{Z}\int_{\overline{\mathcal{A}/\mathcal{G}}} \bar{T}_\alpha(\bar{A})\, e^{-\beta\bar{S}_W}[\bar{A}]\, d\bar{\mu}^0 \\ &= \frac{1}{NZ(a,L_x,L_\tau)}\prod_l \int_G e^{-\beta S_W}\, \mathrm{tr}(\prod_{l\in\alpha} H_l)\, d\mu_H(H_l)\end{aligned} \tag{4.3}$$

for any loop α contained in the lattice. Here $d\mu_H$ is the Haar measure on G and the partition function Z is defined through $\chi(p) = 1$ where, as before, p denotes the trivial hoop. Note that (4.3) is precisely the Wilson integral for computing the vacuum expectation value of the trace of the holonomy. Thus, the space $\overline{\mathcal{A}/\mathcal{G}}$ and the measure $d\bar{\mu}^0$ are tailored to the calculations one normally performs in lattice gauge theory.

We have thus carried out the first four steps outlined above. The last step, taking the continuum limit, is of course the most difficult one. It has been carried out for the general case $G = SU(N)$ [22]. For simplicity, however, in what follows, we will restrict ourselves to the Abelian case $G = U(1)$.

Consider the pattern of areas that the loop α creates on M and select simple loops β_i that enclose these areas. It can be shown that α can be written as the composition of the β_i and the completely horizontal loop c at "future time infinity". Let $k(\beta_i)$ and k be the effective winding numbers of the simple loops β_i, $i = 1, .., , n$ and of the homotopically non-trivial loop c respectively, in the simple loop decomposition of α (that is, the signed number of times that these loops appear in α). Define $|\beta_i|$ to be the number of plaquettes enclosed by β_i. Then, if we set

$$K_n(\beta) := \int_G [\exp -\beta(1 - \mathrm{Re}(g))]\ g^n\ d\mu_H(g), \tag{4.4}$$

the generating functional (on the plane or the cylinder) becomes

$$\chi(\alpha) = \prod_{i=1}^{n} [\frac{K_{k(\beta_i)}(\beta_i)}{K_0(\beta_i)}]^{|\beta_i|} \tag{4.5}$$

for $k = 0$ (and, in particular, for any loop in the the plane), and it vanishes identically otherwise. We can now take the continuum limit. The result is simply:

$$\chi(\alpha) = \lim_{a\to 0} \chi^{a,L_x,L_\tau}(\alpha) = \exp[-\frac{g_0^2}{2}\sum_{i=1}^{n} k(\beta_i)^2 Ar(\beta_i)]. \tag{4.6}$$

if $k = 0$, and $\chi(\alpha) = 0$ if $k \neq 0$, where $Ar(\beta)$ is the area enclosed by the loop β.

A number of remarks are in order. i) We see explicitly the area law in (4.6) (which would signal confinement in $d + 1 > 2$ dimensions). The same is true for non-Abelian groups. ii) As in the case of the $\lambda\phi^4$-model in 2-dimensions, no renormalization of the bare coupling g_0 was necessary in order to obtain a well-defined limit. This is a peculiarity of 2 dimensions. Indeed, in higher dimensions because the bare coupling does not even have the correct physical dimensions to allow for an area law. iii) It is interesting to note that the above expression is completely insensitive to the fact that we have not taken the infinite volume limit, $L_x, L_\tau \to \infty$ on the plane, and $L_\tau \to \infty$ on the cylinder. (The only requirement so far is that the lattice is large enough for the loop under consideration to fit in it.) Thus, the task of taking these "thermodynamic" limits is quite straightforward. iv) In higher dimensions, one can formulate a program for constuction of the measure along similar lines (although other avenues also exist). This may be regarded as a method of obtaining the continuum limit in the lattice formulation. The advantage is that if the limit with appropriate properties exists, one would obtain not only the Euclidean "expectation values" of Wilson loop functionals but also a genuine measure on $\overline{\mathcal{A}/\mathcal{G}}$, a Hilbert sapce of states and a Hamiltonian.

Finally, let us compare our result with those in the literature. First, there is complete agreement with [26]. However our method of calculating the vacuum expectation values, $\chi(\alpha)$, of the Wilson loop observables is somewhat simpler and more direct. Furthermore, it does not require gauge fixing or the introduction of a vector space structure on $\mathcal{A}/\mathcal{G}$ and we were able to treat the cases $M = I\!R^2$ and $M = S^1 \times I\!R$ simultaneously. More importantly, we were able to obtain a closed expression in the $U(1)$ case. Finally, the invariance of $\chi(\alpha)$ under area preserving diffeomorphisms is manifest in this approach; the huge symmetry group of the classical theory transcends to the quantum level. This important feature was not so transparent in the previous rigorous treatments (except for the second paper in [26]).

4.2 A proposal for constructive quantum gauge field theory

In this section, we will indicate how one might be able to arrive at a rigorous, non-perturbative formulation of quantum gauge theories in the continuum. The idea of course is to define a *quantum theory of gauge fields* to be a measure $d\bar{\mu}$ on $\overline{\mathcal{A}/\mathcal{G}}$ –whose support may be considerably smaller than $\overline{\mathcal{A}/\mathcal{G}}$– satisfying the analogs of the Osterwalder-Schrader axioms. These must be adapted to the kinematical non-linearity of gauge theories. We will discuss how the key axioms can be so formulated. As one would expect, they are satisfied in the 2-dimensional example discussed above. Using them, we will arrive at the Hamiltonian framework which turns out to be equivalent to that obtained from canonical quantization.

The two key axioms in the kinematically linear case were the Euclidean invariance and the reflection positivity. These can be extended as follows. Let $E : I\!R^{d+1} \to I\!R^{d+1}$ denote an Euclidean transformation. We require:

$$\chi(E \circ \alpha) = \chi(\alpha) \tag{4.7}$$

for all E in the Euclidean group. The axiom of reflection positivity also admits a simple extension. Let us construct the linear space $\mathcal{R}^+$ of complex-valued functionals $\Psi_{\{z_i\},\{\beta_i\}}$ on $\overline{\mathcal{A}/\mathcal{G}}$ of the form:

$$\Psi_{\{z_i\},\{\beta_i\}}(\bar{A}) = \sum_{i=1}^{n} z_i \ \bar{T}_{\beta_i}[\bar{A}] = \sum_{i=1}^{n} z_i \ \bar{A}[\beta_i] \ , \tag{4.8}$$

where $\beta_i \in \mathcal{H}G$ are independent in the sense of the previous subsection and have support in the positive half space, and z_i are arbitrary complex numbers. Then a measure $d\bar{\mu}$ on $\overline{\mathcal{A}/\mathcal{G}}$ will be said to satisfy the reflection positivity axiom if:

$$(\Psi, \Psi) :=< \theta\Psi, \Psi >:= \int_{\overline{\mathcal{A}/\mathcal{G}}} (\theta\Psi(\bar{A}))^* \Psi(\bar{A}) \ d\bar{\mu}(\bar{A}) \geq 0 \ , \tag{4.9}$$

where θ, as before, is the "time-reflection" operation [4]. The remaining Osterwalder-Schrader axioms can be extended to gauge theories in a similar manner [21, 22], although a definitve formulation is yet to emerge.

Let us now consider the 2-dimensional example for $G = U(1)$. Since the generating functional (4.6) is invariant under all area preserving diffeomorphisms, it is, in particular, invariant under the 2-dimensional Euclidean group on the plane [5] The verification of reflection positivity requires more work.

[4]This formulation is for the case when the gauge group is $U(1)$, or $SU(2)$. For $SU(N)$ $N > 2$, the product of traces of holonomies is not expressible as a linear combination of traces of holonomies. Therefore, the argument of the functionals χ contain $1, 2, ..., N-1$ loops. However, the required extension is immediate.

[5]On the cylinder the Euclidean group, of course, has to be replaced by the isometry group of the metric, that is, the group generated by the Killing vectors ∂_τ and ∂_x.

We will simultaneously carry out this verification *and* the construction of the physical Hilbert space. Let us apply the Osterwalder-Schrader algorithm, outlined in section 2.2, to construct the physical Hilbert space and the Hamiltonian. First, a careful analysis provides us the null space $\mathcal{N}$: its elements are functionals on $\overline{\mathcal{A}/\mathcal{G}}$ of the type:

$$\tilde{\Psi}[\bar{A}] = (\sum_{i=1}^{n} z_i\ \bar{T}_{\beta_i}[\bar{A}]) - (\sum_{i=1}^{n} z_i \chi(\beta_i))\ \bar{T}_p[\bar{A}]\ , \tag{4.10}$$

provided all β_i are homotopically trivial. The physical Hilbert space $\mathcal{H}$ is given by the quotient construction, $\mathcal{H} = \mathcal{R}^+/\mathcal{N}$. On the plane, it is simply the linear span of the trivial vector $T_p = 1$; $\mathcal{H}$ is one-dimensional.

Let us now consider the more interesting case of the cylinder. If one of β_i or β_j contains a homotopically trivial loop and the other does not, $(\theta\beta_i)^{-1} \circ \beta_j$ necessarily contains a homotopically non-trivial loop and the characteristic function (4.6) vanishes at that loop. Finally, if both loops are homotopically non-trivial, then $(\theta\beta_i)^{-1} \circ \beta_j$ necessarily contains a homotopically non-trivial loop unless the effective winding numbers of the non-trivial loops in β_i and β_j are equal. Therefore, the linear spans of T_α's with zero and non-zero k are orthogonal under $(.,.)$. The former was shown to coincide with the null space generated by T_p. To display the structure of the Hilbert space $\mathcal{H} = \mathcal{R}^+/\mathcal{N}$, let us introduce a horizontal loop γ at $\tau = 0$ with winding number one. Then, any loop with winding number $k = n$ can be written as the composition of γ^n and homotopically trivial loops. Therefore $\beta = \gamma^{-n} \circ \alpha$ has zero winding number. Finally, using $\theta\gamma = \gamma$, it is easy to show that $T_{\gamma^n \circ \beta} - \chi(\beta) T_{\gamma^n} \in \mathcal{N}$, implies that the Hilbert space $\mathcal{H}$ is the completion of the linear span of the vectors T_{γ^n} $n \in \mathbb{Z}$. The vectors $\psi_n := T_{\gamma^n}$ form an orthonormal basis in $\mathcal{H}$. Since the quotient of $\mathcal{R}^+$ by the null sub-space $\mathcal{N}$ has a positive definite inner-product, $(.,.)$, it is clear that the generating function χ satisfies reflection positivity. Finally, note that, since the loop γ probes the generalized connections $\bar{A}$ only at "time" zero, the final result is analogous to that for a scalar field where the Hilbert space constuction can also be reduced to to the fields at "time" zero.

Our next task is to construct the Hamiltonian H. By definition, H is the generator of the Euclidean time translation semi-group. Let $\gamma(\tau) := T(\tau)\gamma$ be the horizontal loop at time τ and set $\alpha = \gamma \circ \rho \circ \gamma(\tau)^{-1} \circ \rho^{-1}$, where ρ is the vertical path between the vertices of the horizontal loops. Then, we have $T_{\gamma(\tau)^n} = \chi(\alpha^n) T_{\gamma^n}$ as elements of $\mathcal{H}$, so that

$$\begin{aligned} (\psi_m,\ T(\tau)\psi_n) &= \chi(\alpha^n)\delta_{n,m} = [\exp(-\tfrac{1}{2}n^2 g_0^2 L_x \tau)]\ \delta_{n,m} && (4.11) \\ &=: (\psi_n, \exp(-\tau H)\psi_m)\ . && (4.12) \end{aligned}$$

Finally, the completeness of $\{\psi_n\}_{n\in\mathbb{Z}}$ enables us to write down the action of the Hamiltonian H simply as:

$$H\psi_n = \frac{g_0^2}{2} L_x E^2 \psi_n,\ E\psi_n = -in\psi_n\ . \tag{4.13}$$

Finally, it is clear that the vacuum vector is unique and given by $\Omega(\bar{A}) = 1$. Thus, because the key (generalized) Osterwalder axioms are satisfied by our continuum measure, we can construct the complete Hamiltonian framework.

To conclude, we note that the 2-dimensional model can also be quantized directly using the Hamiltonian methods [25, 22]. The resulting quantum theory is completely equivalent to the one obtained above, starting from the Euclidean theory.

5 Discussion

In this contribution we first pointed out that, in its standard form [3], the basic framework of constructive quantum field theory depends rather heavily on the assumption that the space of histories is linear. Since this assumption is not satisfied in gauge theories (for $d+1 > 2$), a fully satisfactory treatment of quantum gauge fields would require an extension of the framework. We then suggested an avenue towards this goal.

The basic idea was to regard $\mathcal{A}/\mathcal{G}$ as the classical space of histories and to attempt to construct a quantum theory by suitably completing it and introducing an appropriate measure on this completion. To achieve this, one can exploit the "non-linear duality" between loops and connections. More precisely, if one uses loops as probes –the counterpart of test functions in the case of a scalar field– one can follow the general methods used in the kinematically linear theories and introduce the notion of cylindrical functions and cylindrical measures on $\mathcal{A}/\mathcal{G}$. The question is if these can be extended to genuine measures. The answer turned out to be in the affirmative: there exists a completion $\overline{\mathcal{A}/\mathcal{G}}$ of $\mathcal{A}/\mathcal{G}$ such that every cylindrical measure on $\mathcal{A}/\mathcal{G}$ can be extended to a regular, σ-additive measure on $\overline{\mathcal{A}/\mathcal{G}}$. Thus, we have a "non-linear arena" for quantum gauge theories; the discussion is no longer tied to the linear space of tempered distributions. We were able to indicate how the Osterwalder-Schrader axioms can be generalized to measures on the non-linear space $\overline{\mathcal{A}/\mathcal{G}}$. The key open problem is that of singling out *physically appropriate* measures.

The space $\overline{\mathcal{A}/\mathcal{G}}$ is analogous to the algebraic dual $\overline{SH}$ of the Schwarz space one encounters in the kinematically linear case. Therefore, it is almost certainly too big for quantum Yang-Mills theories (although there are indications that it is of the "correct size" for diffeomorphism invariant theories such as general relativity.) That is, although measures which would be physically relevant for Yang-Mills theories could be well-defined on $\overline{\mathcal{A}/\mathcal{G}}$, their support is likely to be significantly smaller. In the kinematically linear case, the Bochner-Milnos theorem provides tools to find physically relevant measures and tells us that their support is the space S' of tempered distributions. The analogous result is, unfortunately, still lacking in our extension to gauge theories. Without such a result, it is not possible to specify the exact mathematical nature of the Schwinger functions of the theory –the "Euclidean expectation values" of the Wilson loop operators. This in turn means that we have no results on the analytic continuation of these functions, i.e., on the existence of Wightman functions. What we *can* formulate, is the notion of reflection positivity and this ensures that the physical Hilbert space, the Hamiltonian and the vacuum exists.

It is clear from the above discussion that our framework is incomplete. We need to introduce appropriately weak topologies on the space of hoops –the probe space– and find generating functionals $\chi(\alpha)$ which are continuous with respect to them. Only then can one have sufficient control on the nature of Schwinger functions. For the moment, $\overline{\mathcal{A}/\mathcal{G}}$ serves only as the "universal home" for the measures we want to explore. As we saw, this strategy was successful in the 2-dimensional Yang-Mills theory.

In higher dimensions, there are reasons to be concerned that $\overline{\mathcal{A}/\mathcal{G}}$ may be too large to play even this "mild" role. That is, one might worry that the elements of $\overline{\mathcal{A}/\mathcal{G}}$ are allowed to be so "pathological" that it would be difficult to define on them the standard operations that one needs in mathematical physics. For instance, $\overline{\mathcal{A}/\mathcal{G}}$ arises only as a (compact, Hausdorff) topological space and does not carry a manifold structure. For physical applications on the other hand, one generally needs to equip the domain spaces of quantum states with operations from differential calculus. Would one not be stuck if one has so little structure? It turns out that the answer is in the negative: Using projective techniques associated with families of graphs, one *can* develop differential geometry on $\overline{\mathcal{A}/\mathcal{G}}$. In particular, notions such as vector fields, differential forms, Laplacians and heat kernels are well defined [15]. Thus, at least for the purposes we want to use $\overline{\mathcal{A}/\mathcal{G}}$, there is no *obvious* difficulty with the fact that it is so large.

A more subtle problem is the following. A working hypothesis of the entire framework is that the Wilson loop operators should be well-defined in quantum theory. From a "raw," physical point of view, this would seem to be a natural assumption: after all Wilson loop functionals are the natural gauge invariant observables. However, technically, the assumption *is* strong. For example, if the connection is assumed to be distributional, the Wilson loop functionals cease to be well-defined. Therefore, in a quantum theory based on such a hypothesis, the connection would not be representable as an operator valued distribution. In particular, in the case of a Maxwell field, such a quantum theory can not recover the textbook Fock representation [27]. More generally, the representations that can arise will be *qualitatively* different from the Fock representation in which the elementary excitations will not be plane waves or photon-like states. Rather, they would be "loopy," concentrated along "flux lines." They would be related more closely to lattice gauge theories than to the standard perturbation theory. In 2 dimensions, this reprentation does contain physically relevant states. Whether this continues to be the case in higher dimensions is not yet clear. It is conceivable that the quantum theories that arise from our framework are only of mathematical interest. However, even if this turns out to be the case, they would still be of considerable significance since as of now there does not exist a single quantum gauge theory in higher dimensions. Finally, if it should turn out that loops are too singular for physical purposes, one might be able to use the extended loop group of Gambini and co-workers [20] as the space of probes. This group has the same "flavor" as the hoop group in that it is also well tailored to incorporate the kinematical non-linearities of gauge theories. However, the hoops are replaced by extended, smoothened objects so that Fock-like excitations are permissible.

To conclude, our main objective here is to revive intertest in manifestly gauge invariant approaches to quantum gauge theories, in which the kinematical non-linearities are met head on right from the beginning. There have been attempts along these lines in the past (see, particularly [21]) which, however, seem to have been abandoned. (Indeed, to our knowledge, none of the major programs for construction of quantum Yang-Mills theories is still being actively pursued.) The specific methods we proposed here are rather tentative and our framework is incomplete in several respects. Its main merit is that it serves to illustrate the type of avenues that are available but have remained unexplored.

Acknowledgements

This work was supported in part by the NSF Grant PHY93-96246 and the Eberly research fund of Penn State Univesity.

References

[1] P. Ramond (1990), *Field theory: a modern primer*, Addison-Wesley, New York.

[2] R.J. Rivers (1987), *Path integral methods in quantum field theory*, Cambridge University Press, Cambridge.

[3] J. Glimm and A. Jaffe (1987), *Quantum physics*, Springer-Verlag, New York.

[4] V. Rivasseau (1991), *From perturbative to constructive renormalization*, Princeton University Press, Princeton.

[5] D. Iagolnitzer (1993), *Scattering in quantum field theories*, Princeton Univ. Press, Princeton.

[6] M. Asorey, P. K. Mitter (1981), *Comm. Math. Phys.* , **80** 43-58;
M. Asorey, F. Falceto (1989), *Nucl. Phys. B* , **327** 427-60.

[7] V. Husain, K. Kuchař (1990), *Phys. Rev. D* , **42** 4070.

[8] I. Gel'fand and N. Vilenkin (1964), *Generalized functions, Vol. IV*, Academic Press, New York.

[9] Y. Yamasaki (1985), *Measures on infinite dimensional spaces*, World Scientific, Singapore.

[10] Yu. L. Dalecky and S.V. Fomin (1991), *Measures and differential equations in infinite dimensional spaces*, Kluwer Ac. Pub., Dordrecht.

[11] A. Ashtekar and C.J. Isham (1992), *Class. Quan. Grav.*, **9** 1069-1100.

[12] A. Ashtekar and J. Lewandowski, 'Representation theory of analytic holonomy $C^\star$ algebras', to appear in *Knots and quantum gravity*, J. Baez (ed), Oxford University Press.

[13] J. Baez, 'Diffeomorphism-invariant generalized measures on the space of connections modulo gauge transformations', hep-th/9305045, to appear in the *Proceedings of the conference on quantum topology*, L. Crane and D. Yetter (eds).

[14] D. Marolf and J. M. Mourão, 'On the support of the Ashtekar-Lewandowski measure' submitted to *Commun. Math. Phys.*

[15] A. Ashtekar and J. Lewandowski, 'Differential calculus in the space of connections modulo gauge transformations', Preprint CGPG.

[16] J. W. Barret (1991), *Int. Journ. Theor. Phys.* ,**30** 1171-1215;
J. Lewandowski (1993), *Class. Quant. Grav.* , **10** 879-904;
A. Caetano, R. F. Picken, 'An axiomatic definition of holonomy', preprint IFM/14-93.

[17] M. Reed, B. Simon (1979), *Methods of Modern Mathematical Physics, vol. 1*, Academic Press, New York.

[18] A. Ashtekar (1987), *Phys. Rev. D* , **36** 1587-1602.

[19] A. Ashtekar (1991), *Non-perturbative canonical quantum gravity*, (Notes prepared in collaboration with R. S. Tate), World Scientific, Singapore.

[20] C. Di Bartolo, R. Gambini, J. Griego, 'The extended Loop representation of quantum gravity', report IFFC/94-13, gr-qc/9406039;
C. Di Bartolo, R. Gambini, J. Griego, J. Pullin , 'Extended Loops : A New Arena for Non-Perturbative Quantum Gravity', CGPG-94/4-3, gr-qc/9404059.

[21] E. Seiler (1982), *Gauge theories as a problem of constructive quantum field theory and statistical mechanics*, Lecture notes in Physics, v. 159, Springer-Verlag, Berlin, New York.

[22] A. Ashtekar, J. Lewandowski, D. Marolf, J. Mourao, T. Thiemann, 'Constructive Quantum Gauge Field Theory in two spacetime dimensions', CGPG preprint, August 94.

[23] D. S. Fine (1990), *Comm. Math. Phys.* , **134** 273-292; D. S. Fine (1991), *Comm. Math. Phys.* , **140** 321-338.

[24] M. Creutz (1983), *Quarks, Gluons and Lattices*, Cambridge University Press, New York.

[25] S. V. Shabanov, Saclay preprint, France, hep-th/9312160.

[26] L. Gross, C. King, A. Sengupta (1989), *Ann. Phys. (NY)*, **194** 65-112; S. Klimek, W. Kondracki (1987), *Comm. Math. Phys.* , **113** 389-402; V. A. Kazakov (1981), *Nucl. Phys. B*, **79** 283.

[27] A. Ashtekar. C. J. Isham (1992), *Phys. Lett.*, **B274**, 393.

On the Hamiltonian formulation of higher dimensional Chern-Simons gravity

Máximo Bañados and Luis J. Garay

Abstract

The analogue of the Ashtekar-Witten approach to (2+1) general relativity is studied in higher odd-dimensional spacetimes. The underlying action is that of Chern-Simons for the anti-de Sitter group.

Gravity in all odd-dimensional spacetimes can be formulated as a Chern-Simons theory for the anti-de Sitter group. In the three-dimensional case, this has been extensively studied. The higher dimensional case was considered in [1], where black hole geometries were found. The aim of this report is to study the Hamiltonian formulation for the action proposed in [1].

We shall start by briefly reviewing the three-dimensional case [2]. The Hamiltonian formulation of this theory is quite simple. The basic variables are the triad e_i^a and the spin connection w_i^{ab}. The action, in a 2+1 decomposition of spacetime, is

$$I = \int_{\Re}\int_{\Sigma}\left[\frac{1}{2}\epsilon_{abc}\epsilon^{ij}e_i^a\dot{w}_j^{bc} - e_0^a P_a - \frac{1}{2}w_0^{ab}J_{ab}\right], \qquad (1)$$

where the constraints P_a and J_{ab} have the form

$$J_{ab} = -\frac{1}{2}\epsilon_{abc}\epsilon^{ij}D_i e_j^c , \qquad (2)$$

$$P_a = -\frac{1}{2}\epsilon_{abc}\epsilon^{ij}(R_{ij}^{bc} + e_i^b e_j^c) . \qquad (3)$$

R_{ij}^{ab} is the curvature associated to the spin connection w_i^{ab} and D_i is its corresponding covariant derivative. From this action, we learn that the basic Poisson bracket is

$$\{e_i^a(x), w_j^{bc}(y)\} = \epsilon_{ij}\epsilon^{abc}\delta(x,y) . \qquad (4)$$

The generators P_a and J_{ab} satisfy the anti-de Sitter algebra [2]. The spin connection w_i^{ab} is the compensation field associated to the Lorentz generator

J_{ab} while e_i^a is the compensation field for the translational part P_a. As it is well known, the above constraints force the anti-de Sitter curvature 2-form to be zero and therefore, the connection is flat. All the degrees of freedom are then associated to global aspects such as holonomies around non-contractible loops [2, 3] and global charges in the case of manifolds with boundaries [4].

An alternative Hamiltonian formulation for the same Lagrangian theory can be obtained by performing a different foliation of spacetime [5]. If we interpret the triad e_μ^a as the non-singular matrix of change of basis from an orthonormal frame $\vec{e}_a$ to a coordinate frame $\vec{e}_\mu$, i.e. $\vec{e}_\mu = e_\mu^a \vec{e}_a$, then a metric structure can be extracted from the triad by the usual formula $g_{\mu\nu} = e_\mu^a e_\nu^b \eta_{ab}$. With respect to that metric, we can define the weight-1 normal to the spatial surface

$$n_a = \frac{1}{2!}\epsilon_{abc}\epsilon^{ij} e_i^b e_j^c \ , \tag{5}$$

and we can decompose the vector ∂_t in its normal and parallel components

$$(\partial_t)^a = N n^a + N^i e_i^a \ . \tag{6}$$

With this foliation, the Hamiltonian form of the Chern-Simons action is

$$I = \int_{\Re}\int_{\Sigma}\left[\frac{1}{2}\epsilon_{abc}\epsilon^{ij}e_i^a \dot{w}_j^{bc} - N^i H_i - N H_\perp - N^{ab}J_{ab}\right] \ , \tag{7}$$

where $N^{ab} = N w_\perp^{ab}$, N and N^i are independent Lagrange multipliers and

$$H_i = e_i^a P_a + \frac{1}{2} w_i^{ab} J_{ab} \ , \tag{8}$$

$$H_\perp = n^a P_a \ . \tag{9}$$

Note that the Poisson bracket remains unchanged.

The generators H_i and $H_\perp$ admit a straightforward interpretation. Indeed, they are the projections of the generator of local translations P_a on to the spatial surface Σ (up to a local rotation represented by the second term in Eqn. (8)) and along the perpendicular direction to that surface, respectively. In physical terms, P_a can be interpreted as the generator of translations in a free-falling system of reference (Lorentzian coordinates) and H_i and $H_\perp$ are then the expression of such generator in the language of surface deformations of general relativity, when an arbitrary foliation of spacetime is performed. The Poisson bracket algebra of these generators is —up to a local Lorentz rotation— the surface deformation algebra with a weight-2 inverse spatial metric given by

$$h^{ij} = \epsilon^{il}\epsilon^{jk} e_l^a e_k^b \eta_{ab}. \tag{10}$$

The above construction is equivalent to the Ashtekar's formulation of 2+1 gravity. Indeed, by making the replacement $E_a^i = \epsilon^{ij}\eta_{ab}e_j^b$ and $A_i^a = \frac{1}{2}\epsilon^a{}_{bc} w_i^{bc}$ one can easily prove that (8) and (9) reduce to the Ashtekar constraints of 2+1 gravity [6, 7].

In this work, we would like to report on the possible extension of the above analysis, and the relation between Witten's and Ashtekar's formulations to higher dimensions. We will make use of the Chern-Simons formulation for gravity in higher dimensions and therefore, we will restrict ourselves to odd-dimensional spacetimes. We will first write the action in terms of the local gauge constraints and then, we will reformulate the theory in the language of surface deformations, i.e. spatial diffeomorphisms and normal deformations.

In $(2n+1)$ dimensions, we consider the Chern-Simons action for the anti-de Sitter group [8], which is equivalent to a Lovelock [9] action with certain fixed coefficients [1]. Note that, since in dimensions higher than three, the Poincaré group has a degenerate Killing metric, it is convenient to take the anti-de Sitter group as the gauge group, so that a negative cosmological constant will be present.

If the spacetime topology is $\Sigma \times \Re$, the Chern-Simons action in $2n+1$ dimensions can be written as

$$I = \int_{\Re}\int_{\Sigma}\left[L^i_{ab}(w,e)\dot{w}^{ab}_i - e^a_0 P_a - \frac{1}{2}w^{ab}_0 J_{ab}\right] , \tag{11}$$

where the constraints P_a and J_{ab} are given by

$$P_a = -\frac{1}{4}\epsilon_{abcde}\epsilon^{ijkl}(R^{bc}_{ij} + e^b_i e^c_j)(R^{de}_{kl} + e^d_k e^e_l) , \tag{12}$$

$$J_{ab} = -\frac{1}{4}\epsilon_{abcde}\epsilon^{ijkl}(R^{cd}_{ij} + e^c_i e^d_j) D_k e^e_l \tag{13}$$

and e^a_0, w^{ab}_0 are non-dynamical Lagrange multipliers. For the sake of simplicity in the notation, all equations that require an explicit realisation for each dimension have been written for a five-dimensional spacetime, although all the results and conclusions obtained from them are valid in any other odd-dimension. The action (11) is the extension to higher dimensions of the action (1). The constraints (12) and (13) are consistent with time evolution, that is, they are preserved in time [8].

The explicit expression of $L^i_{ab}(w,e)$ is not necessary for our purposes but only its exterior derivative in phase space $\Omega = \delta L$,

$$\Omega^{ij}_{ab} = 0 , \tag{14}$$

$$\Omega^{ij}_{abc} = \epsilon_{abcde}\epsilon^{ijkl}(R^{de}_{kl} + e^d_k e^e_l) , \tag{15}$$

$$\Omega^{ij}_{abcd} = \epsilon_{abcde}\epsilon^{ijkl} D_k e^e_l , \tag{16}$$

which is the relevant function that appears in the equations of motion. The identification of the dynamical fields and their Poisson bracket is not direct. The difficulties appear because the form Ω is not invertible and therefore, it does not induce a well-defined Poisson bracket. One can see, however, that the non-invertibility of Ω does not induce further constraints [10] but only makes some of the fields completely arbitrary [8], that is, they are not fixed by the equations of motion. The separation between the dynamical and non-dynamical fields is quite involved and it does not seem possible to perform it without breaking general covariance [11].

In complete analogy to the three-dimensional case, if we perform a $2n+1$ decomposition along the normal to the spatial surface

$$n_a = \frac{1}{4!}\epsilon_{abcde}\epsilon^{ijkl}e_i^b e_j^c e_k^d e_l^e \tag{17}$$

and the tangent vectors e_i^a, we find the higher dimensional analogue of the action (7),

$$I = \int_{\Re}\int_{\Sigma}\left[L^i_{ab}(w,e)\dot{w}_i^{ab} - N^i H_i - N H_\perp - N^{ab}J_{ab}\right] . \tag{18}$$

It is remarkable that the generators H_i and $H_\perp$ have the same form as in three-dimensions, i.e. they are given by Eqns. (8) and (9), where now P_a and J_{ab} have the higher-dimensional form (12) and (13), respectively.

One can check that these constraints are preserved in time. Indeed, the time derivatives of the smeared generators $H(M^i)$ and $H_\perp(M)$ have the form (without loss of generality, we can assume that the smearing functions do not depend on time)

$$\begin{aligned} \dot{H}(M^i) &= H([M,N]^i) + H_\perp(\mathcal{L}_{M^i}N) + J(\mathcal{L}_{M^i}N^{ab}) , & (19)\\ \dot{H}_\perp(M) &= H_\perp(\mathcal{L}_{N^i}M) + H(K^i) + J(K^i w_i^{ab}) , & (20)\end{aligned}$$

where $K^i = h^{ij}(M\partial_j N - N\partial_j M)$ and the inverse weight-2 spatial metric h^{ij} is the natural extension of (10),

$$h^{ij} = \epsilon^{ii_1i_2i_3}\epsilon^{jj_1j_2j_3}e_{i_1}^{a_1}e_{j_1}^{b_1}\ e_{i_2}^{a_2}e_{j_2}^{b_2}\ e_{i_3}^{a_3}e_{j_3}^{b_3}\ \eta_{a_1b_1}\eta_{a_2b_2}\eta_{a_3b_3} . \tag{21}$$

Note that the constraints are polynomial in our basic variables e_i^a and w_i^{ab} as can be seen directly from (8) and (9); so it is the inverse metric h^{ij}.

One could be tempted to interpret this result in a straightforward way by saying that $H(N^i)$ is the generator of spatial diffeomorphisms, and that $H_\perp(N)$ is the generator of normal deformations since, in the right hand side of (19) and (20), we have precisely the terms that appear in the algebra of surface deformations. However, a Poisson bracket has not been specified yet. We remark that, due to the non-invertibility of Ω, not all the fields e_i^a and w_i^{ab} are dynamical and therefore, in the expressions for the time derivatives of the constraints, there appear two kinds of terms. The first ones, which involve the time derivative of the true dynamical fields, can be expressed as the Poisson bracket of those fields with the Hamiltonian. The second ones involve time derivatives of the non-dynamical fields and consequently, cannot be written in terms of Poisson brackets. One can check, however, that the latter terms vanish on the constraint surface. In this sense, the constraints H_i and $H_\perp$ satisfy a first class algebra, although it is not the surface deformation algebra. This issue is under study and will be reported elsewhere [11].

Acknowledgments. We are grateful to C. J. Isham, C. Teitelboim, M. Navarro, J. Mourão, N. Manojlović and J. Zanelli. In particular, we would like to thank M. Henneaux for many crucial comments and ideas. This work was also partially supported by grants 193.0910-93 and 194.0203-94 from FONDECYT (Chile) and by institutional support to the Centro de Estudios Científicos de Santiago provided by SAREC (Sweden) and a group of Chilean private companies (COPEC,CMPC,ENERSIS). L.J.G. was supported by a joint fellowship from the Ministerio de Educación y Ciencia (Spain) and the British Council. M.B. holds a Jack Ewer/Fundación Andes Fellowship.

References

[1] M. Bañados, C. Teitelboim and J. Zanelli (1994), *Phys. Rev.*, **D49** 975.

[2] E. Witten (1988), *Nucl. Phys.*, **B311** 46.

[3] J. E. Nelson and T. Regge (1989), *Nucl. Phys.*, **B328** 190.

[4] The asymptotic structure of (2+1)-gravity in the metric (ADM) representation was studied in J. D. Brown and M. Henneaux (1986), *Commun. Math. Phys.*, **104** 207.

[5] N. Manojlović and A. Miković (1992), *Nucl. Phys.*, **B385** 571.

[6] A. Ashtekar (1991), *Lectures on non-perturbative canonical gravity*, World Scientific, Singapore.

[7] I. Bengtsson (1989), *Phys. Lett.*, **B220** 51.

[8] M. Bañados, this volume, p1.

[9] D. Lovelock (1971), *J. Math. Phys.*, **12** 498.

[10] L. Faddeev and R. Jackiw (1988), *Phys. Rev. Lett.*, **60** 1692.

[11] M. Bañados, L. J. Garay and M. Henneaux, in preparation.

An example of loop quantization

R. Loll

1 Motivation

The background for the investigations to be discussed below is provided by recent studies of canonical quantum gravity in 3+1 dimensions. However, related issues arise in other physical theories that can be formulated on spaces of connections and are invariant under a corresponding group of gauge transformations. We will be interested in the so-called loop approach to the quantization of gravity, and will have a closer look at the (2+1)-dimensional theory, in the hope of gaining further understanding of the general approach.

Let us summarize in a nutshell the ideas that have gone into proposals for a quantization program for general relativity.

* The starting point is Ashtekar's reformulation of 3+1-dimensional Hamiltonian gravity in terms of Yang-Mills variables, namely, an $sl(2,\mathbb{C})$-valued pair (A, E) of a connection one-form and its conjugate momentum [1].
* On the phase space spanned by these variables, define Wilson loop variables $T^0(\gamma) = \mathrm{Tr}\,\mathrm{P}\,\exp \oint_\gamma A$, and momentum-dependent generalizations which together form a closed Poisson bracket algebra of loop functions.
* Next, "quantize" this classical structure by finding representations of the Wilson loop algebra on spaces of wave functions that are themselves labelled by spatial loops γ.
* Rewrite the Hamiltonian in terms of loop variables, regularize it appropriately and look for solutions of the Wheeler-DeWitt equation, i.e. wave functions that are annihilated by the quantum Hamiltonian operator.

Some excitement was generated when in 1990, Rovelli and Smolin claimed that - by working roughly along the lines just described - they had indeed been able to find solutions to the non-perturbative quantum theory [2]. Ever since then, a lot of work has gone into understanding which precise assumptions have led to this conclusion, and to which extent the derivation is unique and can be justified both physically and mathematically. Of central importance in this is the algebraic structure on the set of generalized Wilson loops and what can be said about its representation theory. Given the infinite dimensionality of the problem, these are of course very non-trivial questions, which therefore have led to the study of simpler systems where similar algebraic structures occur. One such model system is given by the theory of (2+1)-dimensional gravity, which also may be formulated as a theory on connection space, and which will be the subject of the remainder of this paper.

2 Introducing 2+1 gravity

The action for 3-dimensional gravity is the Riemann-Hilbert action for the Lorentz metric ${}^3g$,

$$S[{}^3g] = \int_M d^3x \sqrt{-{}^3g}\, R[{}^3g], \tag{2.1}$$

whose form is identical to that of the four-dimensional theory. There is however an alternative action principle leading to essentially equivalent equations of motion, which depends on a co-triad e and an $SO(2,1)$-connection A, namely,

$$S[e, A] = \int_M e \wedge F[A], \tag{2.2}$$

where F denotes the curvature of A. It is in this form that the theory is most obviously exactly soluble [3], although a similar result could be established subsequently in the metric formulation too [4]. We will be interested in the case where the three-dimensional manifold M is a product $M = \mathbb{R} \times \Sigma^g$, with Σ^g a compact, oriented Riemann surface of genus g. The theory is topological in the sense that it does not allow for local field excitations, which leads to drastic simplifications compared with the four-dimensional theory. It becomes non-trivial, with a finite-dimensional physical phase space, through the introduction of a non-trivial topology for M.

One may now perform a Legendre transformation which brings the theory of equation (2.2) into the form of a first-class constrained system with conjugate variable pairs $(A_a^i(x), \tilde{E}_i^a(x))$, and constraints

$$\begin{aligned} \mathcal{D}_a \tilde{E}_i^a &= 0 \\ F_{ab\ i} &= 0, \end{aligned} \tag{2.3}$$

see, for example [5]. Here, a is the spatial and i the internal $SO(2,1)$-index. The first three constraints are the usual Gauss law conditions associated with the invariance under local $SO(2,1)$-rotations, and the remaining three are flatness constraints on the components of the spatial curvature F, and contain the three diffeomorphism constraints of the theory. - For computational simplicity it is often convenient to work in the two-dimensional representation of the gauge group $PSU(1,1)$ (i.e. $SU(1,1)$ with opposite points identified) instead of $SO(2,1)$.

The constraints (2.3) can be solved explicitly which (for $g > 1$) leads to a reduced phase space that is the cotangent bundle over the well-known

Teichmüller space $\mathcal{T}(\Sigma^g)$, which will be the subject of the next section. This latter space is contractible and diffeomorphic to $\mathbb{R}^{6g-6}$. The case $g=1$ is degenerate from a mathematical point of view; its reduced configuration space is diffeomorphic to $\mathbb{R}^2$.

3 The loop formulation

One way of describing points of $\mathcal{T}(\Sigma^g)$ is as follows. Consider the $2g$ generators of the homotopy group $\pi_1(\Sigma^g)$ of the Riemann surface. With each handle of Σ^g we associate a pair (α_i, β_i) of generators, where α_i goes around the i'th hole and β_i around the i'th handle. These generators have to obey the fundamental relation

$$\alpha_1\beta_1\alpha_1^{-1}\beta_1^{-1}\alpha_2\beta_2\alpha_2^{-1}\beta_2^{-1}\dots\alpha_g\beta_g\alpha_g^{-1}\beta_g^{-1}=1. \tag{3.1}$$

We define the holonomy of an element $\gamma\in\pi_1(\Sigma^g)$ as the integral $U_\gamma = \mathrm{P}\exp\oint_\gamma A$. The integral does not depend on the representative in the homotopy class of γ because the physical connections are all flat. A point in the Teichmüller space $\mathcal{T}(\Sigma^g)$ is uniquely labelled by the values of the $2g$ holonomy matrices $U_{\alpha_i}, U_{\beta_i}\in PSU(1,1)$ modulo gauge transformations

$$(U_{\alpha_1},\dots,U_{\beta_g}) \xrightarrow{g} g^{-1}(x_0)\cdot(U_{\alpha_1},\dots,U_{\beta_g})\cdot g(x_0) \tag{3.2}$$

at the common, fixed base point $x_0\in\Sigma^g$ of the generators, and subject to

a) $U_{\alpha_1}U_{\beta_1}U_{\alpha_1}^{-1}U_{\beta_1}^{-1}\dots U_{\alpha_g}^{-1}U_{\beta_g}^{-1}=\mathbb{1}$, and

b) $1-T^0(\gamma)^2<0$, $\forall\gamma\in\{\alpha_i,\beta_i\}$ and $T^0(\gamma):=\frac{1}{2}\mathrm{Tr}\,U_\gamma$.

The first condition is a direct consequence of (3.1), and the second condition picks out (gauge-invariantly) the sector where the holonomy matrices U correspond to boosts about space-like axes [5]. Taking these constraints into account, the counting of the degrees of freedom, $2g\times3-3-3=6g-6$, agrees with the dimensionality of Teichmüller space.

One gets rid of the remaining gauge covariance (3.2) of the holonomies by taking traces, $T^0(\gamma):=\frac{1}{2}\mathrm{Tr}\,U_\gamma$. Doing this for any element of the homotopy group, one obtains an overcomplete set $\{T^0(\gamma),\gamma\in\pi_1(\Sigma^g)\}$ of Wilson loop observables for 2+1 gravity. (Note that just considering Wilson loops of the $2g$ homotopy generators does not lead to a complete set of observables.) Likewise, generalized, momentum-dependent loop variables can be introduced. Thus the kinematical set-up is very much like that for both 3+1 gravity and $SU(2)$ Yang-Mills theory in a canonical loop formulation (see [6] for a review). The difference between those theories lies in the way the physical degrees of freedom are imbedded into the initial configuration (or phase) space, given in terms of the overcomplete set of loop variables. To identify the physical degrees of freedom, one has to solve the so-called

Mandelstam constraints associated with the given gauge group and representation, and also take care of certain inequalities that may exist among the loop variables [7].

Usually there are topological obstructions which prevent us from finding a good global solution to the Mandelstam constraints (and other constraints that come with the theory), i.e. a set of independent Wilson loops that parametrize the physical configuration space globally. It turns out however that the structure of 2+1 gravity is sufficiently simple to allow us to find such solutions explicitly.

We will make use of a particular parametrization for the Teichmüller spaces $\mathcal{T}(\Sigma^g)$, in terms of the so-called Fenchel-Nielsen coordinates [8]. They are associated with the Riemann surface Σ^g equipped with a metric h of constant negative scalar curvature -1. The surface is cut along $3g-3$ simple (non-intersecting) closed geodesics into $2g-2$ "pairs of pants", so that each pants is a piece of the surface, with three circular boundary components.

We associate with the i'th cut a pair (l_i, τ_i) of variables in $\mathbb{R}^+ \times \mathbb{R}$, where l_i measures the length of the boundary component (with respect to h) and τ_i the angle of the relative twist between the two pants that meet along the boundary. The Fenchel-Nielsen coordinates parametrize Teichmüller space globally, leading to the identification $\mathcal{T}(\Sigma^g) \simeq (\mathbb{R}^+)^{3g-3} \times \mathbb{R}^{3g-3}$. A natural symplectic form ω exists on $\mathcal{T}(\Sigma^g)$, the so-called Weil-Petersson form, for which the Fenchel-Nielsen coordinates are canonical pairs, i.e. $\omega = \sum_i dl_i \wedge d\tau_i$. Note that this symplectic form has nothing to do with the canonical symplectic structure on the phase space $T^*\mathcal{T}(\Sigma^g)$ of 2+1 gravity, which is $\Omega = \sum_i dl_i \wedge dp_{l_i} + \sum_i d\tau_i \wedge dp_{\tau_i}$.

The existence of an explicit parametrization of the reduced configuration space simplifies calculations enormously, and allows us to set up a quantization of the reduced system (as opposed to solving the constraints à la Dirac *after* quantizing the original system). Our next step will be to rewrite the Wilson loops associated to certain homotopy group elements in terms of the Fenchel-Nielsen coordinates. This can be done using a result of Okai [9] who constructed an explicit cross section of the principal bundle

$$\begin{array}{c} \mathrm{Hom}(\pi_1(\Sigma^g), PSU(1,1))^{e=2g-2} \\ \downarrow \\ \mathrm{Hom}(\pi_1(\Sigma^g), PSU(1,1))^{e=2g-2}/PSU(1,1) = \mathcal{T}_g, \end{array} \tag{3.3}$$

thus giving explicit expressions U_{α_i}, U_{β_i} for the holonomy matrices in terms of the Teichmüller parameters. The superscript $e = 2g-2$ in (3.3) denotes a technical condition on the space of all homomorphisms which selects exactly the sector we are interested in here. Obviously we can then write all of the gauge-invariant loop observables as $T^0_{(l,\tau)}(\gamma) = \frac{1}{2}\mathrm{Tr}\, U_\gamma(l,\tau)$, and it turns out that a general Wilson loop is a rational function of the hyperbolic functions $\sinh l_i$, $\sinh \tau_i$, $\cosh l_i$ and $\cosh \tau_i$.

Our aim is to construct a loop representation for 2+1 gravity which mimics the corresponding construction for the (3+1)-dimensional theory. The transition from the connection to the loop representation is usually achieved by defining a so-called loop transform between the Hilbert spaces of the two formulations. For example, the loop transform for Yang-Mills theory is given by

$$\tilde{\psi}(\alpha) := \int_{\mathcal{A}/\mathcal{G}} dV([A])\, T^0_A(\alpha)\, \psi([A]). \tag{3.4}$$

The wave functions ψ of the connection representation depend only on the gauge equivalence class $[A]$ of A, and the wave functions $\tilde{\psi}$ in the loop representation on spatial loops α. The transform is to be thought of as a non-linear analogue of a Fourier-type transform. It remains a heuristic device as long as one does not specify an appropriate gauge-invariant measure dV on the connection space. This so far has not been achieved for either gauge theory or general relativity, but in the present case we are more fortunate and can give the explicit form of the loop transform as

$$\tilde{\psi}(\gamma) := \int_{\mathcal{T}(\Sigma^g)} dV(l,\tau)\, T^0_{(l,\tau)}(\gamma)\, \psi(l,\tau), \tag{3.5}$$

where $\gamma \in \pi_1(\Sigma^g)$ and ψ is a square-integrable function in $L^2(\mathcal{T}(\Sigma^g), dV)$. A natural volume element dV on Teichmüller space is the Liouville volume element associated with the symplectic form ω, i.e. $dV = \omega^{3g-3} = dl_1 \wedge d\tau_1 \wedge \ldots \wedge d\tau_{3g-3}$. However, one quickly realizes that with this choice of volume element the integration in (3.5) diverges for general Wilson loops $T^0(\gamma)$. (This problem was first noticed for the $g = 1$ case by Marolf [10].) This happens because $T^0(\gamma)$ depends on hyperbolic sines and cosines which diverge for large values of the Fenchel-Nielsen coordinates l_i and τ_i. This of course could only have occurred because the gauge group for 2+1 gravity is the non-compact group $SO(2,1)$ and the Wilson loops themselves are not bounded functions. A similar problem is not present for $SU(2)$, say, where we have $-1 \leq T^0(\gamma) \leq 1$.

One way of dealing with this problem has been suggested by Ashtekar and the author and consists in modifying the measure in such a way that the integration over Teichmüller space in (3.5) converges for arbitrary Wilson loops [11]. We multiply the volume element by a function $m(l,\tau)$ which is subject to a number of conditions. It turns out that the construction of the loop transform requires m to be of the form $m = \exp \sum_i b_i T^0(\gamma_i)$ of an exponential of a linear combination of some set of Wilson loops $T^0(\gamma_i)$.

Hence the task is to first find functions m that give an appropriate damping behaviour and then establish how the modified measure affects the construction of the loop representation. For the simplest case of $g = 1$ both of these issues were investigated in [11]. One possibility to define m in this case is to set $m = \exp -p(T^0(\vec{q}_1) + T^0(\vec{q}_2))$, where $p > 0$ and $\vec{q}_1$ and $\vec{q}_2$ are two linearly independent vectors in $\mathbb{Z}^2$ (homotopy elements for $g = 1$ are labelled by a pair of integers). As a result one obtains modified expressions for the Wilson loop operators in the loop representation, namely,

$$\begin{aligned}
\left(\hat{T}^0(\vec{k})\tilde{\psi}\right)(\vec{n}) &= \frac{1}{2}\left(\tilde{\psi}(\vec{n}+\vec{k}) + \tilde{\psi}(\vec{n}-\vec{k})\right) \\
\left(\hat{T}^1(\vec{k})\tilde{\psi}\right)(\vec{n}) &= -\frac{i\hbar}{2}(\vec{k}\times\vec{n})\left(\tilde{\psi}(\vec{k}+\vec{n}) - \tilde{\psi}(\vec{k}-\vec{n})\right) + \\
&+ \frac{ip\hbar}{4}\sum_{i=1,2}(\vec{k}\times\vec{q}_i)\Big(\tilde{\psi}(\vec{k}+\vec{n}+\vec{q}_i) - \tilde{\psi}(\vec{k}+\vec{n}-\vec{q}_i) + \\
&\qquad \tilde{\psi}(\vec{k}-\vec{n}+\vec{q}_i) - \tilde{\psi}(\vec{k}-\vec{n}-\vec{q}_i)\Big).
\end{aligned} \tag{3.6}$$

What is new compared to the usual loop representation is the term proportional to p in the expression for the Wilson loop momentum $\hat{T}^1$. It is the loop analogue of the divergence term that has to be added to the expression for $\hat{T}^1$ in the connection representation, in order to make it self-adjoint with respect to the measure $m\,dV$. (Note that the commutator algebra of the loop operators is still exactly the same as in the representation with $m = 1$!)

One may be tempted to take the limit as $p \to 0$ of (3.6) and obtain the expressions of the standard loop representation. However, this limit is not well-defined, which becomes apparent when calculating the scalar products of basic wave functions, like for instance

$$< \tilde{T}^0(\vec{k}), \tilde{T}^0(\vec{n}) >= \int dV\, e^{-p(T^0(\vec{q}_1)+T^0(\vec{q}_2))} T^0(\vec{k}) T^0(\vec{n}). \tag{3.7}$$

We calculated the general scalar product (3.7) for the case $\vec{q}_1 = (1,0)$, $\vec{q}_2 = (0,1)$ and found that the results depend on the modified Bessel functions $K_n(p)$ and inverse powers p^{-n}, and therefore diverge as $p \to 0$. Also, the explicit form of the scalar product is much more complicated than the usual one, which is essentially proportional to $\delta_{\vec{k},\vec{n}}$.

For the higher-genus case, which is algebraically more complicated, we have not yet established suitable damping factors. However, this should be reasonably straightforward, given the recently obtained solution for a set of complete and independent loop invariants for arbitrary genus g [12].

4 Conclusions

We have learned from the preceding discussion that difficulties may arise in the construction of the loop representation whenever the gauge group under consideration is non-compact. In 2+1 gravity and for g=1, we dealt with this problem by introducing an appropriate volume element on the space of connections modulo gauge transformations to make the integration in the loop transform convergent. This in turn led to a new loop representation with unusual expressions for the Wilson loop operators and the scalar product. For the higher-genus case we expect similar results to hold.

This immediately raises the question of whether a similar phenomenon occurs in 3+1 dimensions, where with $SL(2,\mathbb{C})$ we also have a non-compact gauge group. This issue is much harder to address, because of the infinite dimensionality of the theory and the fact that in principle there do not exist good global coordinates on the reduced configuration space. In one sense the (3+1)-dimensional theory may however be simpler, since its gauge group is the complexification of a compact gauge group, which is not true for $SU(1,1)$. So far our knowledge about possible measures for quantum gravity is very limited, although extensive investigations are under way. We may eventually be able to classify inequivalent loop representations, and relate them to classes of suitable measures on the connection space, along the lines proposed above for the "toy model" of 2+1 gravity.

Bibliography

[1] A. Ashtekar (1986), 'New variables for classical and quantum gravity', *Phys. Rev. Lett.*, **57** 2244-7; (1987), 'New Hamiltonian formulation of general relativity', *Phys. Rev. D*, **36** 1578-1602.

[2] C. Rovelli and L. Smolin (1990), 'Loop space representation of quantum general relativity', *Nucl. Phys. B*, **331** 80-152.

[3] E. Witten (1989), '2+1 dimensional gravity as an exactly soluble system', *Nucl. Phys. B*, **311** 46-78.

[4] V. Moncrief (1989), 'Reduction of the Einstein equations in 2+1 dimensions to a Hamiltonian system over Teichmüller space', *J. Math. Phys.*, **30** 2907-14.

[5] A. Ashtekar, V. Husain, C. Rovelli, J. Samuel and L. Smolin (1989), '2+1 gravity as a toy model for the 3+1 theory', *Class. Quant. Grav.*, **6** L185-93.

[6] R. Loll (1993), 'Chromodynamics and gravity as theories on loop space', hep-th/9309056, to appear in Memorial Volume for M.K. Polivanov.

[7] R. Loll (1993), 'Loop variable inequalities in gravity and gauge theory', *Class. Quant. Grav.*, **10** 1471-6.

[8] W. Abikoff (1980), *The real analytic theory of Teichmüller space* Lecture Notes in Mathematics, vol.820, Springer, Berlin.

[9] T. Okai (1992), 'An explicit description of the Teichmüller space as holonomy representations and its applications', *Hiroshima Math. J.*, **22** 259-71.

[10] D.M. Marolf (1993), 'Loop representations for 2+1 gravity on a torus', *Class. Quant. Grav.*, **10** 2625-47.

[11] A. Ashtekar and R. Loll (1994), 'New loop representations for 2+1 gravity', *Class. Quant. Grav.*, to appear.

[12] R. Loll (1994), 'Independent loop invariants for 2+1 gravity', CGPG preprint.

Gauge fixing in constrained systems

J.M. Pons and L.C. Shepley

Abstract

In this paper we show how to eliminate nonphysical degrees of freedom in both the Lagrangian and Hamiltonian formulations of a constrained system. The use of gauge fixing procedures is different in the two cases, but the final result is that the number of degrees of freedom in the two formulations agrees. The two key steps in our method are to use gauge fixing to eliminate ambiguities in the dynamics and to determine the inequivalent initial data. Applications to reparameterization invariant theories are briefly discussed.

1 Introduction

The degrees of freedom in either the Lagrangian or the Hamiltonian formulation of a singular system are reduced in two stages: reduction from using the natural constraints (with consistency requirements) and reduction through a gauge fixing procedure. In general the number of natural constraints in the Lagrangian formulation differs from the number in the Hamiltonian formulation [1]. In this paper we show how to determine the dynamics and fix the gauge in both formalisms. We believe the Lagrangian discussion is new and useful. The result includes a proof that the final number of degrees of freedom in the two formalisms is the same. Important parts of our methods are the two steps of determining the dynamics and determining the independent initial data.

In section 2 we develop a detailed version of the Hamiltonian gauge fixing procedure. In section 3 we study the Lagrangian gauge fixing procedure and show that the number of degrees of freedom in the Hamiltonian and Lagrangian formalisms are equal. In section 4 we describe some details concerning reparameterization invariant theories. (In a future paper, we will expand the discussion and give several examples.)

We assume some regularity conditions: The Hessian matrix of the Lagrangian with respect to the velocities has constant rank, and the rank of the Poisson Bracket matrix of constraints remains constant in the stabilization algorithm (so that a second class constraint can never become first class by adding new constraints to the theory). We always maintain the equivalence between the Lagrangian and the Hamiltonian formalisms and so do not modify the Hamiltonian formalism by adding *ad hoc* constraint terms as Dirac has conjectured [2,3,4,5]. This conjecture has been proved to be completely unnecessary under our regularity conditions [6].

2 Hamiltonian gauge fixing procedure

We start with a singular Lagrangian $L(q^i, \dot{q}^i)$ $(i = 1, \cdots, N)$ $(\dot{q}^i = dq^i/dt)$. The functions $\hat{p}_i(q, \dot{q}) = \partial L/\partial \dot{q}^i$ are used to define the Hessian $W_{ij} = \partial \hat{p}_i/\partial \dot{q}_j$, a matrix with rank less than N (we assume this rank is constant). The Legendre map from velocity space (tangent bundle for configuration space) TQ to phase space (cotangent bundle) T^*Q defined by $p_i = \hat{p}_i(q, \dot{q})$ defines a constraint surface of dimension less than $2N$.

The function $E_L := \hat{p}_i \dot{q}^i - L$ in velocity space (the energy function) is mapped to a function on the primary constraint surface, and in phase space a canonical Hamiltonian H_C may be defined which agrees with this function on the surface. H_C is not unique, and to it may be added a linear combination of constraint functions. The stabilization algorithm uses the time derivatives of the constraints to eliminate some of the arbitrary functions and to add other constraints. Once the stabilization algorithm has been performed, we end up with [3,7,8]:

1: A certain number, M, of constraints. These constraints are arranged into first and second classes. The first class constraints have weakly vanishing Poisson Brackets with all the constraints, and the matrix of the second class constraint Poisson Brackets is nonsingular (there must be an even number of second class constraints). These constraints restrict the dynamics to a constraint surface within T^*Q of dimension $2N - M$.

2: A dynamics (with some gauge arbitrariness) on the constaint surface which is generated, through Poisson Brackets, by the so called Dirac Hamiltonian:

$$H_D := H_{FC} + \lambda^\mu \phi^1_\mu \, .$$

H_{FC} is the first class Hamiltonian, obtained by adding to the canonical Hamiltonian H_C pieces linear in the primary second class constraints. ϕ^1_μ $(\mu = 1, \cdots, P_1)$ are the primary (hence the superscript 1) first class constraints. The secondary and higher first class constraints, obtained from the time derivatives of the ϕ^1_μ, are not used here. The λ^μ are arbitrary functions of time (or spacetime in field theories).

3: A certain number (P_1) of independent gauge transformations generated, through Poisson Brackets,

$$\delta_\mu q^i = \{q^i, G_\mu\} \ , \ \delta_\mu p_i = \{p_i, G_\mu\} \ ,$$

by functions G_μ $(\mu = 1, \cdots, P_1)$ which have the following form [6,9,10,11]:

$$G_\mu = \epsilon_\mu \phi^{K_\mu}_\mu + \epsilon^{(1)}_\mu \phi^{K_\mu - 1}_\mu + \epsilon^{(2)}_\mu \phi^{K_\mu - 2}_\mu + \cdots + \epsilon^{(K_\mu - 1)}_\mu \phi^1_\mu \ ,$$

where ϵ_μ is an arbitrary infinitesimal function of time; $\epsilon^{(r)}_\mu$ is its r-th time derivative; K_μ is the length of the stabilization algorithm for the primary first class constraint ϕ^1_μ; and $\phi^2_\mu, \ldots, \phi^{K_\mu}_\mu$ are secondary through K_μ-ary, first class constraints. It turns out [12] that one can take these gauge generators in such a way that all the first class constraints are involved once and only once in the G_μ, and so their total number equals

$$F := \sum_{\mu=1}^{P_1} K_\mu \ .$$

We distinguish two different steps in the gauge fixing procedure: In the first we fix the laws of evolution; in the second we eliminate the redundancy of initial conditions that are physically equivalent [6].

The arbitrariness in the dynamics is represented by the P_1 functions λ^μ. To get rid of this arbitrariness, we introduce a set of P_1 constraints $\chi^1_\mu \simeq 0$ ($\mu = 1, \cdots, P_1$), defined so that their own stability equations, under dynamical evolution, will determine the functions λ^μ; the matrix

$$C_{\mu\nu} \equiv \{\chi^1_\mu, \phi^1_\nu\}$$

must be non-singular. The conservation in time of this new set of constraints determines λ^ν as

$$\lambda^\nu = -(C^{-1})^{\nu\mu}(\{\chi^1_\mu, H_{FC}\} + \frac{\partial \chi^1_\mu}{\partial t}) \ .$$

The dynamical evolution thus becomes completely determined. The imposition of these constraints causes the dynamics to be further restricted to the $(2N - M - P_1)$-dimensional constraint surface defined by $\chi^1_\mu = 0$.

Even though the dynamics has now been fixed, there is still the possibility of gauge transformations which take one trajectory into another. To check whether these gauge transformations do exist, we need only check their action at a specified time. That is, the points on the set of trajectories at a specified time are unique initial data for the trajectories. If a gauge transformation exists which relates two initial data points, then these two points are physically equivalent. We will obtain the generators of the transformations which take initial points into equivalent ones ("point gauge transformations") and use them to fix the gauge finally.

Consider the gauge generators given above at, say for simplicity, $t = 0$: $G_\mu(0)$. The most arbitrary point gauge transformation at $t = 0$ will be generated by $G(0) = \sum_{\mu=1}^{P_1} G_\mu(0)$. The arbitrary functions ϵ_μ and their derivatives become, at the given time $t = 0$, (infinitesimal) independent arbitrary parameters (there are F in number). We redefine them as

$$\alpha_{\mu,i_\mu} \equiv \epsilon_\mu^{(K_\mu - i_\mu)}(0) \ ; \ (\mu = 1, \cdots, P_1) \ (i_\mu = 1, \cdots, K_\mu) \ .$$

These point gauge transformation generators must be consistent with the new constraints χ^1_μ. This requirement introduces relations among the α_{μ,i_μ}:

$$0 = \{\chi^1_\nu, G(0)\} = \{\chi^1_\nu, \sum_{\mu=1}^{P_1}\sum_{i_\mu=1}^{K_\mu} \alpha_{\mu,i_\mu}\phi^{i_\mu}_\mu\}$$

$$= \sum_{\mu=1}^{P_1}\Big(\sum_{i_\mu=2}^{K_\mu} \alpha_{\mu,i_\mu}\{\chi^1_\nu, \phi^{i_\mu}_\mu\} + \alpha_{\mu,1}C_{\nu\mu}\Big) .$$

Remember the matrix $C_{\nu\mu} = \{\chi^1_\mu, \phi^1_\nu\}$ is nonsingular. These relations imply

$$\alpha_{\rho,1} = -(C^{-1})^{\rho\nu}\sum_{\mu=1}^{P_1}\sum_{i_\mu=2}^{K_\mu} \alpha_{\mu,i_\mu}\{\chi^1_\nu, \phi^{i_\mu}_\mu\} .$$

As a consequence, the independent point gauge generators are

$$\tilde{\phi}^{i_\mu}_\mu \equiv \phi^{i_\mu}_\mu - \phi^1_\rho(C^{-1})^{\rho\nu}\{\chi^1_\nu, \phi^{i_\mu}_\mu\}\ (\mu = 1, \cdots, P_1)\ (i_\mu = 2, \cdots, K_\mu) .$$

Notice that new point gauge generators only exist when there are secondary first-class constraints, that is, when the length K_μ of at least one of the G_μ is greater than one.

Recall that F is the number of first class constraints in the original theory, including primary, secondary, and higher constraints; we conclude that there are $F - P_1$ generators that relate physically equivalent initial conditions. To eliminate the extraneous variables, we will select a unique representative of each equivalence class by introducing a new set of $F - P_1$ gauge fixing constraints, $\chi^{i_\mu}_\mu \simeq 0$ $(\mu = 1, \cdots, P_1)$ $(i_\mu = 2, \cdots, K_\mu)$, such that $\det|\{\chi^{i_\mu}_\mu, \tilde{\phi}^{j_\nu}_\nu\}| \neq 0$, $i_\mu \neq 1$, $j_\nu \neq 1$, in order to prevent any motion generated by $\tilde{\phi}^{j_\nu}_\nu$. The stability requirement is

$$\frac{\partial}{\partial t}\chi^{i_\mu}_\mu + \{\chi^{i_\mu}_\mu, H_D\} \simeq 0 ,$$

which $\simeq 0$ means vanishing on the constraint surface; this requirement simply dictates how the $\chi^{i_\mu}_\mu$ evolve off the initial data surface. Notice that we have explicitly allowed time dependence in the $\chi^{i_\mu}_\mu$ constraints. In fact, time dependence is necessary in the special case when H_{FC} is a constraint (first class, of course).

This ends the gauge fixing procedure. Now we can count the physical number of degrees of freedom: The M constraints left after the stabilization algorithm restricted motion to a $2N - M$-dimensional surface in T^*Q. The gauge fixing constraints needed to fix the evolutionary equations number P_1. Finally there are $F - P_1$ point gauge fixing constraints needed to select physically inequivalent initial points. (The total number of gauge fixing constraints equals the number F of first class constraints in the original theory.) The final number of degrees of freedom is $2N-M-F$. Notice that $M-F$ is the original number of second class constraints and is therefore even. Consequently, $2N-M-F$ is even. This result agrees with the fact that the above procedure makes all constraints into second class ones, and in this case the constraint surface is symplectic [13].

3 Lagrangian gauge fixing procedure

We first use a stabilization algorithm similar to the one used in the Hamiltonian formalism. (In the equations below, we use the summation convention for configuration space indices $i=1,\cdots,N$.) The equations of motion obtained from the Lagrangian L are (assuming for simplicity no explicit time dependence):

$$W_{is}\ddot{q}^s = \alpha_i \ ,$$

where

$$W_{ij} = \frac{\partial^2 L}{\partial \dot{q}^i \partial \dot{q}^j} \ , \ \alpha_i = -\frac{\partial^2 L}{\partial \dot{q}^i \partial q^s}\dot{q}^s + \frac{\partial L}{\partial q^i} \ .$$

If W_{ij} is singular, it possesses null vectors γ^i_ρ, giving constraints $\alpha_i \gamma^i_\rho = 0$. The stabilization algorithm starts by demanding that time evolution preserve the $\alpha_i \gamma^i_\rho$ constraints. Eventually the dynamics is described by a vector field in velocity space

$$X := \frac{\partial}{\partial t} + \dot{q}^i \frac{\partial}{\partial q^i} + a^i(q,\dot{q})\frac{\partial}{\partial \dot{q}^i} + \eta^\mu \Gamma_\mu =: X_0 + \eta^\mu \Gamma_\mu \ ;$$

the a^i are determined from the equations of motion and the stabilization algorithm; η^μ ($\mu = 1,\cdots,P_1$) are arbitrary functions of time; and

$$\Gamma_\mu = {}^{(1)}\gamma^i_\mu \frac{\partial}{\partial \dot{q}^i} \ ,$$

where ${}^{(1)}\gamma^i_\mu$ are a subset of the null vectors of W_{ij}, corresponding to the primary first class constraints found in the Hamiltonian formalism. It is not necessary to use the Hamiltonian technique to find the Γ_μ, but it does facilitate the calculation:

$${}^{(1)}\gamma^i_\mu = \frac{\partial \phi^1_\mu}{\partial p_i}(q,\hat{p}) \ ,$$

where the ϕ^1_μ are the primary first class constraints, and $\hat{p}_i$ stands for the Lagrangian definition of the momenta $\hat{p}_i = \partial L/\partial \dot{q}^i$. The P_1 number of η^μ is the same number as in the Hamiltonian formalism. There are left $M-P_1$ constraints [6].

At this point it is useful to appeal to the Hamiltonian formalism for the computation of the P_1 independent gauge transformations. The result includes the definitions of the functions $\phi_\mu^{j_\mu}$, and then in the Lagrangian formalism we define

$$f^i_{\mu,j_\mu}(q,\dot{q}) := \frac{\partial \phi_\mu^{j_\mu}}{\partial p_i}(q,\hat{p}) \ .$$

These functions give the infinitesimal Lagrangian gauge transformations as

$$\delta_\mu q^i = \sum_{j_\mu=1}^{K_\mu} \epsilon_\mu^{(K_\mu - j_\mu)} f^i_{\mu,j_\mu} \ ,$$

the ϵ_μ being arbitrary functions of time.

As we did in the Hamiltonian formalism, the first step in the gauge fixing procedure will be to fix the dynamics to determine the arbitrary functions η^μ. To this end we introduce P_1 constraints, $\omega_\nu^0 \simeq 0$, such that $D_{\mu\nu} := \Gamma_\mu \omega_\nu^0$ has non-zero determinant: $\det|\Gamma_\mu \omega_\nu^0| = \det|D_{\mu\nu}| \neq 0$. This determines the dynamics as

$$X_F = X_0 - (X_0\omega_\nu^0)(D^{-1})^{\nu\mu}\Gamma_\mu \ .$$

There still remain some point gauge transformations. These transformations may be thought as affecting the space of initial conditions. In fact, we can extract those transformations at $t=0$ that preserve the gauge fixing constraints $\omega_\nu^0 \simeq 0$ from the general gauge transformations. This general transformation is

$$\delta\omega_\nu^0 = \sum_{\mu=0}^{P_1} \delta_\mu \omega_\nu^0 = \sum_{\mu=0}^{P_1} \left(\frac{\partial \omega_\nu^0}{\partial q^i}\delta_\mu q^i + \frac{\partial \omega_\nu^0}{\partial \dot{q}^i}\delta_\mu \dot{q}^i\right) ,$$

where now

$$\delta \dot{q}^i = \frac{d}{dt}\delta q^i = X_F \delta q^i + \frac{\partial \delta q^i}{\partial t} \ .$$

In this expression we use the values at $t=0$:

$$\delta_\mu q^i(0) = \sum_{i_\mu=1}^{K_\mu} \epsilon_\mu^{(K_\mu - i_\mu)}(0) f^i_{\mu,i_\mu} = \sum_{i_\mu=1}^{K_\mu} \alpha_{\mu,i_\mu} f^i_{\mu,i_\mu} \ ,$$

$$\frac{\partial \delta_\mu q^i}{\partial t}(0) = \sum_{i_\mu=1}^{K_\mu} \epsilon_\mu^{(K_\mu - i_\mu + 1)}(0) f^i_{\mu,i_\mu} = \sum_{i_\mu=0}^{K_\mu} \alpha_{\mu,i_\mu} f^i_{\mu,i_\mu+1} \ ,$$

after redefining $\alpha_{\mu,i_\mu} := \epsilon_\mu^{(K_\mu - i_\mu)}(0)$ (we have defined $f^i_{\mu,K_\mu+1} = 0$).

Notice that now, due to the presence of the time derivative of $\delta_\mu q^i$, i_μ runs from 0 to K_μ. This is a key difference with respect to the Hamiltonian case, where the α_{μ,i_μ} parameters had indices i_μ running from 1 to K_μ. We call the result for the independent point gauge transformations (at $t = 0$) $\delta_{(0)}\omega^0_\nu$:

$$\delta_{(0)}\omega^0_\nu = \sum_{\mu=1}^{P_1}\Big(\sum_{i_\mu=1}^{K_\mu}\Big(\frac{\partial\omega^0_\nu}{\partial q^i}f^i_{\mu,i_\mu} + \frac{\partial\omega^0_\nu}{\partial \dot{q}^i}(X_F f^i_{\mu,i_\mu}) + \frac{\partial\omega^0_\nu}{\partial \dot{q}^i}f^i_{\mu,i_\mu+1}\Big)\alpha_{\mu,i_\mu}$$

$$+\alpha_{\mu,0}\Gamma_\mu\omega^0_\nu\Big) ,$$

where $f^i_{\mu,1} = \gamma^i_\mu$.

At this point, recalling that $\det|\Gamma_\mu\omega^0_\nu| \neq 0$, we see that the stability conditions $\delta_{(0)}\omega^0_\nu = 0$ allow the determination of $\alpha_{\mu,0}$ in terms of α_{μ,i_μ} $(\mu = 1,\cdots,P_1)$ $(i_\mu = 1,\cdots,K_\mu)$.

We conclude that the independent point gauge transformations $\delta_{(0)}$ that still remain, relating physically equivalent initial conditions, are parameterized by α_{μ,i_μ} $(\mu = 1,\cdots,P_1)$ $(i_\mu = 1,\cdots,K_\mu)$. Their number equals F, the total number of first class constraints in the Hamiltonian theory. To eliminate these transformations we introduce F new gauge fixing constraints $\omega^{i_\mu}_\mu \simeq 0$ $(\mu = 1,\cdots,P_1)$ $(i_\mu = 1,\cdots,K_\mu)$, with the conditions:

1: The system $\delta_{(0)}\omega^{i_\mu}_\mu = 0$, which is linear in the α_{μ,i_μ} $(\mu = 1,\cdots,P_1)$ $(i_\mu = 1,\cdots,K_\mu)$ has only the solution $\alpha_{\mu,i_\mu} = 0$ (so that no point gauge transformations are left).

2: $X_F(\omega^{i_\mu}_\mu) \simeq 0$ (the requirement of stability under evolution).

Now we have completed the gauge fixing procedure. For reasons similar to the ones raised in the Hamiltonian formalism, there are cases where a time dependent constraint shows up necessarily. Summing up, the gauge fixing constraints introduced in velocity space (that is, in the Lagrangian formalism) are $\omega^{i_\mu}_\mu$ $(\mu = 1,\cdots,P_1)$ $(i_\mu = 0,\cdots,K_\mu)$. Their number is $F+P_1$, and therefore the total number of constraints becomes $(M-P_1)+(F+P_1) = M+F$. The number of degrees of freedom is then $2N-M-F$. Comparison with the results of the previous section shows that we have proved

Theorem. *The number of physical degrees of freedom in constrained Hamiltonian and Lagrangian formalisms is the same.*

Observe that this result, which was expected on physical grounds, is nontrivial. In fact, before introducing the gauge fixing constraints, the dimensions of the constraint surface were different in the two formalisms. This means that the gauge fixing procedure has to make up for this difference—and we see that it does.

4 Comments and conclusions

Reparameterization invariant theories provide interesting cases for the application of the preceeding sections. Here we will simply state two results and give an indication of how they may be proved. A more detailed discussion and examples, including spatially homogeneous cosmologies of Types I and IX, will be treated in a future paper.

The first result is:

Theorem. *The canonical Hamiltonian in a reparameterization-invariant theory (if it is non-zero) is a constraint.*

This comes from the fact that $\delta L \equiv \frac{\partial L}{\partial q^i}\delta q^i + \frac{\partial L}{\partial \dot{q}^i}\delta \dot{q}^i$ is a total derivative: $\delta L = -\frac{d}{dt}(L\delta t)$. From this equality can be found the conserved quantity $G = \frac{\partial L}{\partial \dot{q}^i}\delta q^i + L\delta t$. From the fact that G is a constant of the motion for arbitrary δt in a reparameterization invariant theory, one finds the Theorem.

In general H_C will be a secondary, first class constraint. Then the Dirac Hamiltonian H_D is also a first class constraint and thus a part of $G(0)$, the generator of point gauge transformations. Then, in order to fulfill the two requirements introduced in the second step of the gauge fixing procedure, it is mandatory that at least one of the gauge fixing constraints be time-dependent: Otherwise there is no way to satisfy the gauge fixing conditions. By choosing variables appropriately, we can always end up with only one time-dependent gauge fixing constraint. This is an outline of the proof that:

Theorem. *Reparameterization-invariant theories necessarily require that one of the gauge fixing constraints be time-dependent.*

This result is clearly expected from the physical interpretation of this kind of theory: The existence of reparameterization invariance as a gauge symmetry implies that the evolution parameter —the "time"—is an unphysical variable.

To summarize, in this paper we showed how to determine the dynamics and fix the gauge in both the Hamiltonian and Lagrangian formalisms. Important parts of our methods are the two steps of determining the dynamics and determining the independent initial data: Some constraints are needed to fix the time evolution of the gauge system, which was undetermined to a certain extent. The rest of the constraints eliminate the spurious degrees of freedom that are still present in the setting of the initial conditions of the system. This double role has not been adequately emphasized in the literature. Although we only treat classical systems in this paper, our approach should also be relevant to quantum ones.

We will expand the discussion in a future paper [14] and give examples drawn from spatially homogeneous cosmological models. Also, many of the ideas we discuss may be extended, and we intend to look into various aspects of field theory. For example, the various gauges used in electromagnetic theory, including the Coulomb ($\vec{\nabla} \cdot \vec{A} = 0$), Lorentz ($A^{,\sigma}{}_{\sigma} = 0$), and radiation ($A_0 = 0$) gauges, apply in different formulations, either Lagrangian or Hamiltonian. How they affect the true degrees of freedom of the electromagnetic field may be clarified by our methods.

Acknowledgements

J.M.P. acknowledges support by the Comisión Interministerial para la Ciencia y la Tecnología (project number AEN-0695) and by a Human Capital and Mobility Grant (ERB4050PL930544). He also thanks the Center for Relativity at The University of Texas at Austin for its hospitality.

Bibliography

[1] J.M. Pons (1988), *J. Phys. A: Math. Gen.*, **21** 2705.

[2] P.A.M. Dirac (1950), *Can. J. Math.*, **2** 129.

[3] P.A.M. Dirac (1964), *Lectures on Quantum Mecanics.* Yeshiva Univ. Press, New York.

[4] K. Sundermeyer (1982), "Constrained Dynamics" in *Lecture Notes in Physics.* **169**, Springer Verlag, Berlin.

[5] M. Henneaux, C. Teitelboim and J. Zanelli (1969), *Nucl. Phys.*, **B332** 169.

[6] X. Gracia and J.M. Pons (1988), *Annals of Physics (N.Y.)*, **187** 2705.

[7] C. Batlle, J. Gomis, J.M. Pons and N. Roman (1986), *J. Math. Phys.*, **27** 2953.

[8] E.C.G. Sudarshan and N. Mukunda (1974), *Classical Dynamics: A Modern Perspective.* Wiley, New York.

[9] J.L. Anderson and P.G. Bergmann (1951), *Phys. Rev.*, **83** 1018.

[10] L. Castellani (1982), *Annals of Physics (N.Y.)*, **143** 357.

[11] R. Sugano, Y. Saito and T. Kimura (1986),*Prog. Theor. Phys*, **76** 283.

[12] J. Gomis, M. Henneaux, and J.M. Pons (1990), *Class. Quantum Grav.*, **7** 1089.

[13] M. Gotay, J.M. Nester and G. Hinds (1978), *J. Math. Phys.*, **19** 2388.

[14] J.M. Pons and L.C. Shepley (1994), 'Evolutionary Laws, Initial Conditions, and Gauge Fixing in Constrained Systems', in preparation.

Light-cone formulation of gauge theories

G. Fischer

1 Introduction

This work was initiated by investigations on gauge theories with the method of light-front quantization. Some facts found there force one to consider spinors as the fundamental building blocks in describing physical theories. Because we want to build up theories step by step starting from simple ingredients, we choose the simplest nontrivial spinors. These arise over a $1+1$-dimensional space-time and we call them light-cone (LC) spinors.

For the time being, by LC-formulation we understand a formulation of physical theories in terms of these LC-spinors. In addition, we consider the spinors as quantum states and the spinor formulation as a quantization of space-time manifold. In this picture, the Levi-Civita connection on the space-time is carried over to the Dirac operator.

We treat gauge theories over space-time by extending the spinor space by twisting with appropriate vector bundles what brings in the concept of twisted Dirac operators as quantizations of gauge potentials.

2 Clifford algebra and spinors over $1+1$ dimensional space-time

The $1+1$-dimensional space-time is given by a pseudo-Riemannian two-dimensional manifold $(M_{1,1}; \hat{q})$, where $\hat{q}$ is the metric tensor or space-time metric. The tangent space $V = T_x M_{1,1}$ of $M_{1,1}$ gets a quadratic form q from $\hat{q}$. q has a Lorentzian signature . The Clifford algebra $Cl(V, q)$ corresponding to (V, q) is generated by V and the relation

$$c(v) \cdot c(w) + c(w) \cdot c(v) = -2q(v, w), \tag{2.1}$$

with $c(v)$ being the element of $Cl(V)$ belonging to $v \in V$.

An important reason for considering the two-dimensional space-time as natural is that (V, q) is split. By this we mean that it is the sum of two maximal isotropic subspaces

$$V = V_+ \oplus V_- \tag{2.2}$$

$$q(v,v) = 0 \ , \ v \in V_\pm \ , \ \dim V_\pm = \frac{1}{2} \dim V. \tag{2.3}$$

This fact we call the **natural polarization** of two-dimensional space-time. By this polarization one is given a natural Clifford module by the Graßmann algebra of an isotropic subspace (we decided for V_-) ΛV_-, on which $Cl(V)$ is given as $\mathrm{End}(\Lambda V_-)$ [6].

One arrives at the LC-form of $Cl(V)$ by introducing LC-coordinates $x^+ = \frac{1}{\sqrt{2}}(x^0 + x^1)$ and $x^- = \frac{1}{\sqrt{2}}(x^0 - x^1)$ in V. The Clifford action on ΛV_- is defined via the action of an $v \in V$ as

$$c(v) = c(v^-) + c(v^+) := \sqrt{2}\epsilon(v^-) - \sqrt{2}\iota(v^+), \tag{2.4}$$

where $\epsilon(v^-)$ is exterior multiplication and $\iota(v^+)$ contraction by $q(v^+, \ )$ in ΛV_-. So it is a kind of deformed exterior multiplication on ΛV_- with the contraction term as deformation. In this form, a base for $Cl(V)$ is given by

$$Cl^0 : 1 = \begin{pmatrix} 1 & 0 \\ 0 & 1 \end{pmatrix} \tag{2.5}$$

$$Cl^1 : c_+ = \sqrt{2}\begin{pmatrix} 0 & -1 \\ 0 & 0 \end{pmatrix} , \ c_- = \sqrt{2}\begin{pmatrix} 0 & 0 \\ 1 & 0 \end{pmatrix} \tag{2.6}$$

$$Cl^2 : \Gamma = \begin{pmatrix} 1 & 0 \\ 0 & -1 \end{pmatrix} \tag{2.7}$$

This matrix algebra acts on the real two-dimensional Clifford module $\Lambda V_- \cong \mathbb{R}^2$.

With $Cl(V)$ and its module ΛV_- at hand, we form $\mathbb{Z}_2$-graded vector bundles over the space-time: the Clifford bundle $Cl(M_{1,1})$ assigns to every point of $M_{1,1}$ the Clifford algebra of its tangent space as fibre. This bundle is acting via its pointwise defined action on the spinor bundle S, the fibers of which are given by ΛV_-.

The space-time manifold should be an oriented spin manifold so that S is an associated spinor bundle of the principal bundle $\mathrm{Spin}(M) = P(M, \mathrm{Spin}(1,1))$.

An important fact of this split case is that the homogeneous elements of ΛV_- (the half spinors) are pure spinors (a pure spinor is annihilated under the Clifford action of a maximal isotropic subspace). So there is an isomorphism between the pure spinors (divided by the group $\mathbb{R}^*$) and the isotropic subspaces of V. The isotropic subspaces are the light rays of V and we are forced to call this representation of the light rays by spinors a "quantization of the light rays".

To give this phrase a more definite meaning we show in the following how one could consider spinor formulation as a $\mathbb{Z}_2$-graded analogue of a quantization.

3 Quantization

In the following two columns, we contrast the usual geometric quantization with the spinor formulation as $\mathbb{Z}_2$-graded (super) analogue of it.

usual	**$\mathbb{Z}_2$-graded**
symplectic manifold (N, ω)	"supersymplectic" manifold $(M_{1,1}, \hat{q})$
observables = Poisson algebra $(\mathcal{F}(N), \{\,,\})$ Poisson bracket induced by ω	super-Poisson algebra [1] $\Gamma(\Lambda V(M))$ super-Poisson bracket (Clifford product) induced by $\hat{q}$
prequantum line bundle over N $\longrightarrow$ Hilbert space	spinor bundle over $M_{1,1}$ $\longrightarrow$ $\mathbb{Z}_2$-graded Hilbert space $\mathcal{H} = \mathcal{H}^+ \oplus \mathcal{H}^-$

quantization

$\mathcal{F}(N) \longrightarrow Q(N)$ $Q(\phi) = i\hbar\nabla_{X_\phi} + \phi$	$\Gamma(\Lambda V(M)) \longrightarrow \Gamma(Cl(V))$ $v \longmapsto c(v)$

On the left hand side, the quantization [3] of a symplectic manifold with (antisymmetric) symplectic two-form is sketched, whereas on the right hand side the quantization of a "supersymplectic manifold" with (symmetric) "supersymplectic" two-form is defined. Essentially this quantization is the transition from Graßmann algebra to Clifford algebra. This resembles some kind of deformation of an algebra with the deformation $(\sqrt{2}\iota(v^+))$(2.4) induced by the quadratic form of q.

An important question especially for gauge theories is what role a connection plays in this quantization picture. In this Section we are dealing only with one connection, namely the unique Levi-Civita connection for the pseudo-Riemannian space-time manifold. To every connection

$$\nabla : \Gamma(\mathcal{E}) \longrightarrow \Gamma(T^*M \otimes \mathcal{E}) \tag{3.1}$$

($\mathcal{E}$ vector bundle over $M_{1,1}$) there corresponds a unique covariant derivative along a vector field X_i on $M_{1,1}$

$$\nabla_i := \iota(X_i) \circ \nabla : \Gamma(\mathcal{E}) \longrightarrow \Gamma(\mathcal{E}). \tag{3.2}$$

Hence we define the **quantization of the connection** ∇ by the operator

$$D := c_i \cdot \nabla_i \tag{3.3}$$

Eq. (3.3) is the usual definition of a Dirac operator [2]. D acts as an odd operator in the Hilbert space $\mathcal{H}$ which is formed from appropriate (square integrable) sections of S. By squaring a Dirac operator one gets a generalized Laplace operator

$$H := D^2. \tag{3.4}$$

In cartesian coordinates with constant metric tensor $\hat{q}$ the Christoffel symbols are vanishing (covariant derivatives are equal to partial derivatives) and both D and H get the simple form

$$\begin{aligned} D &= \begin{pmatrix} 0 & -\sqrt{2}\partial_+ \\ \sqrt{2}\partial_- & 0 \end{pmatrix} \\ H &= -2\partial_+\partial_- \quad \text{(wave operator).} \end{aligned} \tag{3.5}$$

With these two operators at hand we have the following $\mathbb{Z}_2$-graded quantum mechanics for the space-time resembling a supersymmetric system:
The Hilbert space $\mathcal{H} = \mathcal{H}^+ \oplus \mathcal{H}^-$ is the state space for even ($\in \mathcal{H}^+$) and odd ($\in \mathcal{H}^-$) particles. The odd part of the Clifford algebra delivers creation ($c_\pm$) and annihilation operators ($c_\mp$) for even (odd) and odd (even) particles. States in $\mathcal{H}^\pm$ are annihilated by $c_\pm$ so the elements of $\mathcal{H}^\pm$ are the vacuum states for even (odd) particles. The products of c_+, c_- yield number operators $N_e = c_+c_-$ and $N_o = c_-c_+$ for the even and odd part, respectively. Furthermore $\Gamma \in Cl^2$ measures the degree of the states in $\mathcal{H}$. The Dirac operator takes over the role of the supersymmetry generator of the system making the transition from even to odd states. Its square H is the Hamilton operator the eigenvalues of which are the energies.

With respect to these two operators one essentially distinguishes two sectors in the statespace $\mathcal{H}$. The first one is formed by states with nonvanishing energy $H\psi \neq 0$. These states are paired by the Dirac operator. Eigenvalue equations in this sector have the form of Dirac equations for massive particles. The second one is the zero-mode sector of D with states satisfying $D\psi = 0$. These states are the supersymmetric ones. Information on the global structure of the system, especially the global topology of the space-time, is contained in this sector. For later investigation of global geometry (vacuum) of gauge theories we will restrict ourselves to the zero-mode sector of D. With the simple form of the Dirac operator (3.5)

$$D\psi = \begin{pmatrix} -\sqrt{2}\partial_+\psi_2 \\ \sqrt{2}\partial_-\psi_1 \end{pmatrix} = 0 \tag{3.6}$$

the states separate into those depending only on x^+ and those depending only on x^-. This fact allows for using here the LC-splitting ($V = V_+ \oplus V_-$) in the following way

$$\Lambda V = \Lambda(V_+ \oplus V_-) \cong \Lambda V_+ \otimes \Lambda V_- \tag{3.7}$$

From the above discussion we know that the Graßmann algebra of a vector space with quadratic form is a super-Poisson algebra [1]. By splitting V we get the product of two super-Poisson algebras what means, in the language of Section 2, that we have split the quantization into two parts. The corresponding classical phase spaces to be quantized are the null-leaves of the space-time manifold

$$M_{1,1} = M_+ \times M_-. \tag{3.8}$$

But on $V_\pm$ the quadratic form controlling the transition from ΛV to $Cl(V)$ is vanishing. Therefore the Clifford algebra and the Graßmann algebra are identical. In the super-quantization setting this means that the quantization is given by an identity map. So we have the "equation"

quantum = classical.

A system for which this equality is realized we call a **free theory**. The motivation for this definition of a free theory lies in the fact that in the future we only want to allow for quadratic forms arising by an interaction (Cartan-Killing forms of Lie algebras of gauge groups) in order to unify the concepts of quantization and interaction.

Now the splitting also is carried over to the Dirac operator which get decomposed into two exterior covariant derivatives on $\Lambda V_\pm$

$$D \begin{array}{l} \nearrow D_+ = \sqrt{2}\epsilon(e^+)\partial_+ \\ \searrow D_- = \sqrt{2}\epsilon(e^-)\partial_-. \end{array} \tag{3.9}$$

Because of this decomposition, the investigation of the Dirac operator can be done separately for the x^+ and x^-–parts.

4 Gauge theories

In the first three Sections, we have sketched the general settings of LC-formulation. Now we want to use it for gauge theories. A gauge theory on space-time is defined by a principal fibre bundle $P(M_{1,1}, G)$ over $M_{1,1}$ with gauge group G. Roughly speaking we want to extend the quantization of $M_{1,1}$ to a quantization of the extended manifold P. The method we apply is twisting [2], what means extending the spinor bundle S by multiplying it with associated vector bundles for P.

$$S_G = S \otimes \mathcal{E}. \tag{4.1}$$

In the former Sections, we were dealing with a single connection (Levi-Civita) whereas in gauge theories we are confronted with an infinite dimensional space $\mathcal{A}$ of connections. Each connection $\alpha \in \mathcal{A}$ on P gives rise to a connection for the twisted spinor bundle

$$\nabla^\alpha(s \otimes e) = (\nabla s) \otimes e + s \otimes (\nabla^{\alpha,\mathcal{E}} e) \text{ with } (s \otimes e) \in S_G \tag{4.2}$$

where in the first term the Levi-Civita connection is acting and in the second one the gauge connection α. With ∇^α we define the **twisted Dirac operator**

$$D^\alpha = c_i \cdot \nabla_i^\alpha \tag{4.3}$$

as in (3.3). This gives a whole family $\mathcal{D}(A)$ of Dirac operators labelled by $\mathcal{A}$. On the space $\mathcal{A}$ of gauge potentials there acts the group of gauge transformations $\mathcal{G}$ as symmetry group ($\mathcal{L}G$ symbols the Lie algebra of $\mathcal{G}$), so there are gauge equivalent elements in $\mathcal{A}$ and $\mathcal{D}(A)$. This makes gauge theory a constrained system which thus needs a reduction in order to get the true degrees of freedom. In the following, we use some known objects and notations of gauge theory [4]: the associated bundle ad(P), the typical fibre of which is the Lie algebra of G, and the spaces of differential-forms over $M_{1,1}$ with values in ad(P), denoted by $\Omega_c^*(M_{1,1}, \mathrm{ad}(P))$ (c indicates compact support). The space of connections is an affine space with tangent space $\Omega_c^1(M_{1,1}, \mathrm{ad}(P))$ [4]. By choosing an origin, we identify $\mathcal{A}$ with this space of Lie algebra valued one-forms. Beyond that $\mathcal{A}$ carries a natural symplectic structure

$$\omega(A, B) = \int_{M_{1,1}} tr(A \wedge B) \ , \ A, B \in \Omega^1(M_{1,1}, \mathrm{ad}(P)) \tag{4.4}$$

and there is a moment map $\mu : \mathcal{A} \longrightarrow \mathcal{L}G^*$ for the $\mathcal{G}$-action. Because of the two-dimensionality of space-time the moment map is the curvature (two-forms dual to zero-forms). The whole story is sketched in the following picture, where σ denotes infinitesimal gauge transformations.

$$\begin{array}{ccccc} \Omega_c^0(M, \mathrm{ad}(P)) & \xrightarrow{d_0^\alpha} & \Omega_c^1(M, \mathrm{ad}(P)) & \xrightarrow{d_1^\alpha} & \Omega_c^2(M, \mathrm{ad}(P)) \\ \| & & \| & & \| \\ \mathcal{L}G & \xrightarrow{\sigma} & \mathcal{A} & \xrightarrow{\mu} & \mathcal{L}G^* \end{array} \tag{4.5}$$

With these identifications we can do a Marsden-Weinstein (MW) like symplectic reduction [5] with one difference: we are dividing out the Lie algebra instead of the symmetry group. So we call the resulting quotient

$$\frac{\mu^{-1}(0)}{\mathcal{L}G} = \frac{\ker d_1^\alpha}{\operatorname{im} d_0^\alpha} = H^1(M_{1,1}, \operatorname{ad}(P)) \tag{4.6}$$

"infinitesimal MW-quotient". It is equivalent to the first cohomology group in the sequence (4.5), which is a complex for flat connections α. Taking into account that we are dealing with a product manifold $M_{1,1} = M_+ \times M_-$ we apply a Künneth formula for the reduction

$$H^1(M_{1,1}, \operatorname{ad}(P)) = H^1(M_+, \operatorname{ad}^+(P)) \oplus H^1(M_-, \operatorname{ad}^-(P)) \tag{4.7}$$

($\operatorname{ad}^\pm(P)$ are the $\operatorname{ad}(P)$-like bundles over $M_\pm$). In this way we get a splitting of the reduction which fits together with the LC-splitting for the zero-mode sector of D. Thus we can also investigate the + and − part of the twisted Dirac operators separately

$$\begin{array}{l} D_+^\alpha = \epsilon(e^+)\nabla_+^{\alpha+} \ ; \ \alpha_+ \in H^1(M_+, \operatorname{ad}^+(P)) \\ D_-^\alpha = \epsilon(e^-)\nabla_-^{\alpha-} \ ; \ \alpha_- \in H^1(M_-, \operatorname{ad}^-(P)) \end{array} . \tag{4.8}$$

As a simple example for this reduction we consider a $U(1)$ gauge theory over the space-time $M_{1,1} = \mathbb{R} \times S^1$. For the twisted spinor bundle we take the tensor product with a trivial complex line bundle $S_{U(1)} = S \otimes (M \times \mathbb{C})$. The $\mathcal{A}$-reduction leads to the first de Rham-cohomology group of the space-time $H^1(M_{1,1}, \mathbb{R}) = H^1(S^1) = \mathbb{R}$. Hence the possible values for the gauge potentials are $\alpha_- \in \mathbb{R}$, $\alpha_+ = 0$ (LC-gauge). But, because of the nonvanishing fundamental group of $M_{1,1}$, there also exist large gauge transformations. These have no corresponding constraints and so they survived the former reduction. To get a full reduction one has to divide the infintesimal MW-quotient by the fundamental group

$$\frac{\mu^{-1}(0)/\mathcal{L}G}{\pi_1(S^1)} = \frac{\mu^{-1}(0)}{\mathcal{G}} = \mathbb{R}/\mathbb{Z} = S^1 \tag{4.9}$$

The reduced $\mathcal{A}$ represents the true degree of freedom. In this case this corresponds to the holonomy of the flat connections on P. For the Dirac operator the reduction restricts the attention essentially to

$$D_-^{\alpha-} = \epsilon(e^-)(i\partial_- + \alpha_-) \tag{4.10}$$

and the investigation of this operator leads to the known results (spectral flow, Θ-vacuum). This example uses LC-formulation in the zero-mode sector. A future work will be to consider the paired sector especially pairings made by curvature in gauge theories.

References

[1] B. Kostant and S. Sternberg (1987), 'Symplectic reduction, BRS cohomology, and infinite-dimensional Clifford algebras', *Ann. Phy. (N.Y.)*, **176** 49.

[2] H.B. Lawson and M.L. Michelsohn (1989), 'Spin geometry', Princeton Mathematical Series **38**, Princeton Univ. Press, Princeton, NJ.

[3] J. Rawnsley (1990), 'Quantization on Kähler manifolds', Lecture Notes in Physics **375**.

[4] M.F. Atiyah and R. Bott (1982), 'The Yang-Mills equations over Riemann surfaces', *Phil. Trans. R. Soc. Lond.* A **308** 523.

[5] J. Marsden and A. Weinstein (1974), 'Reduction of symplectic manifolds with symmetry', *Rep. Math. Phys.*, **5** 121.

[6] N. Berline, E. Getzler and M. Vergne (1992), 'Heat kernels and Dirac operators', *Grundlehren der mathematischen Wissenschaften*, **298**, Springer-Verlag, Berlin.

Hamiltonian constraints and Dirac observables

Luca Lusanna

Presymplectic manifolds underlie all relevant physical theories, since all of them are, or may be, described by singular Lagrangians [1] and therefore by Dirac-Bergmann constraints [2] in their Hamiltonian description. In Galilean physics both Newtonian mechanics [3] and gravity [4] have been reformulated in this framework. In particular one obtains a multi-time formulation of non-relativistic particle systems, which generalizes the non-relativistic limit of predictive mechanics [5] and helps one to understand features unavoidable at the relativistic level, where each particle, due to manifest Lorentz covariance, has its own time variable. Instead, all both special and general relativistic theories are always described by singular Lagrangians. See the review in Ref.[6] and Ref.[7] for the so-called multi-temporal method for studying systems with first class constraints (second class constraints are not considered here).

The basic idea relies on Shanmugadhasan canonical transformations [8], namely one tries to find a new canonical basis in which all first class constraints are replaced by a subset of the new momenta [when the Poisson algebra of the original first class constraints is not Abelian, one speaks of Abelianization of the constraints]; then the conjugate canonical variables are Abelianized gauge variables and the remaining canonical pairs are special Dirac observables in strong involution with both Abelian constraints and gauge variables. These Dirac observables, together with the Abelian gauge variables, form a local Darboux basis for the presymplectic manifold [9] defined by the first class constraints (maybe it has singularities) and coisotropically embedded in the ambient phase space when there is no mathematical pathology. In the multi-temporal method each first class constraint is raised to the status of a Hamiltonian with a time-like parameter describing the associated evolution (the genuine time describes the evolution generated by the canonical Hamiltonian, after extraction from it of the secondary and higher order first class constraints): in the Abelianized form of the constraints these "times" coincide with the Abelian gauge variables on the solutions of the Hamilton equations. These coupled Hamilton equations are the multi-temporal equations: their solution describes the dependence of the original canonical variables on the time and on the parameters of the infinitesimal gauge transformations, generated by the first class constraints. Given an initial point on the constraint manifold, the general solution describes the gauge orbit, spanned by the gauge variables, through that point; instead the time evolution generated by the canonical Hamiltonian (a first class quantity) maps one gauge orbit into another one. For each system the main problems are whether the constraint set is a manifold (a stratified manifold, a manifold with singularities...), whether the gauge orbits can be built in the large starting from infinitesimal gauge transformations, and whether the foliation of the constraint manifold (of each stratum of it) is either regular or singular. Once these problems are understood, one can check whether the reduced phase space (Hamiltonian orbit space) is well defined and which is the most convenient set of gauge-fixings able to yield a realization of it.

Usually Shanmugadhasan canonical transformations are defined only locally, i.e. they are not defined on a whole domain containing the constraint set or at least its stratum under investigation. However, since all relevant physical systems must be defined in Minkowski space-time [with their non-relativistic limit to be understood as a limit in which the velocity of light c is very big with respect to all other velocities but finite; the exact Galilean limit $c \rightarrow \infty$ is a contraction which destroys the genuine relativistic effects due to the geometry of a space-time with Lorentzian signature (+,-,-,-)], and since for all them it is assumed that the Poincaré algebra and group are globally implemented, then it turns out (at least for all systems investigated till now) that there is a special family of Shanmugadhasan canonical transformations globally defined. To find it, one has to take into account the Lorentz signature of Minkowski space-time and the consequent existence of various types of Poincaré orbits with different little groups, and to find canonical variables adapted to the Poincaré group (center-of-mass decompositions), to the geometry of the Poincaré orbits and to the first class constraints. The source of globality may be traced back to the existence of the momentum map for the Poincaré group action on the constraint manifold (or to its different actions on the various strata of the constraint manifold as we shall see). By means of these global Shanmugadhasan canonical transformations one obtains a global symplectic decoupling of the associated Dirac observables from the Abelian constraints and gauge variables: in this way one gets natural realizations of the reduced phase space without the necessity of adding gauge-fixings [10]. The price for this naturalness is that the final canonical Hamiltonian is usually non-linear and non-local in the new variables and one does not know how to quantize in the standard ways.

Therefore a research program was started with the aim to identify all the consequences of the following three hypotheses for special relativistic either pointlike or extended systems: i) Lorentz signature of Minkowski space-time; ii) Hamiltonian description with first class constraints (presymplectic approach); iii) global implementation of the Poincaré group.

With special relativistic systems, for which, by assumption, there is a global implementation of the Poincaré algebra and group, the constraint manifold is a stratified manifold, a disjoint union of strata corresponding to the various types of Poincaré orbits. The main stratum consists of all the configurations of the system with time-like total four-momentum $P^2 > 0$

(with the further subdivision $P_o > 0$ or $P_o < 0$); then there are secondary strata corresponding to $P^2 = 0$ ($P_o > 0$ or $P_o < 0$) and $P_\mu = 0$ (the stratum $P_\mu = 0$ is connected with the classical background of the infrared divergences). The stratum corresponding to space-like orbits $P^2 < 0$ must be absent not to have tachyonic effects. The geometry of the various orbits is different due to the inequivalent little groups. To study the strata $P^2 = 0$ and $P_\mu = 0$ one has to add to the original first class constraints the extra first class ones $P^2 \approx 0$ and $P_\mu \approx 0$ respectively. Since on the main stratum $P^2 > 0$, one has the form $W^2 = -P^2 \vec{\bar{S}}^2$ of the Pauli-Lubanski Casimir of the Poincaré group, this stratum is divided in the two substrata $\vec{\bar{S}} \neq 0$ and $\vec{\bar{S}} = 0$, corresponding to the two inequivalent kinds of orbits of the rotation group generated by the rest-frame Thomas spin $\vec{\bar{S}}$ (again one has to add $\vec{\bar{S}} \approx 0$ to the original constraints to describe this substratum). In the stratum $P^2 = 0$, when the little group is O(2), one has $W^\mu = \lambda P^\mu$, $\lim_{P^2 \to 0} W^2/P^2 = -\lambda^2$, where $\lambda = \vec{P} \cdot \vec{J}/P_o$ is the Poincaré invariant helicity; again one must distinguish $\lambda = 0$ from $\lambda \neq 0$.

Let us now review the one-, two- and many-body systems of relativistic mechanics , which have been described with first class constraints (see Refs.[11] for the bibliography; only the papers of the Firenze group will be quoted, since all them are formulated with a homogeneous formalism). The importance of relativistic mechanics stems from the fact that quantum field theory has no particle interpretation: this is forced on it by means of asymptotic states which, till now, correspond to the quantization of independent one-body systems described by relativistic mechanics [or relativistic pseudoclassical mechanics [12], when one adds Grassmann variables to describe the intrinsic spin]. Besides the scalar particle ($P^2 - m^2 \approx 0$ or $P^2 \approx 0$), one has control on: i) the pseudoclassical electron [13] ($P_\mu \xi^\mu - m\xi_5 \approx 0$ or $P_\mu \xi^\mu \approx 0$, where ξ^μ, ξ_5 are Grassmann variables; $P^2 - m^2 \approx 0$ or $P^2 \approx 0$ are implied; after quantization the Dirac equation is reproduced); ii) the pseudoclassical neutrino [14] ($P_\mu \xi^\mu + \frac{i}{3}\epsilon^{\mu\nu\rho\sigma} P_\mu \xi_\nu \xi_\rho \xi_\sigma \approx 0$, $P^2 \approx 0$, giving the Weyl particle wave equation $P_\mu \gamma^\mu (1 - \gamma_5)\psi(x) = 0$ after quantization); iii) the pseudoclassical photon [15] ($P^2 \approx 0$, $P_\mu \theta^\mu \approx 0$, $P_\mu \theta^{*\mu} \approx 0$, $\theta^*_\mu \theta^\mu \approx 0$, where $\theta^\mu, \theta^{*\mu}$ are a pair of complex Grassmann four- vectors to describe helicity ± 1; after quantization one obtains the photon wave equations $\Box A^\mu(x) = 0$, $\partial_\mu A^\mu(x) = 0$; the Berezin-Marinov Grassmann distribution function allows to recover the classical polarization matrix of classical light and, in quantization, the quantum polarization matrix with the Stokes parameters); iv) the vector particle or pseudoclassical massive photon [16] ($P^2 - \mu^2 + (1-\lambda) P_\mu \theta^{*\mu} P_\nu \theta^\nu \approx 0$, $\theta^*_\mu \theta^\mu \approx 0$, which, after quantization, reproduce the Proca-like wave equation $(\Box + \mu^2) A^\mu(x) - (1-\lambda)\partial^\mu \partial_\nu A^\nu(x) = 0$).

Among the two-body systems, the most important is the DrozVincent-Todorov-Komar model [17] with an arbitrary action-at-a-distance interaction instantaneous in the rest frame as shown by its energy-momentum tensor [18] ($P_i^2 - m_i^2 + V(R_\perp^2)) \approx 0$, i=1,2, $R_\perp^\mu = (\eta^{\mu\nu} - P^\mu P^\nu/P^2)R_\nu$, $R^\mu = x_1^\mu - x_2^\mu$, $P_\mu = P_{1\mu} + P_{2\mu}$). This model has been completely understood both at the classical and quantum level [19] (and references therein). Classically it allowed the discovery of the following sequence of canonical transformations: i) from $[x_i^\mu, p_{i\mu}]$, i=1,2, to a generic set of center-of-mass and relative variables $[X^\mu = x_1^\mu + x_2^\mu, P_\mu, R^\mu, Q_\mu = \frac{1}{2}(p_{1\mu} - p_{2\mu}]$; ii) since the model is defined only for $P^2 > 0$, one can use the standard Wigner boost for these orbits $L^\mu{}_\nu(P, \overset{o}{P}) = \epsilon^\mu{}_\nu(P/\eta\sqrt{P^2}) = \eta^\mu_\nu + 2P^\mu \overset{o}{P}_\nu/P^2 - (P^\mu + \overset{o}{P}{}^\mu)(P_\nu + \overset{o}{P}_\nu)/(P + \overset{o}{P}) \cdot \overset{o}{P}$, where $\eta = \mathrm{sign}\, P_o$ and $\overset{o}{P}_\mu = (\eta\sqrt{P^2}; \vec{0})$, to boost at rest the relative variables R^μ, Q_μ: in this way one gets the canonical basis $[\tilde{X}^\mu, P_\mu, \epsilon_R = P \cdot Q/\eta\sqrt{P^2}, T_R = P \cdot R/\eta\sqrt{P^2}, \vec{\tilde{R}}, \vec{\tilde{Q}}]$, where $\vec{\tilde{R}}, \vec{\tilde{Q}}$ are Wigner spin 1 three-vectors, but $\tilde{X}^\mu$ is not a four-vector; iii) finally, from the center-of- mass variables $[\tilde{X}^\mu, P_\mu]$ one goes to the canonical basis $[\epsilon = \eta\sqrt{P^2}, T = P \cdot \tilde{X}/\eta\sqrt{P^2} = P \cdot X/\eta\sqrt{P^2}, \vec{z} = \eta\sqrt{P^2}(\vec{\tilde{X}} - \vec{P}\tilde{X}^o/P_o), \vec{k} = \vec{P}/\eta\sqrt{P^2}]$, where $\vec{z}$, apart a dimensional factor, is the canonical non-covariant classical analogue of the Newton-Wigner operator in presence of spin (orbital angular momentum here). In this final canonical basis one has: a) the Casimir P^2 is in the basis via ϵ and its conjugate variable is the center-of-mass time in the rest frame; b) suitable combinations of the constraints may be written as $\chi_R = \epsilon_R - (m_1^2 - m_2^2)/2\epsilon \approx 0$, $\chi = \epsilon^{-2}[\epsilon^2 - M_+^2(\vec{\tilde{R}}, \vec{\tilde{Q}})][\epsilon^2 - M_-^2(\vec{\tilde{R}}, \vec{\tilde{Q}})] \approx 0$: the first one determines the relative energy ϵ_R (so that the conjugate gauge variable is the relative time T_R), while the other one determines the four branches of the mass spectrum $[\epsilon = \pm M_\rho = \pm(\sqrt{m_1^2 + \vec{\tilde{Q}}^2 + V(-\vec{\tilde{R}}^2)} + \rho\sqrt{m_2^2 + \vec{\tilde{Q}}^2 + V(-\vec{\tilde{R}}^2)})$, $\rho = \pm$; for equal masses and $P^2 > 0$ there are only two branches $\epsilon = \pm 2\sqrt{m^2 + \vec{\tilde{Q}}^2 + V(-\vec{\tilde{R}}^2)}$, since the two branches with $\epsilon = 0$ have $P^2 = 0]$ and T is the conjugate gauge variable; c) therefore our final canonical basis $[\epsilon, T, \epsilon_R, T_R, \vec{z}, \vec{k}, \vec{\tilde{R}}, \vec{\tilde{Q}}]$ is a quasi-Shanmugadhasan one (by redefining ϵ_R and T, the constraint $\chi_R \approx 0$ may be put in the form $\epsilon_R' \approx 0$) with $\vec{z}, \vec{k}, \vec{\tilde{R}}, \vec{\tilde{Q}}$ playing the role of Dirac observables (with respect to χ_R and T_R) with the evolution determined by $\chi \approx 0$ [i.e. there are four different Hamiltonians $\pm M_\rho(\vec{\tilde{R}}, \vec{\tilde{Q}})$ for the four branches]: in this case, by replacing χ with $\epsilon - (\pm M_\rho(\vec{\tilde{R}}, \vec{\tilde{Q}})) \approx 0$ one could find four Shanmugadhasan bases, one for each branch, and therefore the four sets of final Dirac observables (Jacobi data), but in general a constraint of the kind $\epsilon - H \approx 0$ can be included globally in such bases only if the dynamics generated by the Hamiltonian H is Liouville integrable. At the quantum level, while one has a system of coupled integro-differential Klein-Gordon equations coming from the constraints in the original variables, one gets coupled differential equations in the final ones (in these variables one can define a Cauchy problem). The mass spectrum and the elementary solutions have been found in both the formulations [19] for all potentials for which a complete set of eigenfunctions is known for the operator $-\vec{\nabla}_R^2+V(-\vec{\tilde{R}}^2)$. Moreover, in both cases four scalar products, compatible with both equations (i.e. independent from T and T_R), have been found as generalization of the two existing scalar products of the Klein-Gordon equation: all of them are non-local even in the limiting free case and differ among themselves for the sign of the norm of states on different mass-branches.

The connection with the Bethe-Salpeter equation of the quantized model has been studied in Ref.[20], where it is shown that the constraint wave function can be obtained from the Bethe-Salpeter one by multiplication for a delta function containing the relative energy ϵ_R to exclude the spurious solutions.

The extension of the model to two pseudoclassical electrons and to an electron and a scalar has been done in Ref.[21], and the first was used to get good fits to meson spectra.

While in the literature there are various 3- and n-body non-separable models with first class constraints, see the talk of Longhi[22] for the difficult case of 3-body separable first class constraints.

Of particular importance is the canonical transformation $(x_i^\mu, p_{i\mu}) \mapsto (x^\mu, P_\mu, R_a^\mu, Q_{a\mu}); i = 1,..,n;\ a = 1,..,n-1$, of Ref.[23], which transforms the first class constraints $p_i^2 - m_i^2 \approx 0$ of n free scalar particles into $P \cdot Q_a \approx 0; a = 1,..,n-1$, and an overall mass-shell constraint determining the 2^n branches of the mass spectrum (2^n is a topological number, which is broken when some interaction destroys some mass gap). In analogy to the two-body case for $P^2 > 0$, there is a quasi-Shanmugadhasan canonical transformation from $[x_i^\mu, p_{i\mu}]; i = 1,..,n$ to the base $[\epsilon = \eta\sqrt{P^2}, T = P \cdot X/\eta\sqrt{P^2}, \epsilon_{Ra} = P \cdot Q_a/\eta\sqrt{P^2}, T_{Ra} = P \cdot R_a/\eta\sqrt{P^2}, \vec{z}, \vec{k}, \vec{\tilde{R}}_a, \vec{\tilde{Q}}_a]; a = 1,..,n-1$, with $\vec{\tilde{R}}_a, \vec{\tilde{Q}}_a$ Wigner spin 1 vectors: the n$-$1 constraints $\epsilon_{Ra} \approx 0$ are among the new momenta and their conjugate variables are the n$-$1 relative times T_{Ra}. It is now under investigation with Pauri and Lucenti how to go from the sub-base $[\vec{\tilde{R}}_a, \vec{\tilde{Q}}_a]$ to a sub-base containing $\tilde{S} = \mid \vec{\tilde{S}} \mid$, so that this final canonical basis would contain both the Poincaré Casimirs, besides n$-$1 constraint variables, and therefore would include completely the geometry of the orbits with $P^2 > 0$; this n-body kinematics would contain variables relative not only to the center of mass but also to the total spin (orbital angular momentum) and could be useful in many fields of physics to take into account the effects of the Lorentz signature of Minkowski space-time also at low velocities (quasi-Galilean limit). Another feature which is going to be clarified is the dependence of the inner Lorentz group generated by $S^{\mu\nu}$ $[J^{\mu\nu} = L^{\mu\nu} + S^{\mu\nu},\ L^{\mu\nu} = X^\mu P^\nu - X^\nu P^\mu]$ and of the zitterbewegung of center-of-mass variables from the relative time gauge variables. Next, by replacing the constraints $p_i^2 - m_i^2 \approx 0$ with $p_i \cdot \xi_i - m_i \xi_{5i} \approx 0$ (with ξ_i^μ, ξ_{5i} and ξ_j^μ, ξ_{5j} commuting for $i \neq j$), one should be able to include the intrinsic spin in the final canonical basis.

Both the open and closed Nambu string, after an initial study with light-cone coordinates, have been treated [24] along the lines of the two-body model in the sector $P^2 > 0$. Both Abelian Lorentz scalar constraint and gauge variables have been found and globally decoupled, and a redundant set of Dirac observables $[\vec{z}, \vec{k}, \vec{a}_n]$ has been found. It remains an open problem whether one can extract a global canonical basis of Dirac observables from the Wigner spin 1 vectors $\tilde{a}_n$, which satisfy sigma-model-like constraints; if this basis exists (maybe the previous spin bases could help in this search), it would define the Liouville integrability of the Nambu string and would open the way to quantize it in four dimensions.

When a singular Lagrangian may be reconstructed from the first class constraints of relativistic mechanics, it is reparametrization invariant; to this invariance, via the Noether identities implied by the second Noether theorem, there corresponds the mass-shell constraint, which is quadratic in the momentum for the one-body systems. The meaning of the gauge invariance is that the observer has the freedom to make any choice of what is the "time" with which evolution is described. For many-body systems the other constraints may be chosen linear in the momenta (like $P \cdot Q_a \approx 0$) and the conjugate variables are the relative times: again their gauge nature implies the freedom of the observer to describe the system with any given delay among the pairs of constituents. In the continuum case of the Nambu string, half of the gauge variables may be interpreted as the time and the relative times of the points of the string, while the other half as the freedom for the observer to define what is the "longitudinal space". In all cases the gauge freedom is of the kind of general relativity, in which there is built in the freedom to define what are time and space. Instead in gauge theories, like electromagnetism and Yang-Mills theories, the gauge variables correspond to unobservable degrees of freedom. However in all cases one interprets the first class constraints as restrictions on the Cauchy data of the associated Euler-Lagrange equations: in phase space the Dirac observables are the independent Cauchy data (or Jacobi data, when the canonical Hamiltonian vanishes).

After the Nambu string, this methodology has been applied to classical gauge theories following the pioneering work of Dirac [25] for electromagnetism. By considering a 3+1 splitting of Minkowski space-time and Yang-Mills theory for a trivial principal bundle over the fixed-time Euclidean space R^3 with a semisimple compact, connected, simply connected Lie group as structure group, it was possible to find an exact global Shanmugadhasan canonical transformation to make a symplectic decoupling [26] of the Abelianization of the Gauss' laws and of the conjugate Abelian gauge variables from the Dirac observables, which turn out to be suitable Lie algebra valued transverse vector gauge potentials and transverse electric fields like in the electromagnetic case. This is possible in suitable weighted Sobolev spaces, in which the covariant divergence is an elliptic operator without zero modes [27] and the Gribov ambiguity is absent [otherwise it is the source of a further stratification of the constraint manifold [26] and of the presence of cone over cone singularities due to the stability subgroups of gauge transformations of certain gauge potentials and field strengths]. The discovery of the Green function for the covariant divergence allowed to solve the Gauss law constraints and to find the Green functions of the Faddeev-Popov operator and the square of the covariant derivative in the case of transverse gauge potentials. After the construction of a connection-dependent coordinatization of the trivial principal bundle based on generalized canonical coordinates of first kind on the fibers, the multi-temporal equations for the gauge potential were solved: the gauge potential is decomposed in a pure gauge background connection (the Maurer-Cartan one-form on the group of gauge transformations or BRST ghost) and in a gauge-covariant magnetic gauge potential, whose transversality properties were found by using a generalized Hodge decomposition for one-forms based on the BRST operator interpreted as the vertical derivative on the principal bundle. After an analoguous decomposition of the electric field into transverse and longitudinal parts (the latter containing transverse gauge potential, transverse electric field and Gauss law contributions), the Dirac observables are identified as the restriction to the identity cross-section of the trivial principal bundle of the transverse gauge potential and transverse electric field. Also the gauge invariant Dirac observables (but in this case not physical observables) of Grassmann-valued fermion fields are determined. The physical Lagrangian and Hamiltonian, and the non-Abelian and topological charges are obtained in terms of the previous Dirac observables; the form of the Lagrangian is obtained by means of an explicit realization of the abstract Riemannian metric of Mitter and Viallet built by using the found Green functions. When the structure group is SU(3), one has the classical basis of quantumchromodynamics (QCD). The fundamental role played by the identity cross-section in the determination of global Dirac observables raises the problem whether such observables exist with non-trivial principal bundles and whether the theory of quantum anomalies should be rephrased as the theory of obstructions to the existence of global Dirac observables; if a classical theory does not admit these observables, then either it is already pathological at the classical level or one has to find a physical interpretation of certain gauge degrees of freedom. One has found also the classical background of the quantum superselection rules: since one has Dirac observables Q_a for the non-Abelian charges, one can define a classical superselection sector as the subset of the Dirac observables which have vanishing Poisson brackets with the charges Q_a, i.e. as the subset of scalar observables, and a certain value of the Casimirs like $\sum_a Q_a^2$. In this way only even functions of the "unobservable" Grassmann-valued fermionic Dirac observables are selected. One could think to impose confinement of elementary fermions in a QCD scheme by adding the extra first class constraints $Q_a \approx 0$. It is an open problem whether there is a symplectic structure on the scalar Dirac observables, so that one could quantize only a superselection sector instead of applying the superselection rules after quantization. Having eliminated the gauge degrees of freedom and taken into account the non-localities implied by the implementation of the Poincaré group, the Haag-Kastler program of "local observables" (i.e. localized on compact domains) for gauge theories should start from Dirac observables. Also the role of the center of the group of gauge transformations and of the winding number have been discussed. Instead, it is still open the problem of finding Dirac observables for the standard model of electroweak interactions with its Higgs mechanism for symmetry breaking: in it some of the Gauss laws have to be solved in the momenta of the Higgs fields and not in the Yang-Mills momenta.

Both the Lagrangian and Hamiltonian are non-local and non-polynomial, but without singularities in the coupling constant; as in the Coulomb gauge they are not Lorentz invariant, but the invariance can be enforced on them if one reformulates the theory for the stratum $P^2 > 0$ on space-like hypersurfaces (on light-cones for the stratum $P^2 = 0$) following Dirac [2]. With special relativistic theories, one can restrict the space-like hypersurfaces to hyperplanes orthogonal to P_μ for $P^2 > 0$ (it could be called a Wigner foliation of Minkowski space-time): in this way the transverse gauge potential and electric field become Wigner spin 1 three-vectors and the final dynamics is governed by the first class constraint $\epsilon - H_P \approx 0$, with H_P being the physical Lorentz invariant Hamiltonian. Another byproduct of this construction is the indirect proof of the existence of a center-of-mass decomposition also for classical field theory: namely there is a canonical basis containing center-of-mass variables $\tilde{X}^\mu$, P_μ or $\epsilon, T, \vec{z}, \vec{k}$ plus an infinite number of relative variables with Wigner covariance. Even if this basis has still to be constructed (and even if the independent Dirac observables from the transverse ones have not yet been extracted with a control on Euclidean covariance), the important point is that for all extended special relativistic systems one arrives at a final canonical basis adapted to the Poincaré orbits and to the first class constraints, which contains Dirac observables with Wigner covariance and the non-covariant canonical three-vector $\vec{z}$.

Now in the literature there are three relevant concepts [28] of center-of-mass position all coinciding in the rest frame: i) the canonical non-covariant position (also called center of spin; it is the classical basis of the Newton-Wigner operator in presence of spin), whose role, apart dimensions, is played by $\vec{z}$ in the space of Dirac observables; ii) the covariant non-canonical Fokker center of inertia, obtained from i) by Lorentz transformations; iii) the non-covariant non-canonical Møller center of energy. Only the second one defines an intrinsic world line and it can be shown [28] that all the pseudo-world-lines associated with the other two positions in all possible reference frames fill a world-tube around the Fokker center of inertia, whose intrinsic transverse radius is determined by the Poincaré Casimirs of the relativistic system (assumed in an irreducible Poincaré representation with $P^2 > 0$ and $W^2 \neq 0$): $\rho = |\vec{\bar{S}}|/\sqrt{P^2}c = \sqrt{-W^2}/P^2 c$. This classical intrinsic unit of length, whose quantum counterpart is the Compton wavelength of the configuration multiplied its total spin, has remarkable properties: i) the criticism to classical theories based on the quantum argument of pair production applies only inside the world-tube; ii) the world-tube is the remnant in flat Minkowski space-time of the energy conditions of general relativity [as shown by Møller, if a material body has its radius smaller than ρ, then the classical energy density is not definite positive and the peripheral rotation velocity is higher than the velocity of light]; iii) the classical relativistic theory of position measurements faces the following dilemma which has no analogue in Newtonian theories: a) the measurement of the canonical position [a Dirac observable independent from the gauge-fixings on the relative times] is frame-dependent, namely this position cannot be localized inside the world-tube in a covariant way; b) every other center-of- mass position variable (maybe canonical and covariant as the original naive center-of-mass X^μ, but very often non-canonical) depends on the relative times and its measurement acquires meaning only after a choice of gauge-fixings on them. At the quantum level, the situation is even more complicated due to the theorems of Hegerfeldt [29], according to which, if the Newton-Wigner position is a self-adjoint oerator, then nearly all wave packets will spread in space with a velocity higher than the velocity of light (only wave packets with special power-like tails, living on the boundary of the Hilbert space, do not have this pathology). Waiting for better developed classical and quantum theories of position measurements, we can conclude that there are strong indications that the center-of-mass of extended relativistic systems may not be localized inside the world-tube both at the classical and quantum levels. It could be suggested that in the quantum theory, enriching the Heisenberg undetermination relations with this veto, one obtains an ultraviolet cutoff $c\sqrt{P^2}/S$ for the total energy in the spirit of Dirac and Yukawa. Let us remark that the insertion of the spin Poincaré Casimir in the final canonical variables could introduce further non-covariant variables; it will be interesting to find which is the minimal number and meaning of these non-covariant canonical variables induced by the adaptation to the geometry of the Poincaré orbits and to the first class constraints.

What is not clear is how this ultraviolet cutoff could be used to quantize the non-linear and non-local physical electromagnetic and Yang-Mills Hamiltonians. To try to build a consistent framework with this aim in mind, let us remark that the standard asymptotic Fock spaces used to give a particle interpretation to quantum field theory and to build S matrix theory do not constitute a relativistic Cauchy problem for it: since many-particle states are tensor products of single free particle states, one asymptotic free particle can be in the absolute future of the others. One reflection of this freedom ia given by the spurious solutions of the relativistic two-body bound-state Bethe-Salpeter equation [12], which are excitations in the relative energy ϵ_R conjugate to the relative time T_R. Therefore, one should look for a multi-temporal reformulation of the asymptotic states of quantum field theory by using a canonical basis like the one discussed previously to describe modified asymptotic states of n free particles on a space-like hyperplane orthogonal to their total momentum P_μ, when $P^2 > 0$. This is in the spirit of the Tomonaga-Schwinger [30] formulation of quantum field theory and generalizations of the non-local scalar products described above for two scalar particles (and extended to spin 1/2 and spin 1 particles) should be used in the construction. Formal S matrix theory would be unchanged with the center-of-mass time T replacing the parameter t (there is no evolution in the relative times due to the quantization of the n−1 first class constraints $\epsilon_{Ra} \approx 0$), but then one should formulate a reduction formalism and a perturbative expansion, in which only the total energy $\epsilon = \eta\sqrt{P^2}$ propagates: the relative energies $\epsilon_{Ra}; a = 1, .., n-1$, should not propagate to avoid spurious solutions of the resulting bound-state equations. If this reformulation is possible, one should get a scheme in which the previous ultraviolet cutoff would be natural and maybe there could be the possibility of a further extension of asymptotic states to include permanent bound-states (or only them in theories like QCD). Naturally, the non-covariance of the canonical, probably not self-adjoint, center-of-mass oerator corresponding to the Dirac observables $\vec{z}$ would show up, especially in an attempt to describe this reformulation in a path integral approach, whose construction till now heavily relies on the non-relativistic concept of self-adjoint position operators and on their eigenstates.

Finally the methodology described above should help to solve the first class constraints of general relativity either in the old tetrad gravity formulation [31] or in Ashtekar's approach [32] (where solutions already exist). In tetrad gravity, there are 16 configuration variables in each space-time point and 14 first class constraints in phase space, of which 13 are linear in the momenta (three generators of space diffeomorphisms and the other ten satisfying a Poincaré algebra). One should solve these 13 constraints and find a quasi-Shanmugadhasan canonical basis adapted to them; only for some special class of globally hyperbolic, globally parallelizable, asymptotically flat manifolds this should be possible, if techniques similar to those used for the Yang-Mills Gauss laws will work with space diffeomorphisms. If this can be accomplished, then there should appear in the canonical basis an "energy variable" ϵ (like $\epsilon = \eta\sqrt{P^2}$ for special relativistic systems) such that the time diffeomorphism constraint, which is quadratic in the momenta, could be put in the form $\epsilon - H_P \approx 0$ (like for covariant Yang-Mills theory), with H_P depending only on the two pairs of Dirac observables describing the classical graviton degrees of freedom in each point of space-time; the gauge variable conjugate to ϵ should be the "time variable" for this class of general relativistic manifolds. The eventual globality of the results and the extension to other monifolds are completely open problems at this stage. However, now both Yang-Mills theory with fermions and classical general relativity would be put in the same form and the natural quantization procedure would be the Schrödinger one. If the idea of the ultraviolet cutoff determined by the Poincaré Casimirs would work for special relativistic gauge theories, one could hope to use the Casimirs of the asymptotic Poincaré group to try to regularize quantum gravity. Finally, one should try to solve the constraints of tetrad gravity coupled to matter and Yang-Mills fields and try to face the unsolved problem of how to define elementary particles in the framework of general relativity.

References

[1] L. Lusanna (1979), *Nuovo Cimento*, **B52** 141.
L. Lusanna (1990), *Phys. Rep.*, **185** 1.
L. Lusanna (1991), *Riv. Nuovo Cimento*, **14** 1.
L. Lusanna (1990), *J. Math. Phys.*, **31** 2126.
L. Lusanna (1990), *J. Math.Phys.*, **31** 428.
M. Chaichian, D. Louis Martinez and L. Lusanna (1993), 'Dirac's Constrained Systems: The Classification of Second-Class Constraints', Helsinki Univ. preprint HU-TFT-93-5 .

[2] P.A.M. Dirac (1950), *Can. J. Math.*, **2** 129; (1964),*Lectures on Quantum Mechanics*, Belfer Graduate School of Science, Monographs Series, Yeshiva University.
J.L. Anderson and P.G. Bergmann (1951), *Phys. Rev.*, **83** 1018.
P.G. Bergmann and J. Goldberg (1955), *Phys. Rev.*, **98**, 531 .

[3] G. Longhi, L. Lusanna and J.M. Pons (1989), *J. Math. Phys.*, **30** 1893 .

[4] R. De Pietri, L. Lusanna and M. Pauri (1994), 'Standard and Generalized Newtonian Gravities as 'Gauge' Theories of the Extended Galilei Group: I) The Standard Theory; II) Dynamical Three-Space Theories', Parma Univ. preprints .

[5] L. Bel (1970), *Ann. Inst. H. Poincaré*, **307**; (1971), **14** 189; (1973), **18** 57; (1977), *Mecanica Relativista Predictiva* Report UABFT-34, Universidad Autonoma Barcelona (unpublished) .

[6] L. Lusanna (1992), 'Dirac's Observables: from Particles to Strings and Fields', talk given at the International Symposium on *Extended Objects and Bound States*, Karuizawa, eds. O.Hara, S.Ishida and S.Naka, World Scientific, 1993 .

[7] L. Lusanna (1991), 'Classical Observables of Gauge Theories from the Multitemporal Approach', talk given at the Conference 'Mathematical Aspects of Classical Field Theory', Seattle, in *Contemporary Mathematics* (1992), **132** 531 .

[8] S. Shanmugadhasan (1973), *J. Math. Phys.*, **14** 677.
L. Lusanna (1993), 'The Shanmugadhasan Canonical Transformation, Function Groups and the Extended Second Noether Theorem', *Int. J. Mod. Phys.*, **A8** 4193 .

[9] M.J. Gotay, J.M. Nester and G. Hinds (1978), *J. Math. Phys.*, **19** 2388.
M.J. Gotay and J.M. Nester (1979), *Ann. Inst. Henri Poincarè*, **A30**, 129 and (1980), **A32** 1.
M.J. Gotay and J. Śniatycki (1981), *Commun. Math. Phys.*, **82** 377.
M.J. Gotay (1982), *Proc. Am. Math. Soc.*, **84** 111; (1986), *J. Math. Phys.*, **27** 2051.
G. Marmo, N. Mukunda and J. Samuel (1983), *Riv. Nuovo Cim.*, **6** 1 .

[10] J.M. Arms, M.J. Gotay and G. Jennings (1990), *Adv. Math.*, **79** 43 .

[11] F.M. Lev (1993), *Rivista Nuovo Cim.*, **16,n.2** 1.
J.A. Llosa (ed) (1982), *Relativistic Action-at-a-Distance*, Lecture Notes Phys. n.162, Springer.
G. Longhi and L. Lusanna (eds) (1987), *Constraint's Theory and Relativistic Dynamics*, World Scientific.
L. Lusanna (1981), *Nuovo Cim.*, **65B**,135; (1984), in Proc. VII Seminar on *Problems of High Energy Physics and Quantum Field Theory*, Protvino USSR, vol.I, p.123; (1986) in *Gauge Field Theories*, XVIII Karpacz School, Harwood .

[12] R. Casalbuoni (1976), *Nuovo Cim.*, **33A** 115 and 389.
F.A. Berezin and M.S. Marinov (1977), *Ann. Phys. (N.Y.)*, **104**, 336.
A. Barducci, R. Casalbuoni and L. Lusanna (1977), *Nuovo Cim. Lett.*, **19**, 581; (1977), *Nucl. Phys.*, **B124**, 93; (1981), *Nucl. Phys.*, **B180** [FS2], 141 .

[13] A. Barducci, R. Casalbuoni and L. Lusanna (1976), *Nuovo Cim.*, **35A** 377 .

[14] A. Barducci, R. Casalbuoni, D. Dominici and L. Lusanna (1981), *Phys. Lett.*, **100B** 126 .

[15] A. Barducci and L. Lusanna (1983), *Nuovo Cim.*, **77A** 39 .

[16] A. Barducci and L. Lusanna (1983), *J. Phys.*, **A16** 1993 .

[17] Ph. Droz Vincent (1969), *Lett. Nuovo Cim.*, **1** 839; (1970), *Phys. Scr.*, **2** 129; (1975), *Rep. Math. Phys.*, **8** 79.
I.T. Todorov (1976), *Report Comm. JINR E2-10125, Dubna*, (unpublished); (1978), *Ann.Inst.H.Poincaré*, **28A**,207.
A. Komar (1978), *Phys.Rev.*, **D18** 1881 and 1887 .

[18] A. Barducci, R. Casalbuoni and L. Lusanna (1979), *Nuovo Cim.*, **54A** 340 .

[19] G. Longhi and L. Lusanna (1986) ,*Phys. Rev.*, **D34** 3707 .

[20] H. Sazdjian (1981), *Ann. Phys. (N.Y.)*, **136** 136; (1986), *Phys. Rev.*, **D33** 3401; (1987), *J. Math. Phys.*, **28** 2618 and 1988, **29** 1620; (1989), *Ann. Phys. (N.Y.)*, **191** 52; (1992), in Proc.Int.Symp. *Extended Objects and Bound Systems*, O. Hara, S. Ishida and S. Naka (eds), World Scientific.
J. Bijtebier and J. Brockaert (1992), *Nuovo Cim.*, **A105** 351 and 625; (1992), in Proc.Int.Symp. *Extended Objects and Bound Systems*, O. Hara, S. Ishida and S. Naka (eds), World Scientific .

[21] H.W. Crater and P. Van Alstine (1982), *J. Math. Phys.*, **23** 1697; (1983), *Ann. Phys. (N.Y)*, **148** 57; (1984), *Phys. Rev. Lett.*, **53** 1577; (1984), *Phys. Rev.*, **D30** 2585; (1986), **D34** 1932; (1987), **D36** 3007; (1988), **D37** 1982; (1990), *J. Math. Phys.*, **31** 1998; (1992), *Phys. Rev.*, **D46** 766.
H.W. Crater, R.L. Becker, C.Y. Wong and P. Van Alstine (1992), *Phys. Rev.*, **D46** 5117; in Proc. Int. Symp. *Extended Objects and Bound Systems*, O. Hara, S. Ishida and S. Naka (eds), World Scientific.
H.W. Crater and D. Yang (1991), *J. Math. Phys*, **32** 2374 .

[22] G. Longhi (1994), talk at this Conference .

[23] L. Lusanna (1981), *Nuovo Cim.*, **64A** 65 .

[24] F. Colomo, G. Longhi and L. Lusanna (1990), *Int. J. Mod. Phys.*, **A5** 3347; (1990), *Mod. Phys. Lett.*, **A5** 17.
F. Colomo and L. Lusanna (1992), *Int. J. Mod. Phys.*, **A7** 1705, 4107 .

[25] P.A.M. Dirac (1955), *Can. J. Phys.*, **33** 650 .

[26] L. Lusanna (1994), *Dirac's Observables for Classical Yang-Mills Theory with Fermions* Firenze Univ. preprint DFF 201/1/ 94 .

[27] V. Moncrief (1979), *J. Math. Phys.*, **20** 579 .

[28] M. Pauri (1980), in *Group Theoretical Methods in Physics*,
K.B. Wolf (ed), Lecture Notes Phys. **135**, Springer, Berlin; (1971), *Invariant Localization and Mass-Spin Relations in the Hamiltonian Formulation of Classical Relativistic Dynamics*, Parma Univ. preprint IFPR-T-019 (unpublished) .

[29] G.C. Hegerfeldt (1989), *Nucl. Phys. (Proc.Suppl.)*, **B6** 231 .

[30] S. Tomonoga (1946), *Prog. Theor. Phys.*, **1** 27; (1948), *Phys. Rev.*, **74** 224.
Z. Koba, T. Tati and S. Tomonoga (1947), *Prog. Theor. Phys.*, **2** 101 and 198.
S. Kanesawa and S. Tomonaga (1948), *Prog. Theor. Phys.*, **3** 1 and 101.
J. Schwinger (1948), *Phys. Rev.*, **73** 416 and **74** 1439 .

[31] J.E. Nelson and C. Teitelboim (1978), *Ann. Phys. (N.Y.)*, **116** 86.
M. Henneaux (1983), *Phys. Rev.*, **D27** 986.
J.M. Charap and J.E. Nelson (1983), *J. Phys.*, **A16** 1661 and 3355; (1986), *Class. Quantum Grav.* **3** 1061.
J.M. Charap (1987), in *Constraint's Theory and Relativistic Dynamics*,

G. Longhi and L. Lusanna (eds), World Scientific
J.M. Charap, M. Henneaux and J.E. Nelson (1988), *Class. Quantum Grav.*, **5** 1405 .

[32] A. Ashtekar (1987), *Phys. Rev.*, **D36** 1587; (1986), *New Perspectives in Canonical Gravity*, Bibliopolis; (1991),*Lectures on Non-Perturbative Canonical Gravity*, World Scientific.
M. Henneaux, J.E. Nelson and C. Schomblond (1989), *Phys. Rev.*, **D39** 434.
R. Capovilla, J. Dell and T. Jacobson (1989), *Phys. Rev. Lett.*, **63** 2325.
J.D. Romano (1993), *Gen. Rel. Grav.*, **25** 759 .

Gauging kinematical and internal symmetry groups for extended systems

Roberto De Pietri, Massimo Pauri and Luca Lusanna

Abstract

The possible external couplings of an extended non-relativistic classical system are characterized by *gauging* its maximal symmetry group *at* the *center-of-mass.* The Galilean one-time and two-times harmonic oscillators are exploited as models.

1 Introduction

In this paper we exploit the *gauge* technique to characterize the possible couplings of extended systems. The non-relativistic harmonic oscillator with *center-of-mass* is used as a model. We make the *ansatz* that the essential structural elements and the *extension* of a dynamical system are represented and summarized by its maximal dynamical symmetry group viz. by the algebraic structure of the constants of the motion. Then, we apply the *gauge* procedure to this group by *localizing* it at the *center-of-mass* of the system. We show thereby that the gauge procedure is meaningful also for *dynamical* symmetries besides the usual *kinematical* ones. In spite of the evident paradigmatic and heuristic nature of our *ansatz*, the results obtained here seem notably expressive.

The technical steps of the work are the following: 1) the standard Utiyama procedure for fields is applied to the possible trajectories of the *center-of-mass* as described by a canonical realization of the *extended* Galilei group. This determines the gravitational-inertial fields which can couple to the *center-of-mass* itself. As shown elsewhere [4], the requirement of invariance (properly *quasi-invariance*) of the Lagrangian leads to the introduction of eleven *gauge* compensating fields and their transformation properties. 2) The generalized Utiyama procedure is then applied to the *internal* dynamical U(3) symmetry so that *gauge* compensating fields have to be introduced in connection to the internal angular momentum (*spin*) and the quadrupole momentum. Besides the dynamical nature of the internal transformations, the crucial point is that the Galilei rotations subgroup affects the internal transformations.

The following remarkable results are obtained: **1)** a peculiar form of interaction with the external *gauge* fields; **2)** a modification of the dynamical part of the symmetry transformations, which is needed to take into account the modification of the dynamics itself induced by the *gauge* fields. In particular, the Yang-Mills fields associated to the internal rotations have the effect of modifying the *time derivative* of the internal variables. On the other hand, the Yang-Mills fields associated to the quadrupole momentum have the effect of introducing a sort of *internal metric* in the relative space, together with the modification of the *interaction* mentioned above. Moreover, it is interesting to find that, given their dynamical effect, the internal rotations Yang-Mills fields apparently define a sort of *Galilean spin connection.*

These general features are reminiscent of some peculiar aspects of the *relativistic string* dynamics, in particular the so called *bootstrap* hypothesis according to which the external fields that can couple to the system are all and the same which are already included in the field theory itself.

The determination of the dynamical equations for the Yang-Mills fields remains an open problem.

2 The "one-time" non-relativistic oscillator

The Lagrangian describing the Galilean one-time harmonic oscillator is:

$$\mathcal{L} = \sum_{a=1}^{2} \frac{m_a}{2} \delta_{ij} \dot{x}_a^i \dot{x}_a^j - \frac{k'}{2} \delta_{ij} (x_1^i - x_2^i)(x_1^j - x_2^j) \ . \tag{2.1}$$

Introducing the center-of-mass coordinates:

$$\begin{cases} x^i &= \dfrac{m_1 x_1^i + m_2 x_2^i}{M} \\ r^i &= x_1^i - x_2^i \ , \end{cases} \tag{2.2}$$

the Lagrangian can be rewritten as:

$$\mathcal{L} = \frac{M}{2} \delta_{ij} \dot{x}^i \dot{x}^j + \frac{\mu}{2} \delta_{ij} \dot{r}^i \dot{r}^j - \frac{k'}{2} \delta_{ij} r^i r^j \equiv \mathcal{L}_{CM}(\boldsymbol{x};t) + \mathcal{L}_I(\boldsymbol{r};t) \ , \tag{2.3}$$

where $\mu = \frac{m_1 m_2}{m_1+m_2}$ is the reduced-mass and $M = m_1 + m_2$ is the total mass.

The system has two kinds of symmetries: (1) The Galilei *kinematical* symmetry, whose infinitesimal transformations are given by

$$\begin{cases} \delta x^i &= \epsilon\, \dot{x}^i + \epsilon^i + c_{jk}{}^i \omega^j x^k - t v^i \\ \delta r^a &= \epsilon\, \dot{r}^a + c_{cd}{}^a \omega^b r^c \ , \end{cases} \tag{2.4}$$

under which the Lagrangian is *quasi-invariant*

$$\begin{aligned} \delta\mathcal{L} &= \frac{\partial \mathcal{L}}{\partial x^i}\delta x^i + \frac{\partial \mathcal{L}}{\partial r^a}\delta r^a + \frac{\partial \mathcal{L}}{\partial \dot{x}^i}\frac{d}{dt}\delta x^i + \frac{\partial \mathcal{L}}{\partial \dot{r}^a}\frac{d}{dt}\delta r^a \\ &= \frac{d}{dt}\left[\epsilon\mathcal{L} - M\delta_{ij} v^i x^j\right] \ , \end{aligned} \tag{2.5}$$

and: (2) the U(3) internal *dynamical* symmetry, whose infinitesimal transformations are given by:

$$\begin{cases} \delta x^i &= 0 \\ \delta r^a &= c_{bc}{}^a \theta^b r^c + \dfrac{\mu}{\sqrt{k}} \xi^{ab} \delta_{bc} \dot{r}^c \end{cases} , \tag{2.6}$$

under which the Lagrangian is also *quasi-invariant*:

$$\begin{aligned} \delta L &= \frac{d}{dt}\left[\frac{\mu}{\sqrt{k}} \xi^{ab} \delta_{ac} \delta_{bd} \left(\frac{\mu}{2} \dot{r}^c \dot{r}^d - \frac{k}{2\mu} r^c r^d\right)\right] \\ &\equiv \frac{d}{dt}\left[\frac{\mu}{\sqrt{k}} \xi^{ab} F_{ab}(\dot{\boldsymbol{r}};\boldsymbol{r})\right] . \end{aligned} \tag{2.7}$$

3 Gauging the "*extended* Galilei $\wedge$ U(3) group"

We want to implement an overall description of the dynamical system in terms of *symmetries*: precisely the *space-time* Galilean symmetry of a *point-like structure*, identified with the *center-of-mass*, and the *dynamical* U(3) symmetry interpreted as summarizing the *internal structure*.

We shall proceed as follows. First of all, we observe that Newton's mechanics of a point particle may be described using a trivial configuration bundle defined by the *absolute-time* axis, as *base* manifold, and the configuration space R^3, as *fiber*. The corresponding geometrical structure for a system of two particles having a natural center of mass ($\boldsymbol{x}$) and relative variables ($\boldsymbol{r}$), can still be characterized by a trivial fiber bundle defined by the *absolute-time* axis as *base* manifold. Yet, the *fiber* is now a fiber bundle (Q) by itself, having its *base* defined by a vector space R^3 (Q_x), associated to $\boldsymbol{x}$, and a *fiber* identifiable with the tangent space of the latter, associated to the relative variable $\boldsymbol{r}$ (Q_r). Introducing the dynamics requires that the configuration space Q with coordinates $\{\boldsymbol{x}, \boldsymbol{r}\}$ be replaced by its tangent bundle TQ with coordinates $\{\boldsymbol{x}, \boldsymbol{r}; \dot{\boldsymbol{x}}, \dot{\boldsymbol{r}}\}$. In this way we get a further trivial bundle over the *absolute-time* axis. Therefore, since the dynamical algebra acts only on the $\mathcal{L}_r(t)$ part of the lagrangian (see eq (2.3)) with $\mathcal{L}_r : TQ_r \mapsto R$, in carrying out the *gauging* of the internal dynamical group one is naturally led to replace the arbitrary constants θ^a, ξ^{ab} of the Noether transformations (2.4,2.6) with arbitrary functions $\theta^a(\boldsymbol{x}, t)$, $\xi^{ab}(\boldsymbol{x}, t)$ of the *center-of-mass* coordinates and the *absolute-time*.

In these conditions, one has to expect that, in order the Lagrangian to remain *quasi-invariant* as in eqs.(2.5),(2.7) under the localized U(3) transformations, compensating U(3) Yang-Mills fields depending on the *center-of-mass* coordinates $\boldsymbol{x}$, have to be introduced.

In this way we can *gauge* the kinematical *plus* the dynamical group of the harmonic oscillator at the same time. We look now for a new Action, invariant (*quasi-invariant*) under the transformations:

$$\delta^G x^i = \epsilon(t)\dot{x}^i + \tilde{\eta}^i(\boldsymbol{x},t) \tag{3.1}$$

$$\delta^G r^a = \epsilon(t)\dot{r}^a + c_{bc}{}^a\tilde{\theta}^b(\boldsymbol{x},t)r^c + \frac{\mu}{\sqrt{k}}\xi^{ab}(\boldsymbol{x},t)\delta_{bc}\dot{r}^c \ , \tag{3.2}$$

where

$$\begin{aligned}\tilde{\eta}^i(\boldsymbol{x},t) &= \epsilon^i(\boldsymbol{x},t) + c_{jk}{}^l\omega^j(\boldsymbol{x},t)x^k - tv^i(\boldsymbol{x},t)\\ \tilde{\theta}^b(\boldsymbol{x},t) &= \theta^b(\boldsymbol{x},t) + \omega^b(\boldsymbol{x},t)\ .\end{aligned} \tag{3.3}$$

It is seen that, while the *center-of-mass* coordinates are affected only by the local Galilei transformations (see eq.(3.1)), the *internal* variables get modified by both the Galilei and the local U(3) transformations. Due to these features, we can take advantage of the results already obtained in [4] about the gauging of the *extended* Galilei group. Precisely, in analogy to what we found there, the invariant (*quasi-invariant*) Lagrangian for the *center-of-mass* can be written as

$$\mathcal{L}^G_{CM} = \frac{M}{2\Theta(t)}\left[g_{ij}(\boldsymbol{x},t)\,\dot{x}^i\dot{x}^j + 2A_i(\boldsymbol{x},t)\,\dot{x}^i + 2A_0(\boldsymbol{x},t)\right] \ , \tag{3.4}$$

while the corresponding transformation properties for the *gauge* fields are:

$$\begin{aligned}\delta^G\Theta(t) &= \dot{\epsilon}(t)\Theta(t)\\ \delta^G g_{ij}(\boldsymbol{x},t) &= -\frac{\partial\tilde{\eta}^k(\mathbf{x},t)}{\partial x^i}g_{kj} - \frac{\partial\tilde{\eta}^k(\mathbf{x},t)}{\partial x^i}g_{kj}\\ \delta^G A_0(\boldsymbol{x},t) &= 2\dot{\epsilon}A_0 - A_i\frac{\partial\tilde{\eta}^i}{\partial t} - \Theta\frac{\partial}{\partial t}\left[g_{ij}v^ix^j\right]\\ \delta^G A_i(\boldsymbol{x},t) &= \dot{\epsilon}A_i - A_j\frac{\partial\tilde{\eta}^j}{\partial x^i} - g_{ij}\frac{\partial\tilde{\eta}^j}{\partial t} - \Theta\frac{\partial}{\partial x^i}\left[g_{ij}v^ix^j\right]\ .\end{aligned} \tag{3.5}$$

The *gauging* of the U(3) part is more complicated. Actually, the following changes are needed:

(a) the Yang-Mills fields must include temporal components to account for the functional dependence of the group parameter on *absolute-time*:

(b) Besides local U(3) transformations, the Yang-Mills fields must undergo *kinematical* Galilei transformations in analogy with the gauge fields $A_i(\boldsymbol{x},t)$, $A_0(\boldsymbol{x},t)$.

(c) The same *center-of-mass-localized* U(3) transformations of the relative variable $\boldsymbol{r}$ (see eq.(3.2)) must be modified due to their dynamical nature. Indeed a change of the Lagrangian implies new equations of motion.

Looking at (2.6) and (3.2), we can exploit the natural distinction of the U(3) transformations in a *kinematical* part (the SU(2) transformations associated to *internal* rotations) and a *dynamical* one. We shall consistently denote the Yang-Mills fields as $B_k^{(a)}(\boldsymbol{x},t)$, $B_0^{(a)}(\boldsymbol{x},t)$ (fields associated to the *internal* rotation subgroup) and $B_k^{(ab)}(\boldsymbol{x},t)$, $B_0^{(ab)}(\boldsymbol{x},t)$ (fields associated to the *dynamical* transformations). Then, it can be checked that the following results hold true:

1) By defining an *internal covariant derivative* (associated to the internal rotations) as:

$$Dr^a = \dot{r}^a + \dot{x}^k B_k^{(b)}(\boldsymbol{x},t) c_{bc}{}^a r^c + B_0^{(b)}(\boldsymbol{x},t) c_{bc}{}^a r^c \;, \tag{3.6}$$

and an *internal metric* (associated to the dynamical transformations) as:

$$\Xi^{ab}(\boldsymbol{x},t) = \delta^{ab} - \frac{\mu}{\Theta(t)\sqrt{k}}\left[\dot{x}^k B_k^{(ab)}(\boldsymbol{x},t) - B_0^{(ab)}(\boldsymbol{x},t)\right] \;, \tag{3.7}$$

the effective infinitesimal transformations under which we require invariance of the Action can be written:

$$\delta^G r^a = \epsilon(t)\dot{r}^a + c_{bc}{}^a \tilde{\theta}^b(\boldsymbol{x},t) r^c + \frac{\mu}{\Theta\sqrt{k}}\xi^{ab}(\boldsymbol{x},t)|\Xi^{-1}|_{bc}(\boldsymbol{x},t)\; Dr^c \tag{3.8}$$

2) The *modified internal* Lagrangian $\mathcal{L}_I^G(\boldsymbol{r};t)$ can be written:

$$\mathcal{L}_I^G = \frac{\mu}{2\Theta}|\Xi^{-1}|_{ab}(\boldsymbol{x},t)\; Dr^a\; Dr^b - \frac{k\Theta}{2\mu}\Xi^{ab}(\boldsymbol{x},t)\; \delta_{ac}\delta_{bd} r^c r^d \;. \tag{3.9}$$

3) The transformation properties of the Yang-Mills fields, which include the *kinematical* action of the Galilei group and guarantee invariance (*quasi-invariance*) of the Action, are:

$$\begin{aligned}
\delta^G B_k^{(a)}(\boldsymbol{x},t) &= c_{bc}{}^a \tilde{\theta}^b(\boldsymbol{x},t) B_k^{(c)}(\boldsymbol{x},t) \\
&\quad + \xi^{cd}(\boldsymbol{x},t)[c_{ce}{}^a \delta_{df} B_k^{(ef)}(\boldsymbol{x},t) + c_{de}{}^c \delta_{cf} B_k^{(ef)}(\boldsymbol{x},t)] \\
&\quad - \frac{\partial \tilde{\eta}^i}{\partial x^k} B_i^{(a)}(\boldsymbol{x},t) - \frac{\partial \tilde{\theta}^a(\boldsymbol{x},t)}{\partial x^k} \\
\delta^G B_k^{(ab)}(\boldsymbol{x},t) &= \tilde{\theta}^c(\boldsymbol{x},t)[c_{cd}{}^a B_k^{(db)}(\boldsymbol{x},t) + c_{cd}{}^b B_k^{(ad)}(\boldsymbol{x},t)] \\
&\quad + \frac{1}{4}\xi^{cd}(\boldsymbol{x},t)[c_{ec}{}^a \delta_d^b + c_{ec}{}^b \delta_d^a + c_{ed}{}^a \delta_c^b + c_{ed}{}^b \delta_c^a] B_k^{(e)}(\boldsymbol{x},t) \\
&\quad - \frac{\partial \tilde{\eta}^i}{\partial x^k} B_i^{(ab)}(\boldsymbol{x},t) - \frac{\partial \xi^{ab}(\boldsymbol{x},t)}{\partial x^k} \\
\delta^G B_0^{(a)}(\boldsymbol{x},t) &= c_{bc}{}^a \tilde{\theta}^b(\boldsymbol{x},t) B_0^{(c)}(\boldsymbol{x},t) \qquad\qquad (3.10) \\
&\quad + \xi^{cd}(\boldsymbol{x},t)[c_{ce}{}^a \delta_{df} B_0^{(ef)}(\boldsymbol{x},t) + c_{de}{}^c \delta_{cf} B_0^{(ef)}(\boldsymbol{x},t)] \\
&\quad + \dot{\epsilon} B_0^{(a)}(\boldsymbol{x},t) - \frac{\partial \tilde{\eta}^i}{\partial t} B_i^{(a)}(\boldsymbol{x},t) - \frac{\partial \tilde{\theta}^a(\boldsymbol{x},t)}{\partial t}
\end{aligned}$$

$$\begin{aligned}
\delta^G B_0^{(ab)}(\boldsymbol{x},t) &= \tilde{\theta}^c(\boldsymbol{x},t)[c_{cd}{}^a B_0^{(db)}(\boldsymbol{x},t) + c_{cd}{}^b B_0^{(ad)}(\boldsymbol{x},t)] \\
&+\frac{1}{4}\xi^{cd}(\boldsymbol{x},t)[c_{ec}{}^a\delta_d^b + c_{ec}{}^b\delta_d^a + c_{ed}{}^a\delta_c^b + c_{ed}{}^b\delta_c^a]B_0^{(e)}(\boldsymbol{x},t) \\
&+\dot{\epsilon}B_0^{(ab)}(\boldsymbol{x},t) - \frac{\partial\tilde{\eta}^i}{\partial t}B_i^{(ab)}(\boldsymbol{x},t) - \frac{\partial\xi^{ab}(\boldsymbol{x},t)}{\partial t} .
\end{aligned}$$

4) The variation of the total Action results:

$$\begin{aligned}
\delta^G\mathcal{A} &= \delta^G\int dt\ \mathcal{L}^G_{CM}(\boldsymbol{x};t) + \delta^G\int dt\ \mathcal{L}^G_I(\boldsymbol{r};t) \\
&= \int dt\ \frac{d}{dt}\left[-mg_{ij}(\boldsymbol{x},t)v^i(\boldsymbol{x},t)\ x^j\right] + \int dt\ \frac{d}{dt}\left[\frac{\mu}{\sqrt{k}}\xi^{ab}(\boldsymbol{x},t)\right. \\
&\left.\left(\frac{\mu}{2\Theta}|\Xi^{-1}|_{ac}(\boldsymbol{x},t)\ Dr^c|\Xi^{-1}|_{bd}(\boldsymbol{x},t)\ Dr^d - \frac{k\Theta}{2\mu}\delta_{ac}\delta_{bd}r^cr^d\right)\right] ,
\end{aligned} \tag{3.11}$$

i.e. the Action itself is *quasi-invariant*, as expected.

4 The model in the "two-times" formulation

The present model can be formulated within the multi-temporal framework [3]. In this latter formulation, the classical system is described as relativistic particle system dynamics are. In particular, it turns out that the present model is the direct non-relativistic limit of the Todorov-Komar-Droz-Vincent relativistic harmonic oscillator. The model is described by two 1st-class constraints and has a mutual action-at-a-distance interaction (instantaneous in the center-of-mass frame). Introducing the 16 dimensional phase space with coordinates $t_a, E_a, \boldsymbol{x}_a, \vec{p}_a$ ($\{t_a, E_b\} = -\delta_{ab}$, $\{x^i_a, p_{bj}\} = \delta_{ab}\delta^i_j$), the constraints can be written:

$$\begin{aligned}
&\bar{\chi}_a = E_a - \frac{1}{2m_a}(\delta^{ij}p_{ai}p_{aj} + k\delta_{ij}\rho^i\rho^j) \simeq 0 \quad , \quad a = 1,2 \\
&\{\bar{\chi}_1, \bar{\chi}_2\} = 0 \ ,
\end{aligned} \tag{4.1}$$

where ρ^i is defined by :

$$\rho^i = r^i - \frac{t_1 - t_2}{M}\delta^{ij}P_j \ , \quad (P_i = p_{1i} + p_{2i}) \ . \tag{4.2}$$

Besides eqs.(2.2) and (2.4), we introduce the following definitions:

$$\begin{aligned}
t &= \frac{m_1t_1 + m_2t_2}{M} & t_R &= t_1 - t_2 \\
E &= E_1 + E_2 & E_R &= \frac{m_2E_1 - m_1E_2}{M} .
\end{aligned} \tag{4.3}$$

In order to simplify the discussion, let us now consider the following linear combination of the $\bar{\chi}_a$'s:

$$\bar{\chi}_+ = \bar{\chi}_1 + \bar{\chi}_2 = E - \frac{1}{2M}\delta^{ij}P_iP_j - \frac{1}{2\mu}(\delta^{ij}\pi_i\pi_j + k\delta_{ij}\rho^i\rho^j) \simeq 0 \tag{4.4}$$

$$\bar{\chi}_- = \frac{m_2\bar{\chi}_1 - m_1\bar{\chi}_2}{M} = E_r - \frac{1}{M}\delta^{ij}P_i\pi_j - \frac{m_2 - m_1}{2\mu M}(\delta^{ij}\pi_i\pi_j + k\delta_{ij}\rho^i\rho^j) \simeq 0 \ .$$

Note that for $t_R = 0$, solving the constraint $\bar{\chi}_+ = E - \bar{H} \simeq 0$, the usual Hamiltonian formulation is obtained.

The Galilei generators are:

$$\left\{\begin{array}{lcl} E &=& E_1 + E_2 \\ P_i &=& p_{1i} + p_{2i} \\ J_i &=& c_{ij}{}^k x_1^j p_{1k} + c_{ij}{}^k x_2^j p_{2k} = c_{ij}{}^k x^j P_k + c_{ij}{}^k r^j \pi_k \\ K_i &=& m_1\delta_{ij}x_1^j - t_1 p_{1i} + m_2\delta_{ij}x_2^j - t_1 p_{2i} = M\delta_{ij}x^j - tP_i - t_R\pi_i \ . \end{array}\right. \tag{4.5}$$

These generators have zero Poisson brackets with the constraints and, consequently, with the Dirac Hamiltonian. Furthermore, they satisfy the *extended* Galilei Algebra (for more details see [6]).

On the other hand, the generators of the U(3) dynamical symmetry are:

$$\left\{\begin{array}{lcl} \bar{\tilde{S}}_a &=& c_{ab}{}^c\rho^b\pi_c \\ \bar{\tilde{N}}_{ab} &=& \dfrac{1}{\sqrt{2k}}\pi_a\pi_b + \dfrac{\sqrt{k}}{2}\delta_{ac}\delta_{bd}\rho^c\rho^d \ . \end{array}\right. \tag{4.6}$$

Their Poisson brackets with the two constraints $\bar{\chi}_\pm$ are zero, since $\{\vec{\rho}, E_R - \frac{\vec{p}\cdot\vec{\pi}}{M}\} = 0$. Moreover, we have again the structure of a semi-direct product of the two algebras (4.5)-(4.6). group.

The total symmetry is generated by $\bar{G} = \theta^a\bar{\tilde{S}}_a + \xi^{ab}\bar{\tilde{N}}_{ab}$ as:

$$\left\{\begin{array}{lcl} \delta x^i &=& \dfrac{t_r}{M}\delta^{ij}c_{jk}{}^l\theta^k\pi_l - \dfrac{\sqrt{k}t_r}{M}\xi^{ij}\delta_{jk}\rho^k \\ \delta\rho^a &=& c_{bc}{}^a\theta^b\rho^c + \dfrac{1}{\sqrt{k}}\xi^{ab}\pi_b \end{array}\right. \tag{4.7}$$

Since $\{x^i, \rho^j\} = -\frac{t_R}{M}\delta^{ij} \neq 0$, the *center-of-mass* coordinates that are invariant under the transformation of the dynamical symmetry given, are now given by

$$\tilde{x}^i = x^i - t_R\frac{\delta^{ij}\pi_j}{M} \tag{4.8}$$

These are indeed the variables which have to be used for the gauge procedure according to the program sketched before. Performing the canonical transformation

$$\begin{bmatrix} t & , & E \\ t_R & , & E_R \\ p_i & , & x^i \\ \pi_i & , & r^i \end{bmatrix} \longmapsto \begin{bmatrix} t & , & E \\ t_R & , & \tilde{E}_R = E_R - \frac{\delta^{ij} p_i \pi_j}{M} \\ p_i & , & \tilde{x}^i = x^i - t_R \frac{\delta^{ij} p_j}{M} \\ \pi_i & , & \rho^i = r^i - t_R \frac{\delta^{ij} \pi_j}{M} \end{bmatrix} \qquad [A] \tag{4.9}$$

the canonical generators associated to the Galilei boosts the rotations, become $\bar{K}_i = M\delta_{ij}\tilde{x}^j - tp_i$ and $\bar{J}_i = c_{ij}{}^k \tilde{x}^j p_k + c_{ij}{}^k \rho^j \pi_k$, while the constraint $\bar{\chi}_-$ becomes $\tilde{\bar{\chi}}_- = \tilde{E}_R - \frac{m_1 - m_2}{2M\mu}(\delta^{ab}\pi_a\pi_b + k\delta_{ab}\rho^a\rho^b) \simeq 0$.

Note finally that, as shown in [3], the constraints (4.1) can also be derived from a Lagrangian.

5 Gauging the symmetries in the "two-times" formulation

In the canonical base defined by $((t_a(\tau), t'_a(\tau), \tilde{\boldsymbol{x}}^i(\tau), \tilde{\boldsymbol{x}}'^i(\tau), \boldsymbol{\rho}(\tau), \boldsymbol{\rho}'(\tau))$, the *localized* generator of the *gauge* transformations of the maximal symmetry group is:

$$\begin{aligned} \bar{G}^G &= \epsilon(t)E + \varepsilon^i(\tilde{\boldsymbol{x}}, t)P_i + \omega^i(\tilde{\boldsymbol{x}}, t)\bar{J}_i + v^i(\tilde{\boldsymbol{x}}, t)\bar{K}_i \\ &\quad + \theta^a(\tilde{\boldsymbol{x}}, t)\bar{S}_a + \xi^{ab}(\tilde{\boldsymbol{x}}, t)\bar{N}_{ab} \end{aligned} \tag{5.1}$$

Defining as before $\tilde{\theta}^i(\tilde{\boldsymbol{x}}, t) = \theta^i(\tilde{\boldsymbol{x}}, t) + \omega^i(\tilde{\boldsymbol{x}}, t)$ and $\tilde{\eta}^i(\tilde{\boldsymbol{x}}, t) = \varepsilon^i(\tilde{\boldsymbol{x}}, t) + c_{jk}{}^i \omega^j(\tilde{\boldsymbol{x}}, t)\tilde{x}^j - tv^i(\tilde{\boldsymbol{x}}, t)$, the transformation properties of the configuration variables become:

$$\begin{aligned} \delta^G \tilde{x}^i &= \tilde{\eta}^i(\tilde{\boldsymbol{x}}, t) & \delta^G t &= -\varepsilon(t) \\ \delta^G r^a &= c_{bc}{}^a \tilde{\theta}^b(\tilde{\boldsymbol{x}}, t) r^c + \sqrt{k}\xi^{ab}(\tilde{\boldsymbol{x}}, t)\pi_b & \delta^G t_R &= 0 \end{aligned} \tag{5.2}$$

Clearly, $\delta^G(\delta^{ij}\pi_i\pi_j + k\delta_{ij}\rho_i\rho_j) = 0$, and, since we have also $\bar{\delta}'\tilde{E}_R = 0$, we obtain

$$\delta^G \tilde{\bar{\chi}}_- = \delta^G \left[\tilde{E}_R - \frac{m_1 - m_2}{2M\mu}(\delta^{ij}\pi_i\pi_j + k\delta_{ij}\rho_i\rho_j) \right] = 0 \ . \tag{5.3}$$

Being invariant, $\tilde{\bar{\chi}}_-$ does not couple to nay *gauge* field. Let us consider now the constraints $\tilde{\bar{\chi}}_+$. The non-invariant quantities in the expression of $\tilde{\bar{\chi}}_+$ are:

$$\begin{aligned} \delta^G P_i &= -\frac{\partial \tilde{\eta}^j}{\partial \tilde{x}^i} P_j - \frac{\partial \tilde{\theta}^a}{\partial \tilde{x}^i} \bar{S}_a - \frac{\partial \xi^{ab}}{\partial \tilde{x}^i} \bar{N}_{ab} - M \frac{\partial}{\partial \tilde{x}^i} \left[g_{ij} v^i \tilde{x}^j \right] \\ \delta^G E &= -\frac{\partial \tilde{\eta}^j}{\partial t} P_j - \frac{\partial \tilde{\theta}^a}{\partial t} \bar{S}_a - \frac{\partial \xi^{ab}}{\partial t} \bar{N}_{ab} - M \frac{\partial}{\partial t} \left[g_{ij} v^i \tilde{x}^j \right] \end{aligned} \tag{5.4}$$

Correspondingly, the invariance of $\tilde{\tilde{\chi}}_+$ under the transformations (5.2) can be recovered by simply introducing the minimal coupling

$$\begin{cases} P_i \;\mapsto\; \mathcal{P}_a = \mathbf{H}^i_a(\boldsymbol{x},t)\left[P_i - \frac{M}{\Theta}A_i(\boldsymbol{x},t) - B_i^{(a)}(\boldsymbol{x},t)S_a - B_i^{(ab)}(\boldsymbol{x},t)N_{ab}\right] \\ E \;\mapsto\; \mathcal{E} = \frac{1}{\Theta}\left[E - \frac{M}{\Theta}A_0(\boldsymbol{x},t) - B_0^{(a)}(\boldsymbol{x},t)S_a - B_0^{(ab)}(\boldsymbol{x},t)N_{ab}\right] \end{cases} \tag{5.5}$$

Indeed, if the transformations of the *gauge* fields are given by eqs.(3.5),(3.10) with $\boldsymbol{x}$ replaced by $\tilde{\boldsymbol{x}}$, it follows:

$$\delta^G\tilde{\tilde{\chi}}^G_+ = \delta^G\left[\mathcal{E} - \frac{1}{2M}\delta^{ab}\mathcal{P}_a\mathcal{P}_b - \frac{1}{2\mu}\left(\delta^{ab}\pi_a\pi_b + k\delta_{ab}\rho^a\rho^b\right)\right] = 0 \tag{5.6}$$

Finally, the invariant Lagrangian corresponding to the above constraints, expressed as a function of the the variable $\tilde{\boldsymbol{x}}$, ρ, t, t_R, is

$$\begin{aligned} \mathcal{L}_G^{[1]} \;=\; & \frac{M}{2}\frac{g_{ij}(\tilde{\boldsymbol{x}},t)\tilde{x}'^i\tilde{x}'^j}{\Theta(t)t'} + M\;A_i(\tilde{\boldsymbol{x}},t)\tilde{x}'^i + M\;\Theta(t)\;A_0(\tilde{\boldsymbol{x}},t)t' \\ & + \frac{\mu}{2}\frac{|\Xi^{-1}(\tilde{\boldsymbol{x}},t,t_R)|_{ab}D\rho^a\;D\rho^b}{\Theta(t)t' + \frac{m_2-m_1}{M}T'_R} \\ & - \frac{k}{\mu}\left(\Theta(t)t' + \frac{m_2-m_1}{M}t'_R\right)|\Xi(\tilde{\boldsymbol{x}},t,t_R)|^{ab}\delta_{ac}\delta_{bd}\rho^c\rho^d \end{aligned} \tag{5.7}$$

where in analogy to eqs.(3.6),(3.7), we have defined the *internal covariant derivative* and the *internal metric* as:

$$\begin{aligned} D\rho^a \;&=\; \rho'^a + c_{bc}{}^a\left(\tilde{x}'^k B_k^{(b)}(\tilde{\boldsymbol{x}},t) + t'B_0^{(b)}(\tilde{\boldsymbol{x}},t)\right)\rho^c \\ \tilde{\Xi}^{ab}(\tilde{\boldsymbol{x}},t,t_R) \;&=\; \delta^{ab} + \frac{\mu}{\sqrt{k}}\cdot\frac{\tilde{x}'^k B_k^{(ab)}(\tilde{\boldsymbol{x}},t) + t'B_0^{(ab)}(\tilde{\boldsymbol{x}},t)}{\Theta(t)t' + \frac{m_2-m_1}{M}t'_R}\;, \end{aligned}$$

Clearly, this Lagrangian reduces to the Lagrangian $\mathcal{L}_G$ of eqs.(3.4),(3.9) if $t=\tau$, $t_R=0$. The symmetry transformations under which its invariant (*quasi-invariant*) are given by:

$$\begin{cases} \delta^G\tilde{x}^i \;=\; \tilde{\eta}^i(\tilde{\boldsymbol{x}},t) \\ \delta^G\rho^a \;=\; \epsilon(t)\dot{r}^a + c_{bc}{}^a\tilde{\theta}^b(\tilde{\boldsymbol{x}},t)\rho^c + \frac{\mu}{\sqrt{k}}\xi^{ab}(\tilde{\boldsymbol{x}},t)\frac{|\Xi^{-1}|_{bc}(\tilde{\boldsymbol{x}},t,t_R)D\rho^c}{\Theta(t)t' + \frac{m_2-m_1}{M}t'_R}\;. \end{cases} \tag{5.8}$$

Acknowledgements

This work has been supported partly by the I.N.F.N., Italy (iniziativa specifica FI-2) and partly by the Network *Constrained Dynamical Systems* of the E.U. Programme "Human Capital and Mobility"

References

[1] R. Utiyama (1956), 'Invariant Theoretical Interpretation of Interactions', *Phys. Rev.*, **101** 1597.

[2] A. Barducci, C. Casalbuoni and L. Lusanna (1977), 'Classical spinning particles interacting with external gravitational field', *Nucl. Phys.*,**124 B** 521.

[3] C. Longhi and L. Lusanna (1989), 'On the many-time formulation of classical particle dynamics', *J. Math. Phys.*, **39** 8, 1893.

[4] R. De Pietri, L. Lusanna and M. Pauri (1994), 'Standard and Generalized Newtonian Gravities as "Gauge" Theories of the Extended Galilei Group - I: the Standard Theory', gr-qc/9405046, submitted to *Classical and Quantum Gravity.*

[5] K. Pohlmeyer (1982), 'A group-theoretical approach to the quantization of the free relativistic closed string', *Physics Letters*, **B119** 100;
K. Pohlmeyer and K.H. Rehren (1986), 'Algebraic properties of the invariant charges of the Nambu-Goto theory', *Commun. Math. Phys.*, **105** 593.

[6] M. Pauri and G.M. Prosperi (1966), 'Canonical Realizations of Lie Symmetry Groups', *J. Math. Phys.*, **7**;
M. Pauri and G.M. Prosperi (1968), 'Canonical Realizations of the Galilei Group', *J. Math. Phys.*, **9**.

On the harmonic interaction of three relativistic point particles

G. Longhi

1 Introduction

The problem of N mutually interacting particles in direct interaction has received much attention in recent years [1].

The problem can be stated in the following form: to find N mass constraints which satisfy the following requirements

i) they must be first class constraints in the sense of Dirac [2],[3]

ii) they must have the cluster decomposition property (separability), defined as the possibility of splitting the system in subsystems (clusters) by switching off some of the interactions [5],[6],[7] and [8].

The two requirements (i) and (ii) are difficult to satisfy for more than two bodies. In particular the cluster decomposition property or separability requires the presence of genuine N body forces [9],[10] and [11].

In the present paper we start an analysis of the spinless case, with harmonic interaction. We have chosen this simplified case, since the classical solution is well known, in order to have some insight into the difficulties of the problem. A separable model for three fermions has been proposed in ref.[12].

We will show that it is possible to give the form of the first class constraints in the general case, but defined in implicit form. In our intentions this analysis would be the starting point of a relativistic theory of small vibrations about a stable configuration of some general potential.

In a particular configuration, when two over three of the coupling constants are equal, there is a great simplification, particularly if one uses a suitable gauge fixing. The analysis of this case is in progress.

2 The form of the constraints

The two-body relativistic constraints, for harmonic interaction and equal masses, can be written in the following form (DTK model [13],[7] and [14])

$$\phi_0 = p^2 - 4m^2 + 4q^2 + \omega^2\tilde{\rho}^2, \qquad \phi_1 = (p\cdot q), \tag{1}$$

where $(\,\cdot\,)$ means the scalar product with signature $(+\,-\,-\,-)$, and where

$$p = p_1 + p_2, \qquad q = \frac{1}{2}(p_1 - p_2), \qquad \rho = x_1 - x_2, \qquad \tilde{\rho} = \rho - \frac{(\rho\cdot p)}{p^2}p. \tag{2}$$

x_1, x_2, p_1, p_2 are canonical variables for the two particles, coordinates four vectors and momenta respectively, with Poisson brackets

$$\{p_i^\mu,\ x_j^\nu\} = \delta_{ij}g^{\mu\nu}, \quad (i,j = 1,2,3), \tag{3}$$

where g is the metric. The two constraints ϕ_0, ϕ_1 are first class constraints.

We want to generalize the constraints (1) to three particles, in harmonic interaction with arbitrary coupling constants $k_{ij} = k_{ji} \geq 0$ $(i,j = 1,2,3)$, and equal masses. Let us try to write the three first class constraints we are looking for in the following form

$$\phi_0 = \frac{1}{3}p^2 - 3m^2 + \pi_2^2 + \omega_2^2\tilde{\rho}_2^2 + \pi_3^2 + \omega_3^2\tilde{\rho}_3^2, \tag{4}$$

$$\phi_1 = (\pi_2 \cdot Q_2) - \frac{\mu}{6}(9m^2 - p^2), \tag{5}$$

$$\phi_2 = (\pi_3 \cdot Q_3) - \frac{\nu}{6}(9m^2 - p^2), \tag{6}$$

where p is the total momentum $p = p_1 + p_2 + p_3$, and the variables π_2, π_3 and ρ_2, ρ_3 are relative conjugated variables for three bodies, corresponding to the two normal modes of the system, defined in the following.

In these expressions, the four vectors Q_2 and Q_3 must be chosen in such a way to satisfy several conditions that we will discuss in the following.

The transverse relative coordinates $\tilde{\rho}_2$ and $\tilde{\rho}_3$ are defined as transverse to the corresponding four vectors Q_2 and Q_3 respectively.

The form of ϕ_0 is quite natural. The form of ϕ_2 and ϕ_3 require an explanation. They are defined in such a way to fit the free case, when $\omega_2, \omega_3 = 0$. In this case we define

$$Q_2 = \frac{1}{3}p + \mu\pi_2 - \nu\pi_3, \qquad Q_3 = \frac{1}{3}p - \mu\pi_2 + \nu\pi_3. \tag{7}$$

At this point let us make a digression on the definition of the relative variables in the classical case.

3 The relative variables

Let the classical potential for three particles of equal mass in harmonic interaction be

$$V(\vec{x}_1, \vec{x}_2, \vec{x}_3) = \frac{1}{2}\vec{x}_i \cdot a_{ij}\vec{x}_j, \tag{8}$$

$$a = \begin{pmatrix} k_{12} + k_{31} & -k_{12} & -k_{13} \\ -k_{21} & k_{23} + k_{12} & -k_{23} \\ -k_{31} & -k_{32} & k_{31} + k_{23} \end{pmatrix}, \qquad k_{ij} = k_{ji} \geq 0. \tag{9}$$

If S is an orthogonal matrix which diagonalizes the matrix a, we define

$$S = \begin{pmatrix} \frac{1}{\sqrt{3}} & a_1 & b_1 \\ \frac{1}{\sqrt{3}} & a_2 & b_2 \\ \frac{1}{\sqrt{3}} & a_3 & b_3 \end{pmatrix}, \quad S^T a S = (0, \lambda_2, \lambda_3)_{diag}, \quad S^T S = 1. \tag{10}$$

The eigenvalues λ_2 and λ_3 of the matrix a are such that $0 \leq \lambda_2 \leq \lambda_3$, where

$$\lambda_2 = 0 \quad \text{if and only if} \quad k_{12}k_{23} + k_{23}k_{31} + k_{31}k_{12} = 0,$$

(cluster of two interacting particles and one particle free),

$$\lambda_2 = \lambda_3 \quad \text{if and only if} \quad k_{12} = k_{23} = k_{31}, \quad \text{(isotropic case)},$$

$$\lambda_3 = 0 \quad \text{if and only if} \quad k_{12} = k_{23} = k_{31} = 0, \quad \text{(free case)}$$

There is a useful choice of coordinates in the space of the coupling constants, which corresponds in taking cylindrical coordinates around the axis along the direction $\vec{u} \equiv (1, 1, 1)$ in the first octant.

These coordinates are: the distance R from the origin of a plane orthogonal to $\vec{u}$, and the polar coordinates ρ and η in this plane.

In terms of these coordinates, the elements of the matrix S (with the choice $\det S = 1$) are

$$\begin{aligned} a_1 &= -\frac{1}{\sqrt{6}}(\cos\frac{\eta}{2} + \sqrt{3}\sin\frac{\eta}{2}), \quad a_2 = +\frac{1}{\sqrt{6}}(-\cos\frac{\eta}{2} + \sqrt{3}\sin\frac{\eta}{2}), \\ a_3 &= +\frac{2}{\sqrt{6}}\cos\frac{\eta}{2}, \end{aligned} \tag{11}$$

$$\begin{aligned} b_1 &= +\frac{1}{\sqrt{6}}(\sqrt{3}\cos\frac{\eta}{2} - \sin\frac{\eta}{2}), \quad b_2 = -\frac{1}{\sqrt{6}}(\sqrt{3}\cos\frac{\eta}{2} + \sin\frac{\eta}{2}), \\ b_3 &= +\frac{2}{\sqrt{6}}\sin\frac{\eta}{2}. \end{aligned} \tag{12}$$

Observe that they depend on the angle η only, so the dependence of the matrix S on the coupling constants is only through this angle.In all these equations $\eta \epsilon [0, 2\pi)$.

The elements of the matrix S, a_i, b_i, satisfy a set of useful relations, which are consequence of the orthogonality of S only, which means that these relations hold for any value of the coupling constants k_{ij}:

$$\begin{aligned} a_i^2 &= \frac{1}{3} + \mu a_i - \nu b_i, \quad b_i^2 = \frac{1}{3} - \mu a_i + \nu b_i, \\ a_i b_i &= -\nu a_i - \mu b_i, \quad (i = 1, 2, 3), \end{aligned} \tag{13}$$

$$\mu = 3a_1a_2a_3, \qquad \nu = 3b_1b_2b_3, \qquad \mu^2 + \nu^2 = \frac{1}{6}, \tag{14}$$

$$\mu = \frac{1}{\sqrt{6}}\cos(3\eta/2), \quad \nu = -\frac{1}{\sqrt{6}}\sin(3\eta/2). \tag{15}$$

The new canonical variables are now defined in terms of the matrix S:

$$\begin{aligned} p &= p_1 + p_2 + p_3, \quad \pi_2 = \sum_{i=1,2,3} a_i p_i, \quad \pi_3 = \sum_{i=1,2,3} b_i p_i, \\ x &= \frac{1}{3}(x_1 + x_2 + x_3), \quad \rho_2 = \sum_{i=1,2,3} a_i x_i, \quad \rho_3 = \sum_{i=1,2,3} b_i x_i, \end{aligned} \tag{16}$$

$$\{p^\mu, x^\nu\} = g^{\mu\nu}, \qquad \{\pi_i^\mu, \rho_j^\nu\} = \delta_{ij} g^{\mu\nu}, \qquad (i,j = 1,2,3). \tag{17}$$

Observe that the relations (13) and (16) depend solely on the orthogonality of the matrix S, and do not depend on the values of the coupling constants k_{ij}.

Using the relations (16) we have

$$Q_2 = \sum_{i=1,2,3} a_i^2 p_i, \quad Q_3 = \sum_{i=1,2,3} b_i^2 p_i, \tag{18}$$

which in particular means that, if the p_i are time-like, so are Q_2 and Q_3, being a linear combination of time-like vectors, with positive semi definite coefficients not simultaneously vanishing.

It can be verified that these are indeed the constraints in the free case

$$p_i^2 = m_i^2, \qquad (i = 1,2,3). \tag{19}$$

Let us analyse the definitions of Q_2 and Q_3, with regard to the decomposition of the system in two interacting plus one free particles.

The three possible decompositions in a cluster of two interacting particles plus one free are obtained with $\eta = 0$, $2\pi/3$ or $4\pi/3$, and we have the following situation:

$$\eta = 0, \qquad k_{23} = k_{31} = 0, \qquad \omega_2 = 0, \quad \omega_3^2 = 2k_{12}, \quad \mu = \frac{1}{\sqrt{6}}, \tag{20}$$

$$\nu = 0, \quad Q_2 = \frac{1}{6}(p_1 + p_2 + 4p_3), \qquad Q_3 = \frac{1}{2}(p_1 + p_2); \tag{21}$$

- - - -

$$\eta = \frac{2\pi}{3}, \qquad k_{31} = k_{12} = 0, \qquad \omega_2 = 0, \quad \omega_3^2 = 2k_{23}, \quad \mu = -\frac{1}{\sqrt{6}}, \tag{22}$$

$$\nu = 0, \quad Q_2 = \frac{1}{6}(p_2 + p_3 + 4p_1), \qquad Q_3 = \frac{1}{2}(p_2 + p_3) \tag{23}$$

- - - -

$$\eta = \frac{4\pi}{3}, \qquad k_{12} = k_{23} = 0, \qquad \omega_2 = 0, \qquad \omega_3^2 = 2k_{31}, \quad \mu = \frac{1}{\sqrt{6}}, \tag{24}$$

$$\nu = 0, \quad Q_2 = \frac{1}{6}(p_3 + p_1 + 4p_2), \qquad Q_3 = \frac{1}{2}(p_3 + p_1). \tag{25}$$

From this table we see that the four-vectors Q_3 becomes in each case the total momentum of the subsystem of the two interacting particles.

Observe that Q_2 and Q_3 are constants of the motion, that is that they commute (have vanishing Poisson brackets) with the constraints, and that they are time-like, being a linear combination of the p_i $(i = 1, 2, 3)$, which are assumed to be time-like, with coefficients ≥ 0, not simultaneously vanishing.

4 The definition of Q_2 and Q_3 in the interacting case

Until now we have analysed the free case. To generalize the definition of Q_2 and Q_3 to the interacting case we merely substitute π_2 and π_3 in their expressions, which in the free case are constant, with constant of motion for the harmonic interacting case. This can be done using the fact that the transverse variables (to Q_2 or Q_3) essentially describe two three dimensional harmonic oscillators, with the normal frequencies.

Now, for a classical three dimensional harmonic oscillator, with hamiltonian

$$h = \frac{1}{2}(\vec{p}^2 + \omega^2 \vec{x}^2), \tag{26}$$

(mass = 1), we have that the vector

$$\vec{y} = \alpha(\tau)\vec{p} + \omega\beta(\tau)\vec{x}, \tag{27}$$

with

$$\begin{aligned} &\alpha = \cos(\omega\tau) + \sin(\omega\tau), \qquad \beta = -\cos(\omega\tau) + \sin(\omega\tau), \\ &\cos(2\omega\tau) = 2\omega(\vec{p}\cdot\vec{x})/\sqrt{(...)}, \quad \sin(2\omega\tau) = (\vec{p}^2 - \omega^2\vec{x}^2)/\sqrt{(...)}, \\ &(...) = (\vec{p}^2 - \omega^2\vec{x}^2)^2 + 4\omega^2(\vec{p}\cdot\vec{x})^2. \end{aligned} \tag{28}$$

which satisfies

$$\{\vec{y},\, h\} = 0, \qquad \{y^i,\, y^j\} = 0, \qquad (i,j = 1,2,3). \tag{29}$$

This suggests to perform the following substitution in the expressions of Q_2 and Q_3

$$\pi_2 = \tilde{\pi}_2 + \underline{\pi}_2 \rightarrow Y_2 + \underline{\pi}_2 \qquad \pi_3 = \tilde{\pi}_3 + \underline{\pi}_3 \rightarrow Y_3 + \underline{\pi}_3, \tag{30}$$

where $Y_{2,3}$ are defined in analogy with $\vec{y}$, being essentially three-dimensional

$$Y_2 = \alpha_2 \tilde{\pi}_2 + \omega_2 \beta_2 \tilde{\rho}_2, \qquad Y_3 = \alpha_3 \tilde{\pi}_3 + \omega_3 \beta_3 \tilde{\rho}_3, \tag{31}$$

and $\underline{\pi}_k$ is the longitudinal part of π_k with respect to Q_k, $(k = 2,3)$.

We get

$$\begin{aligned} Q_2^\mu &= \frac{1}{3} p^\mu + \mu(Y_2^\mu + \underline{\pi}_2^\mu) - \nu(Y_3^\mu + \underline{\pi}_3^\mu), \\ Q_3^\mu &= \frac{1}{3} p^\mu - \mu(Y_2^\mu + \underline{\pi}_2^\mu) + \nu(Y_3^\mu + \underline{\pi}_3^\mu). \end{aligned} \tag{32}$$

In this way we get the following result: the Q_2 and Q_3 are still constants of the motion, that is they commute with the constraints $\phi_{0,1,2}$, the constraints are still first class, and moreover the decomposition in cluster is maintained.

The demonstration that Q_2 and Q_3 commute with the constraints is a bit complicated, but it essentially requires the use of the implicit function theorem.

The main problem is to find the explicit form of the four-vectors Q_2 and Q_3, which are defined in an implicit way only by the equations (32). But, as a matter of fact, it is not strictly necessary to calculate them, we may eliminate all the scalar quantities involving them in order to get the explicit form of the constraints in terms of the particles variables. This is an elimination algebraic problem, which is known to have always a solution.

There is one particular case in which we may get a great simplification. Namely when two of the three coupling constants are equal. This situation will correspond to the values of $\eta = l\pi/3$, with $l = 0, 1, ..., 6$. In all these cases or μ or ν will vanish. In any of these cases, if we introduce the gauge fixing $(\rho_2 \cdot Q_2) = 0$ and $(\rho_3 \cdot Q_3) = 0$, we may get the explicit expression of Q_3^2, and we may explicitly verify that Q_2 and Q_3 are indeed time-like.

5 The cluster decomposition

The structure of Q_2 and Q_3 in the free case shown in equations (21), (23) and (25) is conserved in the interacting case.

Since the model have the cyclic symmetry, let us verify the cluster decomposition in the case $k_{23} = k_{31} = 0$, $\eta = 0$, which means $\alpha_2 = 1$ and $\beta_3 = -1$.

The relative variables become

$$\rho_2 = \frac{1}{\sqrt{6}}(2x_3 - x_1 - x_2), \qquad \rho_3 = \frac{1}{\sqrt{2}}(x_1 - x_2), \tag{33}$$

$$\pi_2 = \frac{1}{\sqrt{6}}(2p_3 - p_1 - p_2), \qquad \pi_3 = \frac{1}{\sqrt{2}}(p_1 - p_2). \tag{34}$$

The constraints become

$$\phi_0 = \frac{1}{2}\left[p_{12}^2 - 4m^2 + 2(p_3^2 - m^2) + 4q_{12}^2 + \omega_3^2(r_{12}^2 - \frac{(r_{12}\cdot p_{12})^2}{p_{12}^2})\right], \tag{35}$$

$$\phi_1 = \frac{3}{2\sqrt{6}}(p_3^2 - m^2), \qquad \phi_2 = \frac{1}{\sqrt{2}}(p_{12}\cdot q_{12}), \tag{36}$$

where

$$p_{12} = p_1 + p_2, \qquad q_{12} = \frac{1}{2}(p_1 - p_2), \qquad r_{12} = x_1 - x_2, \tag{37}$$

which are equivalent to the constraints of the DTK model [13],[7], [14],for two bodies.

6 The non relativistic limit

The non relativistic limit can be easily get, and the result is that, at equal times, the constraint ϕ_0, becomes

$$\phi_0 = 2m\left[\varepsilon - \sum_{i=1,2,3}\frac{\vec{p}_i^2}{2m} - \frac{1}{2m}\sum_{i,j=1,2,3}\vec{x}_i\cdot a_{ij}\vec{x}_j\right], \tag{38}$$

where ε is the n.r. energy of the center of mass. For what regards the two other constraints ϕ_1 and ϕ_2, they merely give two equations for the relative energies.

7 Concluding remarks

The model we have discussed is one possible approach to the problem of small vibrations of a relativistic three body system around a stability point of some potential.

It must be stressed that it is heavily based on the complete integrability of the corresponding classical system. In particular it is based on the classical harmonic oscillator treatment of the Appendix B, where it is crucial the use of the canonical observable τ, as a kind of an "internal clock" of the system, in place of the usual external parameter time, which cannot have any meaning in the present context.

Only due to the complete integrability of the model it was possible to find the involutive set of constants of the motion y^i.

This circumstance in turn manifests itself in the constancy of the four vectors Q_2 and Q_3, which are a linear combination of the two constants of motion Y_2 and Y_3, which are intrinsically three dimensional, and the two scalar $(\pi_2 \cdot Q_2)$ and $(\pi_3 \cdot Q_3)$, which are constants of motion too.

By adding to these the four constants p^μ we have twelve involutive constants of motion, so that the model is Liouville integrable. Since there are three constraints, we get indeed nine constants of motion, the right number for three particles.

As mentioned in the Introduction, when two over three of the coupling constants are equal (which means, for instance, to put $\eta = 0$ or π, in which case we get $k_{23} = k_{31}$), we get a great simplification. In this case the explicit form of Q can be calculated, and a further simplification may be obtained by the choice of the gauge fixing $(\rho_k \cdot Q_k) = 0$, $(k = 2, 3)$ (relative times determined).

References

[1] F.M. Lev (1993), *Riv. N. Cimento*, **16** 1; see also ref. [4].

[2] P.A.M. Dirac (1950), *Can. J. Math.*, **2** 129.
P.A.M. Dirac (1964), *Lectures on Quantum Mechanics.* Yeshiva University.
J.L. Anderson and P.G. Bergmann (1951), *Phys. Rev.*, **83** 1018.
P.G. Bergmann and J. Goldberg (1955), *Phys. Rev.*, **98** 531.

[3] E.C.G. Sudarshan and Mukunda (1974), *Classical Mechanics: a Modern Perspective.* Wiley.
A.J. Hanson, T. Regge and C. Teitelboim (1975), *Constrained Hamiltonian Systems.* Accademia Nazionale dei Lincei, n. 22, Roma.
K. Sundermeyer (1982), *Constraint Dynamics with Applications to Yang-Mills Theory, General Relativity, Classical Spin, Dual String Model.* Lecture Notes in Physics **169**, Springer, Berlin.

[4] G. Longhi and L. Lusanna (eds.) (1986), *Proceedings of the Firenze Workshop.*, World Scientific, Singapore.

[5] L.L. Foldy (1962), *Phys. Rev.*, **122** 275; (1977), *Phys. Rev.*, **15** 3044.
L.L. Foldy and R.A. Krajcik (1974), *Phys. Rev.*, **32** 1025; (1975), *Phys. Rev.*, **D10** 1777; **D12** 1700.

[6] S.N.Sokolov (1978), 'Theory of relativistic direct interactions', Serpukov report IHEP-OFT 78-125; (1977), *Dokl. Akad. Nauk. SSSR*, **233** 575, Engl. trans.: (1977), *Sov. Phys. Dokl.*, **22** 198; (1978), *Teor. Mat. Fiz.*, **36** 193, Engl. trans.: (1978), *Theor. Math. Phys.*, **36** 682.
S.N. Sokolov and Shatny (1979), *Theor. Math. Phys.*, **37** 1029.

[7] I.T. Todorov (1976), JINR Report E2-10125, Dubna.
I.T.Todorov (1978), *Ann. Inst. H. Poincaré*, **A28** 207; (1973), in *Properties of the Fundamental interactions*, A. Zichichi (ed.), Ed. Compositori, Bologna.

[8] H. Sazdjian (1989), *Ann. Phys. (NY)*, **191** 52.
Ph. Droz-Vincent (1980), *Ann. Inst. H. Poincaré*, **A32** 317; (1983), *Lett. Nuovo Cimento*, **38** 177; (1985), *Phys. Lett.*, **B159** 393.
V. Iranzo, J. Llosa, F. Marquez and A. Molina (1981), *Ann. Inst. H. Poincaré*, **35A** 1; (1984), **40A** 1.

[9] See I.T. Todorov, in ref. [4] and the references quoted therein.

[10] J. Llosa (ed.) (1982), *Relativistic Action-at-a-Distance, Classical and Quantum Aspects.*, Lecture Notes on Physics **162**, Springer.

[11] H. Sazdjian (1981), *Ann. Phys. (NY)*, **136** 136.
F. Rohrlich (1981), *Phys. Rev.*, **D23** 1305.

[12] J. Bijtebier (1990), *Il Nuovo Cim.*, **103A** 669.

[13] Ph. Droz-Vincent (1975), *Rep. Math. Phys.*, **8** 79.

[14] A. Komar (1978), *Phys. Rev.*, **D18** 1881, 1887.

Non existence of static multi-black hole solutions in 2+1 dimensions

O. Coussaert and M. Henneaux

Abstract

It is shown that there is no static pure multi-black-hole solution of the (2+1)-Einstein equations with negative cosmological constant. The result is extended to the stationnary case and to the Einstein-Maxwell theory, for which the absence of pure multi-black hole solutions with freely specifiable positions is also established.

1 Introduction

Four-dimensional Einstein-Maxwell theory admits remarkable static multi-black-hole metrics in which the electrostatic repulsion exactly balances the gravitational attraction [1, 2, 3]. These solutions have the following important properties.

(i) each black hole contained in the solutions is of the extreme Reissner-Nordstrom type, i.e., has its mass equal to its charge;

(ii) the relative positions of the individual black holes are free parameters that can be specified independly of their masses (the exact balance between the electromagnetic repulsion and the gravitational attraction holds for any configuration); and

(ii) all the multi-black hole solutions have some exact supersymmetries [4, 5, 6]; actually, they are the only black hole solutions with this property.

Recently, three-dimensional metrics describing single black holes have been found in the context of the Einstein theory with a negative cosmological constant [7]. Even though the single (2+1)- black hole solutions occur in a much simpler setting, they share global causal properties quite similar to their (3+1)-parents [8].

The purpose of this note is to investigate whether one can construct static multi-black-hole solutions in 2+1 dimensions. We show that this is not possible, in spite of the fact that multiparticle solutions are known to exist for (2+1)-Einstein theory with zero cosmological constant [9]. We find that any attempt to put two (or more) black holes at rest in the same space necessarily produces unwanted naked conical singularities at which the Einstein equations break down. These singularities are identical with the inconsistent singularities discussed -and rejected- in [10] for multiparticle solutions in a de Sitter background. The problems can be traced to the non vanishing of the cosmological constant.

2 General static solution of the vacuum Einstein equations with negative cosmological constant in 2+1 dimensions

In 2+1 dimensions, the vacuum Einstein theory is locally trivial. That is, any point in a solution of the vacuum Einstein equations (with a negative cosmolgical contant) possesses an open neighbourhood that is isometric to an open subset of the anti-de Sitter space. Thus, an uncharged black hole has no influence on the local behaviour of geodesics. The geodesic deviation equation is the same as in the absence of the black hole.

By contrast, the three-dimensional Maxwell field is non trivial. Electric sources produce a non zero Coulomb field which does modify the trajectories of charged particles. There is no perfect matching any more between the gravitational and the electrostatic interactions. For this reason - and because there is no local gravitational field proper to the black-hole to be balanced- we shall first look for neutral multi-black-hole solutions.

The most general static metric can be brought to the form

$$ds^2 = -N^2dt^2 + \phi(dx^2 + dy^2) \tag{2.1}$$

where the lapse N and the conformal factor ϕ are functions of x and y only. The general solution of the vacuum Einstein equations with particle sources has been given, for the metric (2.1), in references [9] and [10]. A slight and straighforward modification ot the techniques of [10] covers the black hole case with negative cosmological constant $\Lambda = -l^{-2} < 0$.

One finds that the general solution is paramerized by one meromorphic differential $\omega = V^{-1}(z)dz$, where $z = x + iy$. In terms of $V(z)$, the lapse fulfills

$$\partial_\zeta N - N^2 = \epsilon \tag{2.2}$$

where the real variable ζ is conformally invariant and is given by

$$\zeta = \frac{1}{2}\left[\int^z \omega + \int^{z^*} \omega^*\right]. \tag{2.3}$$

Here ϵ is a constant. The case $\epsilon < 0$ yields point particles without horizons and is treated in [10]. So we consider here only $\epsilon \geq 0$.

We start with the simplest case $\epsilon = 0$. The solution of (2.2) with $\epsilon = 0$ reads

$$N = -\frac{1}{\zeta} \qquad (\epsilon = 0) \tag{2.4}$$

while the conformal factor is given by

$$\phi = \frac{1}{\zeta^2}\frac{l^2}{|V|^2} \qquad (\epsilon = 0) \tag{2.5}$$

The range of the coordinates x and y is limited by the zeros and poles of $\ln \zeta$ and of V.

To gain insight into the solutions (2.4-2.5) , let us first consider the case where the differential has one pole, say at the origin, $V^{-1} = 1/z$. This choice yields $\zeta = \ln \rho$, with $\rho = \sqrt{x^2 + y^2}$. The metric is then

$$ds^2 = -\frac{1}{(ln \rho)^2}dt^2 + \frac{l^2}{\rho^2(\ln \rho)^2}(d\rho^2 + \rho^2 d\varphi^2) \tag{2.6}$$

The change of variables $\ln \rho = -\frac{l}{r}$ brings it to the form

$$ds^2 = -\left(\frac{r}{l}\right)^2 dt^2 + \left(\frac{l}{r}\right)^2 dr^2 + r^2 d\varphi^2 \tag{2.7}$$

which is just the exterior metric of a single zero mass black hole [7]. The horizon-singularity is at the origin $r = 0 = \rho$ where both V^{-1} and ζ have a pole. The circle at infinity $r = \infty$ is mapped on the circle $\rho = 1$, where ζ vanishes. Thus, $-\infty < \zeta < 0$ and $0 < \rho < 1$; the exterior black hole space is the punctured unit disk. In that range of x and y, V^{-1} does not vanish and there is no singularity of the metric. Note that the metric (6) with $1 < \rho < \infty$ represents also the exterior black hole metric. This is because the metric (6) is invariant under the inversion $\rho \rightarrow 1/\rho$. In that range of ρ, infinity is at $\rho = 1$ and the horizon-singularity is at $\rho = \infty$. Finally, note that a modification of the residue of V^{-1} at $z = 0$ ($V^{-1} \rightarrow \alpha V^{-1}$) can be absorbed in a rescaling of t.

The zero mass black hole is the black hole solution with the maximum number of exact supersymmetries [11]. It appears thus as the best candidate for constructing a multi-black-hole solution. Even in that case, however, one cannot put many black holes together at rest in the same universe, as we now show.

To generate a multi-black hole solution, one would have to take the function V^{-1} to have many poles, one for each hole,

$$V^{-1} = \sum_k \frac{\alpha_k}{z - z_k} \quad , k = 1, ..., N \tag{2.8}$$

This leads to

$$\zeta = \sum_k \ln(\rho_k)^{\alpha_k} \quad , \rho_k = \sqrt{(x-x_k)^2 + (y-y_k)^2} \tag{2.9}$$

The metric describe then N black holes since it is of the form (6) to leading order in the vicinity of each pole $z = z_k$. However , in order to have a true multi-black-hole solution, at least two of these black holes should be in the same universe. Namely, the line at infinity, given by $\zeta = 0$, should not isolate the black holes from one another. This would happen if one were to take, say, $V^{-1} = \frac{1}{z} + \frac{1}{(z-100)}$. There are then two black holes, but each of them is in a separate universe since the line $\zeta = 0$ has two components, each component surrounding one of the black holes.

Furthemore, there should be no zero of V^{-1} in the allowed range of x and y. This is because a zero of V^{-1} with $\zeta \neq 0$ yields a singularity in the metric (2.1) since ϕ vanishes there. This singularity is easily seen to be a naked conical singularity with angle excess equal to a multiple of 2π , i.e., it is a branch point of a multi-covering of the plane. This branch point is not protected by a horizon and is thus visible from infinity. So, the solution is not a pure multi-black-hole solution. Worse than that, the branch point does not follow a geodesic. The Einstein equations fail to hold at the branch point, which is for that reason unacceptable [10].

It is these two requirements - (i) the black holes should be in the same universe; (ii) there should be no zero of V^{-1} - that are in conflict and eliminate the multi-black-hole solution.

Indeed one has:

Theorem: let U be a connected component of the metric (2.1), (2.8), (2.4), (2.5), (2.9) containing at least two black holes. Then, there is a point in U where V^{-1} vanishes.

Proof: the coefficients α_k of the black holes in the region U have the same sign since $\zeta \to +\infty$ close to the holes with negative α's and $\zeta \to -\infty$ close to the holes with positive α's , and the frontier of U has $\zeta = 0$. Assume for definiteness $\alpha_k > 0$. The case $\alpha_k < 0$ is treated similarly. Now, in the region U , the lines $\zeta = const$ have as many components as there are black holes for ζ negative enough.

But by assumption, the line $\zeta = 0$ has a single component, namely, the boundary of U ("infinity"). This implies that $d\zeta$ vanishes somewhere inside U , for the value(s) of ζ where two components of $\zeta = const$ meet. [This may occur for infinite values of the coordinates (x, y), but can be brought to finite values by a conformal transformation.] But if $d\zeta = 0$, then, V^{-1} vanishes.

This is because ζ is the real part of a holomorphic function $\Psi \equiv \zeta + i\eta = \int dz V^{-1}$. By the Cauchy-Riemmann equations, $d\zeta = 0$ implies that $d\eta = 0$, and thus $V^{-1} = 0$. This proves the theorem.

The same argument applies to the case with $\epsilon > 0$, which describes massive black holes. This time, the function N is given by

$$N = \sqrt{\epsilon} \tan \sqrt{\epsilon}(\zeta - \zeta_0) \tag{2.10}$$

and ϕ reads

$$\phi = \frac{\epsilon}{V(z)V^*(z^*) \cos^2 \sqrt{\epsilon}(\zeta - \zeta_0)} \tag{2.11}$$

The poles of V^{-1} correspond to the black hole singularities. The (x, y)-plane is divided into the regions $n\frac{\pi}{\sqrt{\epsilon}} < \zeta - \zeta_0 < \zeta = \zeta_0 + (2n+1)\frac{\pi}{2\sqrt{\epsilon}}$, each of which defines a different exterior solution. The horizons are given by $\zeta = \zeta_0 + n\frac{\pi}{\sqrt{\epsilon}}$. The lines $\zeta = \zeta_0 + (2n+1)\frac{\pi}{2\sqrt{\epsilon}}$ are the lines at infinity. There are again unavoidable branch points singularities where V^{-1} vanishes (the lines defining the horizons should have as many components as there are black holes; the line at infinity should have a single component if the black holes are to be in the same universe). Trying to push the zeros of $d\zeta$ beyond infinity forces the black holes to be in distinct unverses. Trying to hide the zeros of $d\zeta$ inside the horizons does not work either because the horizon must then have a single component and the solution is a single-black-hole solution.

3 Comments

(i) The absence of static multi-black-hole solutions contrasts with the existence of multi-particle solutions known to hold for $\Lambda = 0$ [9]. The origin of the difficulties can be traced to the non vanishing of the cosmological constant and can be heuristically understood as follows. The negative cosmological constant may be viewed as providing a long range attractive force, in the sense that there is no geodesic at rest in the standard static coordinate system of the anti-de-Sitter spacetime, except at the origin. Any particle initially at rest in the static coordinate system and not at the origin will subsequently "fall" on the origin. Now, the black holes should somehow follows geodesics. Thus, once one has "put" a black hole at the origin, there is no other place where one can put a second one at rest, unless one considers multi-coverings of the spatial sections. Each sheet of the multi-covering would be provided with a geodesic at rest where one can "put" another black hole. But the branch points of the covering cannot be at the origin (where the black holes are put) and thus, do not follow themselves geodesics, leading to a breakdown of the Einstein equations [10]. The difficulty with constructing multi-black-hole solutions in 2+1 dimensions is accordingly not that one should balance the individual gravitational field of the black holes but rather, that one should balance the effect of the cosmological constant.

(ii) The absence of multi-black hole solutions is rather surprising from the point of view of the Chern-Simons formulation of 2+1 gravity. In the context of the Chern-Simons theory, the exterior black hole metric appears as a flat SO(2,2)-connexion with non trivial holonomy around one point. In the absence of angular momentum, that holonomy is equivalent to the element $mP_{(2)}$[1] of SO(2,2) [8, 13]. In a particular gauge, the corresponding SO(2,2)-connexion is given by:

$$A = -mJ_{12}d\varphi_1 \tag{3.1}$$

where φ_1 is the angle around the point (x_1, y_1) where the black hole sits. The gauge transformation

$$\begin{aligned} U = & \quad \exp(tJ_{03})\exp(\alpha(r)J_{02}) \\ \alpha(r) = & \quad \ln\left[\tan(\frac{1}{2}m\ln r)\right] \end{aligned} \tag{3.2}$$

brings the connexion to the form

$$\begin{aligned} A = P_{(1)}d\alpha(r) + \cosh\alpha(r)P_{(2)}d\varphi_1 + \sinh\alpha(r)P_{(0)}dt \\ + \sinh\alpha(r)J_{(0)}d\varphi_1 + \cosh\alpha(r)J_{(2)}dt, \end{aligned} \tag{3.3}$$

which agrees with the black hole metric with mass m (the $P_{(a)}$-components coincide with the triad components; the $J_{(k)}$-components coincide with the $SO(2,1)$-connection components). In this light, it would appear that a multi-black-hole solution with n black holes should simply be a flat SO(2,2)- connexion with non trivial holonomy around n distinct points. Such a connexion is easily constructed. For instance:

$$A = \sum_k m_k P_{(2)} d\varphi_k \tag{3.4}$$

is a solution. Here, the φ_i are the angles around the n distinct points. By a gauge transformation similar to (3.2) where $\alpha(r)$ is now defined as

$$\alpha(r) = \ln\left[\tan(\frac{1}{2}\sum_k m_k ln\rho_k)\right], \tag{3.5}$$

one can bring (3.4) to the form

$$\begin{aligned} A = P_{(1)}d\alpha(r) + \cosh\alpha(r)P_{(2)}\sum_k m_k d\varphi_k + \sinh\alpha(r)P_{(0)}dt \\ + \sinh\alpha(r)J_{(0)}\sum_k m_k d\varphi_k + \cosh\alpha(r)J_{(2)}dt. \end{aligned} \tag{3.6}$$

[1]The generators of the SO(2,2)-lie algebra have the following brackets: $[J_{(a)}, J_{(b)}] = \varepsilon_{(a)(b)(c)}J^{(c)}, [P_{(a)}, P_{(b)}] = \varepsilon_{(a)(b)(c)}J^{(c)}, [J_{(a)}, P_{(b)}] = \varepsilon_{(a)(b)(c)}P^{(c)}$, One has $J_{01} = J_{(0)}, J_{13} = J_{(1)}, J_{03} = J_{(2)}, J_{23} = P_{(0)}, J_{12} = -P_{(2)}, J_{02} = P_{(1)}$.

If one computes the triad associated with (3.6), one gets the metric (2.1) describing n massive black holes but also the unavoidable branch point singularities where the Einstein equations break down. Thus, while the solution (3.6) is perfectly regular as a Chern-Simons solution, it is singular as a solution of the metric formulation of 2+1 gravity. The fact that the Chern-Simons theory is blind to the branch-point singularity follows from the fact that the holonomies around these points are unity.

It would be of interest to explore further this question and to understand under which conditions exact global equivalence between the Chern-Simons and metric formulations holds.

4 Charge and angular momemtum

In the previous discussion, the charge and angular momemtum of the black hole were set equal to zero from the outset. One may wonder wether non zero charges or non zero angular momemta could not bypass the above difficulties. It is easy to see that this is not the case if the multi-black-hole solution is required to fulfill the central demand that the positions of the holes can be prescribed independly of the black hole parameters, as in 3+1 dimensions.

Indeed, let $g_{ij}(m_k, J_k, Q_k; r_k)$ be the would-be multi-black-hole metric. Take the limit $J_k \rightarrow 0, Q_k \rightarrow 0$ while keeping the black hole positions fixed. In the limit, one would get a multi-black-hole solution with masses $m_k \geq 0$, zero angular momemtum and zero charge. But such a multi-black-hole solution does not exist, even if $m_k = 0$. Hence, there is no charged or rotating, static or stationnary multi-black-hole solution fulfilling the condition that the positions of the holes are free parameters.[Recall futhermore that the charged black holes are not asymptotic to anti-de Sitter spacetime in (2+1) dimensions and are from this point of view topological [11]].

5 Conclusions

We have shown in this note that contrary to what happens in 3+1 dimensions, there are no static multi-black-hole solutions in 2+1 dimensions. The impossibility is due to the cosmological constant, which is necessary for the existence of event horizons [7]. This suggest that one should look for dynamical multi-black-hole solutions, along the four-dimensional lines of [14]. The "multi-center" solutions of [15] are in this context of great interest. However, they fail to be true black hole solutions since they necesserily contain additional branch point singularities (which follow geodesics) whenever they are more than one black hole.

6 Acknowledgements

We are both grateful to John Charap for the invitation to participate at the Meeting *Geometry of Constrained Dynamical Systems* held at the Isaac Newton Institute. Fruitful discussions with Max Bañados, Gérard Clément, Claudio Teitelboim and Jorge Zanelli are gratefully acknowledged. This work has been supported in part by a research grant from the F.N.R.S. (Belgium) and by research contracts with the Commission of the European Communities. O.C. is Chercheur IRSIA.

References

[1] A. Papapetrou (1947), *Proc. R. Irish Acad.*, **A51** 191.

[2] S.D. Majumdar (1947), *Phys. Rev.*, **72** 930.

[3] J.B. Hartle and S.W. Hawking (1972), *Comm. Math. Phys.*, **26** 87.

[4] P.C. Aichelburg and R. Guven (1981), *Phys. Rev.*, **D24** 2066.

[5] G.W. Gibbons (1985), in *Supersymmetry and Related Topics*, edited by F. del Aguila, J, de Azcárraga and L. Ibáñez, World Scientific, Singapore; p.147;
G.W. Gibbons and C.M. Hull (1982), *Phys. Lett.*, **109B** 190.

[6] K.P. Tod (1983), *Phys. Lett.*, **B121** 241.

[7] M. Bañados, C. Teitelboim and J. Zanelli (1992), *Phys. Rev.Lett.*, **69** 1849.

[8] M. Bañados, M. Henneaux, C. Teitelboim and J. Zanelli (1993), *Phys. Rev.*, **D48** 1849.

[9] S. Deser, R. Jackiw and G. 't Hooft (1984), *Ann. Phys. (N.Y.)*, **152** 220 and reference therein.

[10] S. Deser and R. Jackiw (1984), *Ann. Phys. (N.Y.)*, **153** 405.

[11] O. Coussaert and M. Henneaux (1994), *Phys. Rev. Lett.*, **72** 183.

[12] E. Witten (1988/1989), *Nucl. Phys.*, **B311** 46.

[13] D. Cangemi, M. Leblanc and R.B. Mann (1993), *Phys. Rev.*, **D48** 3606.

[14] D. Kastor and J. Traschen (1992), UMHEP-380, hepth9212035.

[15] G. Clement (1993), hepth9402013, GCR94/02/01.

Spherically Symmetric Gravity and the Notion of Time in General Relativity

H.A. Kastrup and T. Thiemann

Abstract

It is shown - in Ashtekar's canonical framework of General Relativity - that the constraints of spherically symmetric (Schwarzschild) gravity in 4 dimensional space-time can be solved completely yielding two canonically conjugate observables for asymptotically flat space-times, namely mass and - surprisingly - time. The emergence of the time observable is a consequence of the Hamiltonian formulation and its subtleties concerning the slicing of space and time and is not in contradiction to Birkhoff's theorem. Our results can be expressed within the ADM formalism, too, and their relation to the equivalent ones Kuchař obtained recently are briefly discussed. Quantization of the system and the associated Schrödinger equation depend on the allowed spectrum of the masses.

1 Introduction

The issue 'time' is perhaps the most crucial one in canonical - especially quantum - gravity and a number of different approaches have been pursued in recent years (see the excellent reviews[1, 2]). Whereas the discussions of general aspects are certainly essential, one might possibly learn a lot by the analysis of a single - even very simple - system for which the corresponding quantisation can be carried through completely and in which the quantity time appears as a classical and quantum gauge invariant 'observable'.

Such a system is spherically symmetric pure gravity. In view of Birkhoff's theorem which seems to eliminate completely the notion of time for such systems this assertion may appear to be quite surprising. However, if one accepts a canonical (Hamiltonian) framework for this system it "creates" the observable "time" all by itself! This is not so surprising anymore if one observes that 'mass' is certainly an observable for such a system and if it is possible to reduce the original phase space with its numerous gauge degrees of freedom to a nontrivial physical one consisting of observables only then the quantity canonically conjugate to the mass should be 'time'. This is exactly what happens. We found this[3][4][5] by using Ashtekar's canonical framework[6] and recently Kuchař[7] obtained essentially the same results within the ADM-formalism.

Solving the constraints of classical spherically symmetric pure gravity one obtains two canonically conjugate observables, mass M and time T, the mass as a constant of integration m and the 'observable' T as the integral

$$T[q_{xx}, R] = 2\int_{\Sigma} dx(1-2m/R)^{-1}\sqrt{(dR/dx)^2 - q_{xx}(1-2m/R)} \tag{1.1}$$

over a 1-dimensional 'radial' Cauchy surface Σ, where R and q_{xx} are defined by the line element

$$ds^2 = -(N(x,t)\,dt)^2 + q_{xx}(x,t)(dx + N^x(x,t)\,dt)^2 + R^2\,(d\theta^2 + \sin^2\theta d\phi^2)\,. \tag{1.2}$$

The (local) variable x is assumed to coincide asymptotically ($x \to \infty$) with the usual Euclidean radial variable r. The functional T obeys the following 'eqs. of motion':

$$dT/dt = \{T, H_{tot}\}_{\bar{\Gamma}} = \{T, H_{red}\} = 2N^{(\infty)}(t), \quad H_{red} == 2m\,N^{(\infty)}(t), \tag{1.3}$$

where H_{tot} means the total Hamiltonian consisting of the (nonvanishing) constraints and the surface terms (see eq. (2.3) below), $\{.,.\}$ denotes the Poisson bracket, H_{red} means the value of H_{tot} on the constraint surface $\bar{\Gamma}$ where the constraints vanish and $N^{(\infty)}(t)$ is the lapse function $N(t)$ in the spatially asymptotic region under consideration (there may be several of such regions each one having its own lapse function).

This means that both Hamiltonians generate time translations as symmetry transformations - in contrast to gauge transformation - in asymptotically flat regions. The dependence of the "observable" $T(t)$ on the "unobservable" asymptotic time parameter t is determined by the gauge dependent quantity $N^{(\infty)}(t)$, ($N^{(\infty)} = 1$ for space-times which are asymptotically Minkowski flat). The factor 2 in eqs. (1.3) is due to the normalization of the Hamiltonians adopted here. For compact Σ we have $N^{(\infty)}(t) = 0$ and then T commutes with H_{tot} and H_{red}!

Quantisation of the (1+1)-dimensional canonical system formed by the observables m and T depends on the spectrum of m, namely whether it covers the whole real axis or whether it is bounded from below.

2 The model

In the spherically symmetric case the basic canonical variables in Ashtekar's approach[1] to quantum gravity, namely the connection coefficients $A_a^i(x)$ as configuration variables and the densitized triads $\tilde{E}_i^a(x), a = 1,2,3, i = 1,2,3$ as momentum variables, can be expressed by 6 functions $A_I(t,x)$, and $E^I(t,x)$, $I = 1,2,3$, where the E^I are real and the A_I are complex. Here t is a "time" variable and x is a (local) spatial variable which becomes the usual Euclidean radial variable r at spatial infinity.

The metric (q_{ab}) on the 3-dimensional (Cauchy) surfaces $\Sigma^3 = \Sigma \times S^2$ takes the form

$$(q_{ab}) = \text{diag}(\frac{E}{2E^1}, E^1, E^1 \sin^2\theta), \quad E = (E^2)^2 + (E^3)^2, \tag{2.1}$$

Integrating the Einstein-Ashtekar action over the unit sphere in $\Sigma^3 = S^2 \times \Sigma$, where Σ is an appropriate 1-dimensional manifold (see below)and dropping an overall factor $4\pi/(8\pi G)$, we get the following effective (1+1)-dimensional action for spherically symmetric gravity[4]

$$S = \int_R dt[\int_\Sigma dx(-iE^I \dot{A}_I) - H_{tot}], \tag{2.2}$$

with the total Hamiltonian

$$H_{tot} = \int_\Sigma dx(i\Lambda G - iN^x(V_x - A_1 G) + \underset{\sim}{N} C) + \tag{2.3}$$

$$+Q_r + P_{ADM} + E_{ADM}, \quad \text{where}$$

$$G = (E^1)' + A_2 E^3 - A_3 E^2 \; : \tag{2.4}$$

Gauss constraint function;

$$V_x = B^2 E^3 - B^3 E^2 \; : \tag{2.5}$$

vector constraint function;

$$C = (B^2 E^2 + B^3 E^3) E^1 + \frac{1}{2} E B^1 \; : \tag{2.6}$$

scalar constraint function;

$$(B^1, B^2, B^3) = (\frac{1}{2}((A_2)^2 + (A_3)^2 - 2), A_3' + A_1 A_2, -A_2' + A_1 A_3) : \tag{2.7}$$

"magnetic" fields;

Λ : Lagrange multiplier for the Gauss constraint; (2.8)

N^x : "shift" (see eq. (1.2)) (2.9)
and Lagrange multiplier for the vector constraint;

$\underset{\sim}{N} = N/\sqrt{(EE^1/2)}$: Lagrange multiplier for the (2.10)
scalar constraint, N : lapse (see eq. (1.2));

[1]for further details see refs.[3][4][6]

$$Q_r = -i\int_{\partial\Sigma} \Lambda E^1 : \tag{2.11}$$

charge of the remaining $O(2)$-symmetry ;

$$P_{ADM} = \int_{\partial\Sigma} iN^x(A_2E^2 + (A_3 - \sqrt{2})E^3) : \tag{2.12}$$

ADM-momentum in x-direction;

$$E_{ADM} = \int_{\partial\Sigma} \underset{\sim}{N}(A_2E^3 - (A_3 - \sqrt{2})E^2)E^1 \ : \text{ADM-energy} \tag{2.13}$$

A dot means d/dt and a prime d/dx.

The (equal "time") Poisson brackets are

$$\{A_I(x), E^J(y)\} = i\delta_I^J\delta(x,y)\ ,\ \{A_I(x), A_J(y)\} = \{E^I(x), E^J(y)\} = 0. \tag{2.14}$$

The normalization of the energy E_{ADM} is such that

$$E_{ADM} = 2m\, N^{(\infty)}, N^{(\infty)} = N(t, x \in \partial\Sigma) \tag{2.15}$$

for the Schwarzschild mass m (with Newton's constant $G = 1, c = 1$).

The Hamiltonian H_{tot} generates gauge transformations and motions:

$$\delta A_I = \{A_I, H_{tot}\}\epsilon, \quad \delta E^I = \{E^I, H_{tot}\}\epsilon, \tag{2.16}$$

where ϵ is a corresponding infinitesimal parameter[2].

3 Cauchy surfaces Σ and asymptotic properties at spatial infinity

For spherically symmetric systems the topology of the Cauchy 3-manifold Σ^3 is necessarily of the form $\Sigma^3 = S^2 \times \Sigma$ where Σ is a 1-dimensional manifold. Here we are only interested in topological situations where Σ asymptotically flat, that is it has the form (see refs. [3][4])

$$\Sigma = \Sigma_n\ ,\ \Sigma_n \cong K \cup \bigcup_{A=1}^{n} \Sigma_A\ , \tag{3.1}$$

i.e. the hypersurface is the union of a compact set K (diffeomorphic to a compact interval) and a collection of ends (each of which is diffeomorphic to the positive real line without the origin) i.e. asymptotic regions with outward orientation and all of them are joined to K. This means, we have n positive

[2]for more details see ref.[4]

real lines, including the origin, but one end of each line is common to all of them, i.e. these parts are identified.

As an example consider the Kruskal-extended Schwarzschild-manifold where we have *two* ends Σ_1 and Σ_2 each of which belongs to the asymptotic region $x \to \infty$ with $N_2^{(\infty)} = -N_1^{(\infty)} = -N^{(\infty)}$. This is, of course, the most interesting situation, however the case $n > 2$ is a theoretical possibility, too.

We fix[4] the appropriate asymptotic behaviour of the quantities E^I and A_I by requiring
i) that the integrands of the Liouville form

$$\Theta_L = -i \int_\Sigma E^I dA_I \tag{3.2}$$

and the symplectic form

$$\Omega = -i \int_\Sigma dE^I \wedge dA_I \tag{3.3}$$

behave asymptotically as $O(1/x^2)$ in order for the integrals to converge, and that
ii) the Gauss constraint function (2.4) vanishes at least as $O(1/x^2)$ at spatial infinity.

In addition we assume the following asymptotic behaviour of the Lagrange multipiers:

$$\Lambda \to o(1/x^2), \quad N^x \to O(1/x^2), \quad N^{(\infty)} = O(1), \Rightarrow \underset{\sim}{N} \to O(1/x^2). \tag{3.4}$$

4 Symplectic reduction and the observables of the model

Ashtekar's canonical framework allows for degenerate metrics, too (see eq. (2.1)). Here we only discuss the nondegenerate case $E^1 E \neq 0$. Then the vanishing of the scalar and vector constraint functions V_x and C (eqs. 2.5 and 2.6) is equivalent to the vanishing of the functions

$$C_2 = B^2 E^1 + \frac{1}{2} B^1 E^2, \quad C_3 = B^3 E^1 + \frac{1}{2} B^1 E^3, \tag{4.1}$$

because

$$E^1 V_x = E^3 C_2 - E^2 C_3, \quad C = E^2 C_2 + E^3 C_3\ . \tag{4.2}$$

The vanishing of C_2 and C_3 implies

$$E^2 = gB^2,\ E^3 = gB^3,\ g = -2\frac{E^1}{B^1}\ . \tag{4.3}$$

Here we exclude the trivial case $B^1 = 0$ (see ref.[4]). In addition we have

$$A_3C_2 - A_2C_3 = E^1(B^1)' + \frac{1}{2}B^1((E^1)' - G)\ , \tag{4.4}$$

$$A_2C_2 + A_3C_3 = A\,\gamma\,E^1 + \frac{1}{2}B^1(A_2E^2 + A_3E^3)\ , \tag{4.5}$$

where G is the Gauss constraint function (2.4) and

$$A \equiv (A_2)^2 + (A_3)^2,\ \ \gamma \equiv A_1 + \alpha' = \frac{A_2B^2 + A_3B^3}{A}\ , \alpha = \arctan\frac{A_3}{A_2}\ . \tag{4.6}$$

We here assume $A \neq 0$ (recall that A_2, A_3 are complex). As to $A = 0$ see below.

From eq. (4.4) we get the constraint

$$K_1 \equiv \frac{1}{2}B^1(E^1)' + (B^1)'E^1 = 0\ ,\ \text{or}\ (E^1)' = g(B^1)'\ . \tag{4.7}$$

Integration yields

$$\sqrt{E^1}B^1 = const. = -2m\ . \tag{4.8}$$

(The identification of the constant of integration with the mass parameter -2m follows from the comparison of the asymptotic behaviour of E^1 and B^1 with that of the Schwarzschild solution[4].) This is our first observable. That this is so follows from

$$(\delta(\sqrt{E^1}B^1))|_\Gamma = \{(\sqrt{E^1}B^1), H_{tot}\}|_\Gamma = 0, \tag{4.9}$$

Defining (see eq. (2.1))

$$R = \sqrt{E^1} \tag{4.10}$$

we have (see eq. (2,7))

$$B^1 = -\frac{2m}{R}\ ,\quad A = 2(1 - \frac{2m}{R})\ . \tag{4.11}$$

We have $B^1 = -1$ and $A = 0$ on the horizon $R = 2m$. We shall come back to this coordinate singularity later on.

For the further discussion it is convenient to make use of the $O(2)$-symmetry in the (2,3)-"plane" of the variables (A_2, A_3) and (E^2, E^3) by introducing cylindrical coordinates

$$(A_2, A_3) = \sqrt{A}(\cos\alpha, \sin\alpha),\ \ (E^2, E^3) = \sqrt{E}(\cos\beta, \sin\beta)\ . \tag{4.12}$$

They imply the relations

$$G = (E^1)' - \sqrt{AE}\sin(\alpha - \beta) \tag{4.13}$$

and

$$E^I dA_I = \pi_\gamma d\gamma + \pi_1 dB^1 + \pi_\alpha d\alpha - \frac{d}{dx}(E^1 d\alpha) , \tag{4.14}$$

where

$$\pi_\gamma = E^1 , \quad \pi_1 = \sqrt{\frac{E}{A}}\cos(\alpha - \beta) , \quad \pi_\alpha = G . \tag{4.15}$$

We see that the change of variables is tantamount to a canonical transformation where one of the new momenta is the Gauss constraint function G!

In addition, the r.h.s of eq. (4.5) now takes the form

$$A(\pi_\gamma \gamma + \frac{1}{2}\pi_1 B^1) , \tag{4.16}$$

implying the constraint

$$K_2 \equiv 2\pi_\gamma\gamma + \pi_1 B^1 = 0, \quad \text{or } \pi_1 = g\gamma . \tag{4.17}$$

Combining the results obtained so far we can reduce the Liouville form by partial integration as follows

$$\Theta_L = -i\int_\Sigma E^I dA_I = i\int_\Sigma \sqrt{\pi_\gamma}\, B^1 d(\frac{\pi_1}{\sqrt{\pi_\gamma}}) + d\hat{S} = m\, dT + d\hat{S} , \tag{4.18}$$

where eqs. (4.8) and (4.17) have been used and where

$$T = T[\pi_1, \pi_\gamma] = \int_\Sigma \lambda, \quad \lambda \equiv -2i\frac{\pi_1}{\sqrt{\pi_\gamma}} . \tag{4.19}$$

The quantity T is our second observable:
It follows from the falloff properties imposed on A_I and E^I that $\lambda \to O(1/x^2)$ for large x and so the integral T exists as far as the ends are concerned. As to the horizon $R = 2m$ see below.

Furthermore, the relations (2.16) imply[4]

$$\begin{aligned} \delta\lambda &= \{\lambda, H_{tot}\}\epsilon \qquad\qquad (4.20) \\ &= \frac{d}{dx}\left(\lambda N^x + \frac{32m^3(B^1)'}{A(B^1)^4}\underset{\sim}{N}\right)\epsilon = \frac{d}{dx}\left(\lambda N^x + \frac{2R'N}{(1-\frac{2m}{R})\sqrt{q_{xx}}}\right)\epsilon . \end{aligned}$$

The expression in brackets of eq. (4.20) approaches the value $2N^{(\infty)}$ for large x. Thus, we get the important result

$$\delta T = \{T, H_{tot}\}\epsilon = 2\sum_{A=1}^{n} N_A^{(\infty)}\epsilon \ , \tag{4.21}$$

where we have allowed for different lapses $N_A^{(\infty)}$ at different ends Σ_A (see eq. (3.1)) and where we have assumed that there are no contributions from possible inner boundary points and that each end "sees" the same mass m. The occurrence of the lapses on the r.h.s. of δT is due to our allowance for time translations as a *symmetry transformation* at spatial infinity ([8]), as opposed to a gauge transformation for which $N^{(\infty)} = 0$.

If spacetime is Minkowski-like at spatial infinity we have $N^{(\infty)} = 1$ and - interpreting the parameter ϵ as the (infinitesimal) proper time of an asymptotic observer and considering one end only- we get

$$\dot{T} = 2, \quad T = 2t + const. \ . \tag{4.22}$$

Thus, *the observable T is to be interpreted as a time variable!* (The factor 2 is a consequence of our normalization of the energy - see eq. (2.15)).

The mass m is canonically conjugate to the time T: Let us define the mass operator M by

$$M = -\frac{1}{2}\int_\Sigma \sqrt{\pi_\gamma}B^1\chi, \quad \int_\Sigma \chi = 1 \ , \tag{4.23}$$

where χ is a suitable smooth test function the support of which can again be concentrated on a given end Σ_A. From eqs. (3.2) and (4.14) we infer the Poisson brackets

$$\{B^1(x), \pi_1(y)\} = i\delta(x,y), \ \{\gamma(x), \pi_\gamma(y)\} = i\delta(x,y), \ \{\pi_1(x), \pi_\gamma(y)\} = 0 \text{ etc.} \tag{4.24}$$

for the new canonical variables. We therefore have

$$\begin{aligned} \{T, M\} &= \int_\Sigma dx \int_\Sigma dy\, i\{(\frac{\pi_1}{\sqrt{\pi_\gamma}})(x), (\sqrt{\pi_\gamma}B^1)(y)\chi(y)\} \\ &= \int_\Sigma dx \int_\Sigma dy\, i\sqrt{\frac{\pi_\gamma(y)}{\pi_\gamma(x)}}\{\pi_1(x), B^1(y)\}\,\chi(y) = \int_\Sigma dy\,\chi(y) = 1 \ . \end{aligned} \tag{4.25}$$

We see that we can interpret spherical symmetric gravity as a (1+1)-dimensional completely integrable canonical system where the mass M is the action and the time T the angle variable.

This is in complete agreement with the structure of the reduced Hamiltonian $H_{red} = H_{tot}|_\Gamma$ which according to eqs. (2.3) and (2.15) takes the form (again considering one end only)

$$H_{red} = 2\,M\,N^{(\infty)}, \tag{4.26}$$

because the constraints G, V_x and C vanish now and we have $P_{ADM} = 0$ from our choice of the asymptotic properties of the fields and of N^x. In addition we have $Q_r = 0$ because of the decay (3.4) of Λ. Eq. (4.25) then yields

$$\dot{T} = \{T, H_{red}\} = 2N^{(\infty)}\ , \tag{4.27}$$

in complete agreement with eq. (4.21) above.

5 Relations to the ADM-formalism

We first express T as functional of the metric coefficients q_{xx} and R : From eqs. (4.13) and (4.15) we get - using the constraint $G = 0$ -

$$\pi_1^2 = \frac{E}{A} - \frac{((E^1)')^2}{A^2}\ . \tag{5.1}$$

Dividing by E^1 and taking the square root yields

$$\lambda = 2\frac{1}{A\sqrt{E^1}}(((E^1)')^2 - A\,E)^{1/2} = 2(1-2m/R)^{-1}((R')^2 - q_{xx}(1-2m/R))^{1/2}\ , \tag{5.2}$$

where the relations

$$q_{xx} = \frac{E}{2E^1}\ ,\ R = \sqrt{q_{\theta\theta}} = \sqrt{E^1}\ ,\ A = 2(1+B^1) = 2(1-\frac{2m}{R})$$

have been used. We then get

$$\begin{aligned} T[q_{xx}, R] &= \int_\Sigma 2(1-2m/R)^{-1} w(q_{xx}, R)\ , \\ w(q_{xx}, R) &= ((R')^2 - q_{xx}(1-2m/R))^{1/2}\ . \end{aligned} \tag{5.3}$$

For the Schwarzschild solution

$$q_{xx} = (1 - \frac{2m}{x})^{-1}\ ,\quad R = x\ , \tag{5.4}$$

the quantity T vanishes. Such a vanishing T is compatible with eq. (4.21) because the Schwarzschild-Kruskal manifold has *two* ends for $x \to \infty$ with $N_2^{(\infty)} = -N_1^{(\infty)}$ so that the sum on the r.h.s. of eq. (4.21) vanishes! It is amusing that a second end is required here in order to obtain consistency of the formalism! Nonvanishing values of T will be discussed below.

The important eq.(4.21) can be derived in the ADM framework, too[4][7]: The observable T is related in a very simple way to the functional

$$\begin{aligned} S[q_{xx}, R; m) &= \int_\Sigma u(q_{xx}, R, m) , \qquad (5.5)\\ u(q_{xx}, R, m) &= 2R(R' \cosh^{-1}\left(\frac{R'}{\sqrt{q_{xx}(1-\frac{2m}{R})}}\right) - w) ,\\ \cosh^{-1}\left(\frac{R'}{\sqrt{q_{xx}(1-\frac{2m}{R})}}\right) &= \ln(R'+w) - \frac{1}{2}\ln(q_{xx}(1-\frac{2m}{R})) , \end{aligned}$$

derived in ref.[3], which provides an exact Hamilton-Jacobi solution of the (classical) Wheeler-DeWitt eq.

$$[q_{xx}^2 p_x^2 - q_{xx} R p_x p_R + R^2(2RR'' + (R')^2 - \frac{q'_{xx} RR'}{q_{xx}} - q_{xx})] = 0 \qquad (5.6)$$

for spherical symmetric gravity where p_x and p_R are the canonical momenta conjugate to q_{xx} and R, respectively.

Inserting

$$\begin{aligned} p_x &= \frac{\delta S}{\delta q_{xx}} = \frac{\partial u}{\partial q_{xx}} = -\frac{wR}{q_{xx}}, \qquad (5.7)\\ p_R &= \frac{\delta S}{\delta R} = \frac{\partial u}{\partial R} - \frac{d}{dx}\left(\frac{\partial u}{\partial R'}\right) \qquad (5.8)\\ &= -2w - \frac{1}{w}(2RR'' - \frac{q'_{xx}RR'}{q_{xx}} - \frac{2m}{R}q_{xx}) \qquad (5.9) \end{aligned}$$

into the Wheeler-DeWitt eq. (5.6) solves that equation identically. In the same way the ADM diffeomorphism constraint

$$q_{xx}p_x - (2q_{xx}p_x)' + R'p_R = 0 \qquad (5.10)$$

is fulfilled.

The quantities S and T are related as follows: Eq. (4.18) gives

$$\Theta_L = -T\,dm + d(\hat{S} - mT) . \qquad (5.11)$$

On the other hand we have

$$dS[q_{xx}, R; m) = \int_\Sigma (\frac{\delta S}{\delta q_{xx}} dq_{xx} + \frac{\delta S}{\delta R} dR) + \frac{\partial S}{\partial m} dm = \Theta_L + \frac{\partial S}{\partial m} dm \qquad (5.12)$$

Comparing the last two equations we can identify $\hat{S} - mT = S$ and so we have

$$T = \frac{\partial S}{\partial m} \ . \qquad (5.13)$$

This important relation can be verified by an explicit calculation!

We now briefly comment on the relation of our results to those of Kuchař[7]. They are equivalent and can be translated by observing the following relations:
He has $q_{xx} = \Lambda^2$ and R as above. Furthermore, Kuchař's P_Λ and P_R are related to our p_x and p_R by $p_x = -P_\Lambda/\Lambda$, $p_R = -2P_R$. Furthermore $\lambda = 2(1-2m/R)^{-1}w = 2P_M = -2T'(r)$ (notice that $\Lambda P_\Lambda = Rw$ and compare Kuchař's eq.(80).) Our Hamilton-Jacobi function S from above is related to that of Kuchař bei $S = -2\omega$ (his eq. (97)) . Kuchař considers the mass m as the canonical variable and the time T as the canonically conjugate momentum, we do it the other way round.

We now come to the discussion of configurations for which the observable T does not vanish. Our approach is to start with given values of the observables T and m and ask for those values of N^x and N which are compatible with them and with the remaining gauge degrees of freedom:
Consider the special gauge $E^1 = x^2$ so that $\delta E^1 = 0$! Eq. (2.16) for E^1 implies the condition[4]

$$N^x + \frac{1}{2}x^2(1 - \frac{2m}{x})\lambda \underset{\sim}{N} = 0 \qquad (5.14)$$

for N^x. It shows that N^x has to be nonvanishing for a nonvanishing λ!
The usual choice for the Schwarzschild parametrization is $N^x = 0$ so that $\lambda = 0$ and $T = 0$ in that case.
We see that a nonvanishing T is associated with a slicing (foliation) of space-time which necessarily has a nonvanishing lapse N^x!

Next we ask what gauge freedom is left in general for the gauge dependent functions N^x and $\underset{\sim}{N}$ once the observables m and T are given:
As the first gauge dependent quantity we may take R. As the second we take a function f with the property[4]

$$\lambda = Tf' \ , \ \int_{\partial\Sigma} f = 1 \ , \ \text{i.e.} \lim_{x\to\infty} f = 1 \ . \qquad (5.15)$$

Expressed in terms of the observables m and T and the gauge quantities R and f' the metric coefficient q_{xx} takes the form

$$q_{xx} = \frac{(R')^2}{(1 - \frac{2m}{R})} - \frac{1}{4}(Tf')^2(1 - \frac{2m}{R}) \ . \qquad (5.16)$$

We here finally see how we can deal with the horizon $R = 2m$ and the apparent associated singularity of the integrand of T: We can interpret the eq.(5.16) such that it determines the behaviour of q_{xx} in the neighbourhood of $R = 2m$ once (Tf') is given!

For the shift N^x and the lapse N in terms of R and f we get the equations[4]:

$$R'N^x + (1 - \frac{2m}{R})\frac{Tf'}{\sqrt{q_{xx}}}N = \dot{R} , \qquad (5.17)$$

$$Tf'N^x + (1 - \frac{2m}{R})^{-1}\frac{2R'}{\sqrt{q_{xx}}}N = 2N^{(\infty)}f + T\dot{f} . \qquad (5.18)$$

For the special gauge $R = x$ we have

$$N = \frac{2N^{(\infty)}f + T\dot{f}}{2\sqrt{q_{xx}}} , \quad N^x = -\frac{(1 - 2m/x)Tf'(2N^{(\infty)}f + T\dot{f})}{2q_{xx}} . \qquad (5.19)$$

6 Quantisation

As we have only two physical degrees of freedom, M and T, quantisation is easy. It depends, however, on the spectrum of the Mass operator $\hat{M}$, namely whether it is bounded from below or not:

1. $-\infty < m < +\infty$.

We have the commutation relations

$$[\hat{M}, \hat{T}] = -i . \qquad (6.1)$$

Choosing the representation

$$\hat{T}\phi(T) = T\phi(T) , \hat{M}\phi(T) = -i\frac{d}{dT}\phi(T) \qquad (6.2)$$

with the scalar product

$$(\phi_1, \phi_2) = \int_{-\infty}^{+\infty} dT \phi_1^*(T)\phi_2(T) \qquad (6.3)$$

we get from eq. (4.26) the Schroedinger eq. for the wave function $\psi(T,t)$:

$$i\partial_t\psi(T,t) = H\psi(T,t) = 2N^{(\infty)}\hat{M}\psi(T,t) = -i2N^{(\infty)}\frac{\partial}{\partial T}\psi(T,t) , \qquad (6.4)$$

which has the plane wave solutions

$$e_m(T,t) = \frac{1}{\sqrt{2\pi}}e^{-im(2\tau(t) - T)}, \quad \text{where } \dot{\tau} = N^{(\infty)}(t) . \qquad (6.5)$$

Notice that the Schroedinger eq. implies the continuity equation

$$\partial_t(\psi^*\psi) + 2N^\infty \partial_T(\psi^*\psi) = 0 \quad \text{or} \quad \partial_\tau(\psi^*\psi) + 2\partial_T(\psi^*\psi) = 0. \tag{6.6}$$

In view of the positive energy theorem for isolated gravitational systems the assumption that the spectrum of $\hat{M}$ is unbounded from below appears to be unphysical and should be changed:
2. $0 < m < \infty$.
Here, if the commutation relation (6.1) still holds, $\hat{T}$ cannot be selfadjoint - and therefore not diagonalizable - because otherwise $\exp(i\mu\hat{T})$, μ real, would be a unitary operator which generates translations by μ in m-space violating the spectral condition $m > 0$. The problem has been discussed in detail by Klauder et al.[9] and by Isham et al.[10]:
If one defines the operator

$$\hat{S} = \frac{1}{2}(\hat{M}\hat{T} + \hat{T}\hat{M}) \ , \tag{6.7}$$

then we get the Lie algebra commutator

$$[\hat{S}, \hat{M}] = i\hat{M} \tag{6.8}$$

of the affine group in one dimensions, i.e. $\hat{S}$ generates scale transformations of the spectrum of $\hat{M}$:

$$e^{-i\beta\hat{S}}\hat{M}e^{i\beta\hat{S}} = e^{\beta}\hat{M} \ , \beta \text{ real} \ . \tag{6.9}$$

In the space of functions $\chi(m)$ we may choose the operator representations

$$\hat{M} = m \ , \hat{S} = \frac{i}{2}(m\frac{d}{dm} + \frac{d}{dm}m) = i(m\frac{d}{dm} + \frac{1}{2}) \tag{6.10}$$

which are selfadjoint with respect to the scalar product

$$(\chi_1, \chi_2) = \int_0^{+\infty} dm\, \chi_1^*(m)\chi_2(m) \tag{6.11}$$

and where

$$(e^{i\beta\hat{S}}\chi)(m) = e^{-\frac{1}{2}\beta}\chi(e^{-\beta}m) \tag{6.12}$$

leaves the scalar product invariant.
The Schoedinger eq. here has the form

$$i\partial_t\chi(m,t) = 2N^{(\infty)}m\,\chi(m,t) \ , \tag{6.13}$$

with the solutions

$$\chi(m,t) = e^{-2im\tau(t)}g(m) \ , (g,g) < \infty \ . \tag{6.14}$$

7 Comments

One of the most amazing properties of the Hamiltonian approach to spherically symmetric gravity is that it 'generates' its own time observable $T[\Sigma]$ as a functional of the intrinsic metric coefficients q_{xx}, R or equivalent data on the Cauchy surface Σ. The interpretation relies on the spatial asymptotic flatness of spacetime and the existence of a corresponding Minkowski eigentime. Nevertheless T exists for compact Σ, too, but then it commutes with H_{tot} and its interpretation is less obvious.

As to the quantum theory of the system one has the (old) problem that the operator $\hat{T}$ is canonically conjugate to a quantity ($\hat{M}$) the spectrum of which is bounded from below so that $\hat{T}$ cannot be self-adjoint! Otherwise the quantum theory is simple like that of a free particle.

An important question is, of course, what happens if one couples the system to matter, e.g. to a massless scalar field. Then the minisuperspace model becomes a midisuperspace one and the difficulties in solving the constraints increase considerately[5]. Nevertheless one would like to derive the Hawking radiation, at least approximately, if the black hole in a state with mass m undergoes a (radiation) transition to a state with mass $m - \delta m$!

References

[1] C. Isham (1993), 'Canonical quantum gravity and the problem of time', in *Integrable Systems, Quantum Groups and Quantum Field Theories* L.A. Ibort and M.A. Rodriguez (eds.), Kluwer Academic Publishers, London; p157.

[2] K. Kuchař (1992), 'Time and interpretations of quantum gravity', in *Proc. of the 4th Canadian Conference on General Relativity and Astrophysics*, G. Kunstatter et al. (eds.), World Scientific, Singapore; p211.

[3] T. Thiemann and H.A. Kastrup (1993), 'Canonical quantization of spherically symmetric gravity in Ashtekar's self-dual representation', *Nucl. Phys.* , **B399** 211.

[4] H.A. Kastrup and T. Thiemann (1993), 'Spherically symmetric gravity as a completely integrable system', RWTH Aachen preprint PITHA 93-35 (Nov. 93), gr-qc 9401032, to be publ. in *Nucl.Phys.B.*

[5] T. Thiemann (1993), 'On the quantisation of gravity in the framework of the Ashtekar-formalism', PhD Thesis RWTH Aachen.

[6] A. Ashtekar (1987), 'New Hamiltonian formulation of general relativity', *Phys. Rev.* , **D36** 1587;
A. Ashtekar (1991), *Lectures on Non-Perturbative Canonical Gravity*, World Scientific, Singapore;
C. Rovelli (1991), 'Ashtekar formulation of general relativity and loop-space non-perturbative quantum gravity: a report', *Class. Quantum Grav.*, **8** 1613;
C.J. Isham (1991), 'Conceptual and geometrical problems in quantum gravity', in *Recent Aspects of Quantum Fields (Proc. Schladming, Austria 1991)*, H. Mitter and H. Gausterer (eds.), Springer-Verlag, Berlin; p123;
J.D. Romano (1993), 'Geometrodynamics vs. connection dynamics', *Gen.Rel.Grav.* , **25** 759.

[7] K. Kuchař (1994), 'Geometrodynamics of Schwarzschild Black Holes', University of Utah preprint, gr-qc 9403003. See also D. Louis-Marinez, J. Gegenberg and G. Kunstatter, Phys.Lett. B 321(1994)193; D. Louis-Martinez and G. Kunstatter, Phys.Rev. D(1994)5227

[8] R. Beig and N. ó Murchada (1987), 'The Poincaré Group as the Symmetry Group of Canonical General Relativity', *Ann.Physics*, **174** 463.

[9] J.R. Klauder and E.W. Aslaksen (1970), 'Elementary Model for Quantum Gravity', *Phys. Rev. D*, **2** 272 and refs. therein.

[10] C.J. Isham and A.C. Kakas (1984), 'A group theoretical approach to the canonical quantisation of gravity', *Class. Quantum Grav.* , 1 621 and 633;
C.J. Isham, 'Topological and Global Aspects of Quantum Theory', in *Relativity, Groups and Topology II*, B.S. DeWitt and R. Stora (eds.), North Holland, Amsterdam; p1059.

Canonical decomposition of Belinskiĭ-Zakharov one-soliton solutions

Nenad Manojlović, Greg Stephens and Tatijana Vukašinac

Abstract

We perform the standard canonical (3+1) decomposition of the Belinskiĭ-Zakharov one-soliton solution. Our starting point is the general Bianchi I solution which we obtain by applying the symmetry transformation to the Kasner metric. We then construct the symplectic form for the Bianchi I model and on the way show that we have all the physical degrees of freedom for our starting solution. The Belinskiĭ-Zakharov soliton transfomation requires two degrees of freedom in the general Bianchi I solution to be frozen. Although integration of the linearized system, in the non-diagonal case, is a non-trivial step, the particular form of our starting solution simpifies the calculation and reduces the problem effectively to the diagonal case. Therefore, in our case, it is straightforward to obtain the one-soliton metric. Finally, we point out the problems related to the fact that the one-soliton solution is defined only in a certain region of the co-ordinate chart.

1 Introduction

The Einstein field equations for space-times that admit a two-dimensional Abelian group of isometries which acts orthogonally and transitively on non-null orbits are non-linear partial differential equations in two variables [Kramer et al. 1980]. Since the pioneering work of Geroch, it has been known that the field equations in the stationary axisymmetric case admit an infinite dimensional group of symmetry transformations [Geroch 1971, 1972]. This result has encouraged the research in solution-generating methods, the main idea being that the complete class of solutions can then be generated from a particular solution, such as flat space [Cosgrove, 1980-1982]. Subsequently, several solution-generating techniques have been developed, such as the Kinnersley-Chitre transformations [Kinnersley 1977, Kinnersley and Chitre 1977-1978], the Hauser-Ernst formalism [Hauser and Ernst 1979-1981], Harrison's Bäcklund transformations [Harrison 1978,1980] and the Belinskiĭ and Zakharov inverse scattering method [Belinskiĭ and Zakharov 1978].

The main step in the Belinskiĭ-Zakharov formalism is to write down linear eigenvalue equations whose integrability conditions are the given non-linear system. The methods of functional analysis can be applied to generate new solutions of the linear system from old, and hence new solutions of the original system from old [Zakharov and Shabat 1974, 1979, Zakharov 1978]. The particular solutions that can be generated are the soliton solutions. The soliton solutions share a number of common properties with classical particles, namely, they are localized solutions that propagate energy, have a particular velocity of propagation and some persistence of shape which is maintained even when two solitons collide [Verdaguer 1993]. As shown by Belinskiĭ and Zakharov, the soliton transformation needs to be generalized when applied in the context of the Einstein equations with a two-parameter Abelian group of motions. The generalization is that the stationary poles are substituted by the pole trajectories [Belinskiĭ and Zakharov 1978, Verdaguer 1993].

Our goal is to identify physicsl degrees of freedom for the general one-soliton solution and to construct the corresponding physical phase space for the model. As the starting solution we choose the general Bianchi type I solution and we also fix the spatial manifold Σ to be the three torus T^3. We first show how to obtain the general Bianchi type I solution by applying the symmetry transformation to the Kasner metric [Manojlović and Mena Marugán 1993]. The standard canonical (3+1) decomposition of the metric, for the homogeneous foliation, then yields the symplectic structure for the Bianchi type I model. The Belinskiĭ-Zakharov soliton transfomation, which is presently defined only locally in one co-ordinate neighborhood on Σ, requires two degrees of freedom in the general Bianchi type I solution to be frozen. Although integration of the linearized system, in the non-diagonal case, is a non-trivial step, the particular form of our starting solution simpifies the calculation and reduces the problem effectively to the diagonal case. Therefore, in our case, it is straightforward to obtain the one-soliton metric. However, this metric is defined only in a certain region of the co-ordinte chart that does not cover the whole Σ [Belinskiĭ and Zakharov 1978, Verdaguer 1993]. Although it is possible to continue the one-soliton metric with the unperturbed Bianchi type I, it is not difficult to see that the first derivatives of the metric suffer discontinuities on certain light-cones [Belinskiĭ and Zakharov 1978, Verdaguer 1993]. In particular, for the case when the spatial topology is compact, one can show that is not possible to decompose the one-soliton metric in such a way that the corresponding conjugate momenta does not suffer from discontinuities on the relevant light-cones.

This report is organized as follows. In setion two we constrict the general Bianchi type I solution starting from the Kasner metric and also the symplectic structure for this model. In section three we turn to the Belinskiĭ-Zakharov soliton transfomation. We first obtain the one-soliton metric with the maximal number of integrational constants, for the Bianchi type I seed. Then, we point out the problems with the discontinuities of the first derivatives of the solution and we show how these are related to the problems of defining the canonical theory for this system. In section four we present our conclusions.

2 The Bianchi type I

The Bianchi type I model is spatially homogeneous spacetime which admits a three dimensional isometry Lie group $SO(2) \times SO(2) \times SO(2)$ that acts simply transitively on each leaf Σ of the homogeneous foliation [Jantzen 1979, Manojlović and Mena Marugán 1993]. As a consequence, there exists a set of three left-invariant vector fields L_I on Σ which form the Lie algebra of the group $SO(2) \times SO(2) \times SO(2)$

$$[L_I, L_J] = 0 \ . \tag{2.1}$$

Dual to the vector fields L_I, one can introduce a set of three left-invariant one-forms χ^I which satisfy the Maurer-Cartan equations

$$d\chi^I = 0 \ . \tag{2.2}$$

If we begin by considering a certain set of left-invariant one-forms χ^I, for which the three-metric is given by γ_{IJ}, then any other set of left-invariant forms $\tilde{\chi}^I$ will be related to χ^I by a transformation $\tilde{\chi}^I = M^I{}_J \chi^J$ that maintains the symmetries of the model. The inverse (M^{-1}) must always exist, since χ^I is a set of three linearly independent one-forms. The structure constants must remain unchanged under the transformation defined by $M^I{}_J$. But since for Bianchi type I, $C^I{}_{JK} = 0$ it follows that, in this case, any invertible matrix $M \in GL(3, \mathbb{R})$ defines a permissible transformation [Jantzen 1979, Manojlović and Mena Marugán 1993]. If the topology of Σ is fixed to be the three torus T^3, then it is possible to choose the coordinates that are adopted to the group action

$$\chi^1 = dy^1 \quad \chi^2 = dy^2 \quad \chi^3 = dy^3 \ . \tag{2.3}$$

We further simplify the notation by writing $y^0 = t$. The general Bianchi I solution can be written in the form [Manojlović and Mena Marugán 1993]

$$ds^2 = -dt^2 + \gamma_{IJ} dy^I dy^J \ , \tag{2.4}$$

here

$$\gamma_{IJ} = (M^T)_I{}^P \tilde{\gamma}_{PQ} M^Q{}_J \ , \tag{2.5}$$

with $(M^T)_I{}^J$ is the transpose of the matrix $M^I{}_J$ and $\tilde{\gamma}_{IJ}$ is the spatial part of the Kasner metric [Verdaguer 1993]

$$\tilde{\gamma}_{IJ} = \begin{pmatrix} t^{2q_1} & 0 & 0 \\ 0 & t^{2q_2} & 0 \\ 0 & 0 & t^{2q_3} \end{pmatrix} \ , \tag{2.6}$$

with

$$q_1 + q_2 + q_3 = {q_1}^2 + {q_2}^2 + {q_3}^2 = 1 \ . \tag{2.7}$$

We use the Iwasawa decompostion for the general element of the group $GL(3, \mathbb{R})$ into the product of elements of diagonal, nilpotent and rotational subgroup. Hence the parametization for the matrix $M^I{}_J$

$$M = M_D M_N R$$

$$= (\det M)^{\frac{1}{3}} \begin{pmatrix} e^{C_3+\sqrt{3}C_4} & 0 & 0 \\ 0 & e^{C_3-\sqrt{3}C_4} & 0 \\ 0 & 0 & e^{-2C_3} \end{pmatrix} \begin{pmatrix} 1 & C_5 & C_6 \\ 0 & 1 & C_7 \\ 0 & 0 & 1 \end{pmatrix} R, \tag{2.8}$$

where M_D is diagonal matrix, M_N is the upper triangular matrix, R is the rotational matrix, *i.e.* $R \in SO(3)$ and C_3, C_4, C_5, C_6, $C_7 \in \mathbb{R}$. In addition, we write $(\det M) = {R_0}^3$ and solve eqns. (2.7)

$$R_0 = ({C_1}^2 + {C_2}^2)^{\frac{1}{6}} \quad q_1 = \frac{1}{3}\Big(1 - \frac{C_1 + \sqrt{3}C_2}{\sqrt{{C_1}^2 + {C_2}^2}}\Big) \ , \tag{2.9a}$$

$$q_2 = \frac{1}{3}\Big(1 - \frac{C_1 - \sqrt{3}C_2}{\sqrt{{C_1}^2 + {C_2}^2}}\Big) \quad q_3 = \frac{1}{3}\Big(1 + \frac{2C_1}{\sqrt{{C_1}^2 + {C_2}^2}}\Big) \ , \tag{2.9b}$$

where C_1, $C_2 \in \mathbb{R}$. There is an additional symmetry that is not included in the transformations eqn. (2.5) and that is the transformation of the form $t \to t + const.$

The first step in the canonical (3+1) decomposition is the choice of foliation. We choose the homogeneous foliation X^μ

$$X : \Sigma \times \mathbb{R} \to \mathcal{M} \quad , \quad X^\mu(x^I, t) = y^\mu \ , \tag{2.10}$$

where Σ is the three torus, in our case, with the coodinates x^I, $I = 1, 2, 3$, t is the coordinate on $\mathbb{R}$ and $\mathcal{M}$ is the four dimensional spacetime with the coordinates y^μ, $\mu = 0, 1, 2, 3$, such that,

$$X^\mu{}_{,\alpha} = \delta^\mu_\alpha \ , \tag{2.11}$$

where $X^\mu{}_{,\alpha} = \partial_\alpha X^\mu$, α is a coordinate index on $\Sigma \times \mathbb{R}$, *i.e.*, $\alpha = \{t, I\}$. The normal vector n_μ is then determined by the equations

$$n_\mu X^\mu{}_{,I} = 0 \ , \tag{2.12a}$$

$$g^{\mu\nu} n_\mu n_\nu = -1 \ . \tag{2.12b}$$

Using eqn. (2.4) and eqn. (2.11) we obtain $n_0 = 1$ and $n_I = 0$, $I = 1, 2, 3$.

The lapse function N and the shift vector N^I are then define by

$$\dot{X}^\mu = N g^{\mu\nu} n_\nu + N^I X^\mu{}_{,I} \ . \tag{2.13}$$

From eqn. (2.11) and eqns. (2.12) it follows that

$$N^I = 0 \quad , \quad N = -1 \ . \tag{2.14}$$

The metric on the three dimensional manifold Σ is given by

$$\gamma_{IJ} = g_{\mu\nu} X^\mu{}_{,I} X^\nu{}_{,J} \ , \tag{2.15}$$

and with the foliation (2.11) γ_{IJ} agrees with eqn. (2.5). The extrinsic curvature K_{IJ} is defined by

$$K_{IJ} = \frac{1}{2N}\left(-(\partial_t\gamma)_{IJ} + (L_{\vec{N}}\gamma)_{IJ}\right) \ . \tag{2.16}$$

Using eqn. (2.5) and eqn. (2.14) the extrinsic curvature can be written in the following form

$$K_{IJ} = \frac{1}{2}(\partial_t\gamma)_{IJ} = \frac{1}{2}(M^T)_I{}^P(\partial_t\tilde{\gamma})_{PQ} M^Q{}_J \ . \tag{2.17}$$

The momenta π conjugate to γ are defined by

$$\pi^{IJ} = -\frac{\sqrt{\det\gamma}}{k^2}\left(\gamma^{IM}K_{MN}\gamma^{NJ} - \gamma^{IJ}\gamma^{MN}K_{MN}\right) \ , \tag{2.18}$$

where $k \equiv \frac{16\pi G}{c^2}$. Substituting eqn. (2.5) and eqn. (2.17) into eqn. (2.18) we obtain the expressions for the conjugate momenta

$$\pi^{IJ} = R_0{}^3(M^{-1})^I{}_P\tilde{\pi}^{PQ}(M^{-1\,T})_Q{}^J \ , \tag{2.19}$$

here

$$\tilde{\pi}^{IJ} = \frac{1}{k^2}\begin{pmatrix} (1-q_1)t^{-2q_1} & & \\ & (1-q_2)t^{-2q_2} & \\ & & (1-q_3)t^{-2q_3} \end{pmatrix} \ . \tag{2.20}$$

The symplectic structure is defined by

$$\Omega = d\pi^{IJ} \wedge d\gamma_{IJ} \ . \tag{2.21}$$

We substitute the expressions (2.5,19) for the three metric γ_{IJ} and its conjugate momenta π^{IJ} into the symplectic structure eqn. (2.21) and evaluate the exterior differential with respect to the co-ordinates on the space of

Bianchi type I solutions. Using the properties of the exterior differential and some obvious identities the symplectic form can be expressed as

$$\Omega = \Omega_1 + \Omega_2 \ , \tag{2.22}$$

where

$$\Omega_1 = \frac{2}{k^2}\Big(dC^1 \wedge dC^3 + dC^2 \wedge dC^4\Big) \ , \tag{2.23}$$

and

$$\Omega_2 = -\frac{1}{k^2}\, d\Big(l_I \omega^I\Big) \ , \tag{2.24}$$

with

$$l_I = \epsilon_{IJK}\Big({M_N}^{-1}({R_0}^3\tilde{\pi}\tilde{\gamma})M_N - {M_N}^T({R_0}^3\tilde{\pi}\tilde{\gamma}){M_N}^{-1\,T}\Big)^{JK} \ , \tag{2.25}$$

and the one-forms ω^I are defined by

$$\omega^I = \frac{1}{2}\epsilon^I{}_{JK}\big(dRR^T\big)^{JK} \ , \tag{2.26}$$

here $\epsilon^I{}_{JK} = \delta^{IM}\epsilon_{MJK}$. It is not difficult to see that the one-forms ω^I satisfy the Maurer-Cartan equations for the group $SO(3)$. From eqn. (2.24), it is clear that Ω_2 is the natural symplectic structure on $T^*(SO(3))$, the cotangent bundle of the group $SO(3)$. We have obtained (locally) the physical phase space for Bianchi I model. To obtain the global structure of the phase space for this model we need additonal coordinate transformation for Ω_1 [Manojlović and Mena Marugán 1993]. But since our main goal is to obtain the physical degrees of freedom for the Belinskiĭ-Zakharov one-soliton we will now proceed with the soliton transformation.

3 The Belinskiĭ-Zakharov one-soliton transformation: Bianchi type I

We apply the Belinskiĭ-Zakharov one-soliton transformation on the general Bianchi I solution eqn. (2.4). In this section we limit ourselves to the local considerations in one co-ordinate neighborhood on Σ. The soliton transformation requires the starting solution to be in the following form [Belinskiĭ and Zakharov 1978]

$$ds^2 = f^{(0)}\left(-d\tau^2 + dz^2\right) + {g^{(0)}}_{ab}dy^a dy^b \ , \tag{3.1}$$

here, in general case, $f^{(0)} = f^{(0)}(\tau, z)$, ${g^{(0)}}_{ab} = {g^{(0)}}_{ab}(\tau, z)$, and $a,\ b = 1$, 2. One way to transform the metric (2.4) into the form (3.1) would be to write (2.4) as

$$ds^2 = -dt^2 + g_{ab}dy^a dy^b + f(dy^3)^2 \ , \tag{3.2}$$

in our case $g_{ab} = g_{ab}(t)$ and $f = f(t)$, and then the second step would be the co-ordinate transfomation $\{t, y^3\} \to \{\tau, z\}$ which is such that the metric (3.2) transforms into the metric (3.1). The first step requires some components of the matrix M to be set to zero. Since our aim is the consider the soliton solutions as the dynamical system, we have to be sure that the metric (3.2) represents a consistent symplectic sub-manifold within the general Binachi I model. To show this we will use the following parmetrization for the rotational matrix

$$R_{IJ} = \cos r\,(\delta_{IJ} - \widehat{\theta}_I\widehat{\theta}_J) + \widehat{\theta}_I\widehat{\theta}_J + \sin r\,\epsilon_{IJK}\widehat{\theta}_K\,, \tag{3.3}$$

with $\widehat{\theta}_I = \frac{\theta_I}{r}$, $r = \sqrt{\sum_{K=1}^{3}\theta_K^2}$, and $I = 1, 2, 3$. After a straightforward calculation, we obatin the expessions for the left-invariant one-forms ω^I

$$\begin{aligned}\omega^I &= \omega^I{}_J\, d\theta^J = \frac{1}{2}\,\epsilon_{IMN}\Big(\partial_J R\, R^T\Big)^{MN} d\theta^J \\ &= \Big(\frac{\epsilon_{IJK}}{r}\,\widehat{\theta}_K(1-\cos r) + \frac{\sin r}{r}(\delta_{IJ} - \widehat{\theta}_I\widehat{\theta}_J) + \widehat{\theta}_I\widehat{\theta}_J\Big)\, d\theta^J\,. \end{aligned}\tag{3.4}$$

From the definiton of l_I, eqn. (2.25), the above expressions for the one-forms ω^I, eqn. (3.4), and the symplectic form Ω_2, eqn. (2.24), it follows that it is consistent to set simultaneously $C_6 = C_7 = 0$ and $\theta_1 = \theta_2 = 0$. Hence the matrix M is

$$M = R_0 \begin{pmatrix} e^{C_3+\sqrt{3}C_4} & 0 & 0 \\ 0 & e^{C_3-\sqrt{3}C_4} & 0 \\ 0 & 0 & e^{-2C_3} \end{pmatrix} \times$$

$$\begin{pmatrix} 1 & C_5 & 0 \\ 0 & 1 & 0 \\ 0 & 0 & 1 \end{pmatrix} \begin{pmatrix} \cos\theta & \sin\theta & 0 \\ -\sin\theta & \cos\theta & 0 \\ 0 & 0 & 1 \end{pmatrix} \tag{3.5}$$

here $\theta \equiv \theta_3$. Consequently, we have the metric (2.4) in the form (3.2) with

$$g_{ab} = (\mathcal{M}^T)_a{}^c\, \tilde{g}_{cd}\, \mathcal{M}^d{}_b\,, \quad f = {R_0}^2\, e^{-4C_3} t^{2q_3}\,, \tag{3.6}$$

here

$$\mathcal{M} = \begin{pmatrix} 1 & C_5 \\ 0 & 1 \end{pmatrix} \begin{pmatrix} \cos\theta & \sin\theta \\ -\sin\theta & \cos\theta \end{pmatrix}\,, \tag{3.7a}$$

$$\tilde{g} = {R_0}^2 \begin{pmatrix} e^{2(C_3+\sqrt{3}C_4)} t^{2q_1} & 0 \\ 0 & e^{2(C_3-\sqrt{3}C_4)} t^{2q_2} \end{pmatrix}\,, \tag{3.7b}$$

and R_0, q_I, $I = 1, 2, 3$, are defined in eqn. (2.9). The next step is the co-ordinate transformation $\{t, y^3\} \to \{\tau, z\}$

$$\tau = \frac{e^{2C_3}}{R_0(1-q_3)} t^{(1-q_3)} \quad , \quad z = y^3 \ . \tag{3.8}$$

Finally, we have the metric in the form (3.1) with

$$f^{(0)} = (R_0 e^{-2C_3})^{2(1+p_3)}(q\tau)^{2p_3} \quad , \quad g^{(0)}{}_{ab} = (\mathcal{M}^T)_a{}^c \tilde{g}^{(0)}_{cd} \mathcal{M}^d{}_b \quad , \tag{3.9}$$

here $\mathcal{M}^a{}_b$ is defined in eqn. (3.7a),

$$\tilde{g}^{(0)} = \begin{pmatrix} n_1\tau^{2p_1} & 0 \\ 0 & n_2\tau^{2p_2} \end{pmatrix} \quad , \tag{3.10}$$

with

$$n_1 = R_0^{2(1+p_1)} q^{2p_1} e^{2((1-2p_1)C_3+\sqrt{3}C_4)} \quad , \tag{3.11a}$$

$$n_2 = R_0^{2(1+p_2)} q^{2p_2} e^{2((1-2p_2)C_3-\sqrt{3}C_4)} \quad , \tag{3.11b}$$

and $q \equiv q_1 + q_2$, $p_I = q^{-1}q_I$, $I = 1, 2, 3$. Our starting Bianchi I solution is now in the required form with the maximal number of degrees of freedom compatible with the Belinskiĭ-Zakharov formalism.

The crucial step in the Belinskiĭ-Zakharov soliton transformation is to obtain the two-by-two matrix function $\psi^{(0)}(\tau, z, \lambda)$ that satisfies the linearized system

$$D_z\psi^{(0)} = U^{(0)}\psi^{(0)} , \tag{3.12a}$$

$$D_\tau\psi^{(0)} = V^{(0)}\psi^{(0)} , \tag{3.12b}$$

where λ is a complex parameter independent of the co-ordinates $\{\tau, z\}$, the differential operators D_z and D_τ are defined as

$$D_z = \partial_z \ - \frac{2\lambda}{\lambda^2 - \alpha^2}\Big(\alpha\,\alpha_z + \lambda\,\alpha_\tau\Big)\,\partial_\lambda \ , \tag{3.13a}$$

$$D_\tau = \partial_\tau \ - \frac{2\lambda}{\lambda^2 - \alpha^2}\Big(\lambda\,\alpha_z + \alpha\,\alpha_\tau\Big)\,\partial_\lambda \ , \tag{3.13b}$$

and two-by-two matrix fields $U^{(0)}$ and $V^{(0)}$ are given by

$$U^{(0)}(t, z, \lambda) = \frac{1}{2}\Big(\frac{A^{(0)}}{\lambda - \alpha} + \frac{B^{(0)}}{\lambda + \alpha}\Big) , \tag{3.14a}$$

$$V^{(0)}(t, z, \lambda) = \frac{1}{2}\Big(\frac{A^{(0)}}{\lambda - \alpha} - \frac{B^{(0)}}{\lambda + \alpha}\Big) , \tag{3.14b}$$

in our case the real matrices $A^{(0)}$ and $B^{(0)}$ are equal

$$A^{(0)} = B^{(0)} = -\alpha\, \partial_\tau g^{(0)}\, g^{(0)-1} \quad , \tag{3.15}$$

Also, we require that the matrix function $\psi^{(0)}(\tau, z, \lambda)$ is related to the metric $g^{(0)}(\tau, z)$, by

$$g^{(0)}(\tau, z) = \psi^{(0)}(\tau, z, \lambda)\big|_{\lambda=0} \,. \tag{3.16}$$

This step is non-trivial in the general case for the non-diagonal $U^{(0)}$ and $V^{(0)}$ [Belinskiĭ and Francaviglia 1982]. However, in our case the system (3.12) can be simplified in the following way. We define $\tilde{\psi}^{(0)}$ by

$$\psi^{(0)} = \mathcal{M}^T\, \tilde{\psi}^{(0)} \mathcal{M} \,. \tag{3.17}$$

It is easy to see that the corresponding linearized system for $\tilde{\psi}^{(0)}$ is diagonal. Therefore, it is straightforward to obtain $\tilde{\psi}^{(0)}$ such that system (3.12) is satisfied [Belinskiĭ and Zakharov 1978, Jantzen 1980], explicitly

$$\tilde{\psi}^{(0)} = \begin{pmatrix} K\,(\alpha^2 + 2\beta\lambda + \lambda^2)^{p_1} & 0 \\ 0 & \frac{1}{K}\,(\alpha^2 + 2\beta\lambda + \lambda^2)^{p_2} \end{pmatrix} \quad , \tag{3.18}$$

here $K = {n_1}^{p_2} {n_2}^{-p_1}$ and $\beta = \sqrt{n_1 n_2}\, z$.

Finally, the Belinskiĭ-Zakharov one-soliton metric can be written in the form

$$ds^2 = f^{(1)}\,(-d\tau^2 + dz^2) + {g^{(1)}}_{ab} dy^a dy^b \,, \tag{3.19}$$

with

$$f^{(1)} = \mathcal{N} {R_0}^{2p^2-3} q^{2p^2-2} \tau^{2p^2-1}\, \frac{\cosh(\mathrm{p\,r} + \mathcal{D})}{\sinh(\frac{\mathrm{r}}{2})} \quad , \tag{3.20a}$$

here $\mathcal{N} = \mathcal{C}\, \frac{m_1 m_2}{2\omega}\, e^{-(4p^2+3)C_3}$ and

$$g^{(1)}{}_{ab} = (\mathcal{M}^T)_a{}^c\, \tilde{g}^{(1)}_{cd}\, \mathcal{M}^d{}_b \quad , \tag{3.20b}$$

with

$$\tilde{g}^{(1)}_{ab} = \frac{1}{\cosh(\mathrm{p\,r} + \mathcal{D})} \times$$

$$\begin{pmatrix} \cosh(\mathrm{p_1 r} + \mathcal{D})\, \mathrm{n_1}\, \tau^{2\mathrm{p_1}} & -\sinh(\frac{\mathrm{r}}{2})\, \sqrt{\mathrm{n_1 n_2}}\, \tau \\ -\sinh(\frac{\mathrm{r}}{2})\, \sqrt{\mathrm{n_1 n_2}}\, \tau & \cosh(\mathrm{p_2 r} - \mathcal{D})\, \mathrm{n_2}\, \tau^{2\mathrm{p_2}} \end{pmatrix} \tag{3.21}$$

here $p \equiv \frac{1}{2}(p_1 - p_2)$, $r = 2\ln\left(\frac{\mu}{\tau}\right)$ with $\mu = \omega - z - \sqrt{(\omega - z)^2 - \tau^2}$ and $\mathcal{D} = \ln\left(\sqrt{\frac{n_1}{n_2}}\frac{m_2}{m_1}(2\omega)^{(p_1-p_2)}\right)$. Notice that $\mathcal{C}$ and ω are real constants, and m_a is a real constant vector. Therefore, four new real constants are introduced with the Belinskiĭ-Zakharov soliton transformation.

An important remark is that the one-soliton solution eqn. (3.19) is defined only for $(\omega - z)^2 - \tau^2 \geq 0$ [Belinskiĭ and Zakharov 1978, Verdaguer 1993]. It is possible to continue the one-solution metric in the region $(\omega - z)^2 - \tau^2 < 0$ with the unperturbed Bianchi I solution eqn. (3.1,9). However, the first derivatives suffer discontinuities on the light-cone $(\omega - z)^2 - \tau^2 = 0$ [Belinskiĭ and Zakharov 1978, Verdaguer 1993]. At the moment we are not sure if the discontinuities of the first derivatives indicate a genuine singularity, or the pathology of the co-ordinte system we are using. We are investigating this point currently [Manojlović, G. Stephens and T. Vukašinac]. Here we can only argue that if the singularity is genuine, then the Belinskiĭ-Zakharov procedure has to be modified so that it is globally compatible with the topology of the spatial manifold Σ, which is the three torus T^3 in this case. Namely, Gowdy's result, based on the general consideration of the global structure of the spacetimes admitting the two-dimensional Abelian group of isometries which acts orthogonally and transitively on non-null orbits, tells us that the T^3 universes always begin with a singularity at $\tau = 0$ and then expand forever [Gowdy 1974]. Therefore, if the singularity on the light-cones are genuine then the one-soliton solution (3.19) cannot be compatible with the T^3 global topology of the spatial manifold Σ.

In addition, it is possible to see that if the z co-ordinate is compact then any point on the given spatial slice will reach the light-cone $(\omega - z)^2 - \tau^2 = 0$ in finite proper time. [1] A particular consequence of this is the fact that, if the singularity is genuine, then it is not possible find a foliation such that the momenta conjugate to the spatial metric stay finite on the light-cones. To illustrate this point we choose the homogeneous foliation X^μ, eqn. (2.10). The normal vector n_μ is then determined by the equations

$$n_\mu X^\mu{}_{,I} = 0 \ , \qquad (3.22a)$$

$$g^{(1)\ \mu\nu} n_\mu n_\nu = -1 \ , \qquad (3.22b)$$

here $g^{(1)\ \mu\nu}$ is the inverse of the one-soliton metric eqn. (3.19). Using eqn. (2.11) we obtain $n_0 = \sqrt{f^{(1)}}$ and $n_I = 0$, $I = 1, 2, 3$.

The lapse function N and the shift vector N^I are then given by

$$N^I = 0 \quad , \quad N = -\sqrt{f^{(1)}} \ . \qquad (3.23)$$

[1] The proper time T is defined to be such that the metric (3.19) takes the form $ds^2 = -dT^2 + g^{(1)}{}_{ab}dy^a dy^b + f^{(1)}dz^2$

The metric on the three dimensional manifold Σ is

$$\gamma^{(1)}{}_{IJ} = \begin{pmatrix} \mathcal{M}^T & \\ & 1 \end{pmatrix} \begin{pmatrix} \tilde{g}^{(1)} & \\ & f^{(1)} \end{pmatrix} \begin{pmatrix} \mathcal{M} & \\ & 1 \end{pmatrix} , \tag{3.24}$$

here the matrix $\mathcal{M}$, the function $f^{(1)}$ and the metric $\tilde{g}^{(1)}$ are defined in eqns. (3.7a,20a,21) respecitvely. The extrinsic curvature $K^{(1)}{}_{IJ}$ is given by

$$K^{(1)}{}_{IJ} = \frac{1}{2\sqrt{f^{(1)}}} \begin{pmatrix} \mathcal{M}^T & \\ & 1 \end{pmatrix} \begin{pmatrix} \partial_\tau \tilde{g}^{(1)} & \\ & \partial_\tau f^{(1)} \end{pmatrix} \begin{pmatrix} \mathcal{M} & \\ & 1 \end{pmatrix} , \tag{3.25}$$

and the momenta $\pi^{(1)}$ conjugate to $\gamma^{(1)}$ can be expressed as

$$\pi^{(1)\ IJ} = \begin{pmatrix} \mathcal{M}^{-1} & \\ & 1 \end{pmatrix} \begin{pmatrix} \tilde{\pi}^{(1)\ ab} & \\ & \tilde{\pi}^{(1)\ 33} \end{pmatrix} \begin{pmatrix} \mathcal{M}^{-1T} & \\ & 1 \end{pmatrix} , \tag{3.26}$$

here

$$\tilde{\pi}^{(1)\ ab} = \frac{\alpha}{k^2 \tau} \times$$

$$\left(\frac{\mathrm{p}_2 \cosh(\mathrm{p}_1 \mathrm{r} + \mathcal{D}) + \mathrm{p}_1 \cosh(\mathrm{p}_2 \mathrm{r} - \mathcal{D})}{2 \cosh(\mathrm{p}\, \mathrm{r} + \mathcal{D}) \sinh^2\left(\frac{\mathrm{r}}{2}\right)} \tilde{g}^{(0)-1\ ab} + p^2\, \tilde{g}^{(1)-1\ ab} \right) \tag{3.27a}$$

and

$$\tilde{\pi}^{(1)\ 33} = \frac{\alpha}{k^2 \tau\, f^{(1)}} , \tag{3.27b}$$

where the metrics $\tilde{g}^{(0)}_{ab}$, $\tilde{g}^{(1)}_{ab}$ and the function $f^{(1)}$ are given in the eqns. (3.10,21,20a) respectively. Crucial observation is that $\tilde{\pi}^{(1)\ ab}$, and therefore $\pi^{(1)\ ab}$, contain the factor $\frac{1}{\sinh^2\left(\frac{r}{2}\right)}$ which blow up on the light-cone $(\omega - z)^2 - \tau^2 = 0$ and this is the problem for the canonical theory.

4 Conclusions and directions for future research

We considered the Belinskiĭ and Zakharov one-soliton solutions as a dynamical system. In order to identify the physical degrees of freedom we started with the general Bianchi type I solution in the case when the spatial manifold Σ is fixed to be the three torus. Using the homogeneous foliation we constructed the symplectic structure for the Bianchi type I model and showed that we have all the degrees of freedom for this model. Our next step was the Belinskiĭ-Zakharov soliton transformation, which is presently define only in one co-ordinate neighborhood on Σ. We showed that in this case it is necessary to freeze the two physical degrees of freedom, in the general

Bianchi type I solution. The crucial step in the soliton transformation was the integration of the linearized system. In our case the linearized system is non-diagonal, but the particular form of our starting solution enabled us to reduce the problem to the diagonal case. Consequently, it was straightforward to calculate the one-soliton metric. This metric is defined only in the region where $(\omega - z)^2 - \tau^2 \geq 0$. Although it is possible to continue the metric in the interior of the light-cone $(\omega - z)^2 - \tau^2 = 0$, it turns out that the first derivatives suffer discontinuities on the light-cone [Belinskiĭ and Zakharov 1978, Verdaguer 1993]. At the moment we are trying to see if this is the consequence of a pathological co-ordinate system, or if it is a genuine singularity [Manojlović, G. Stephens and T. Vukašinac]. If the singularity is genuine, then the Belinskiĭ-Zakharov soliton transformation has to be modified in order to accommodate the globally non-trivial spatial topologies. In addition, it can be see that in the case when the z co-ordinate is compact, as in our case, it is not possible to decompose the metric in such way that the conjugate momenta do not suffer from discontinuities on the light-cone. We have illustrated this point for the case of homogeneous foliation. Namely, in this case the momenta have the factor $\frac{1}{\sinh^2\left(\frac{r}{2}\right)}$ which blows up on the light-cone $(\omega - z)^2 - \tau^2 = 0$. Therefore the problems with the possible singularity and the non-trivial spatial topologies have to be resolved before we can discuss the phase space structure for the Belinskiĭ-Zakharov one-solton [Manojlović, G. Stephens and T. Vukašinac].

5. Acknowledgments

We would like to thank A. Ashtekar, R. Loll, G. Mena Marugán, J. Mourao and R. Picken for discussions. N.M. was supported by the European Union HCM Network on Constrained Dynamical Systems, contract Nr. ERBCHRXCT930362.

6 References

D. Kramer, H. Stephani, M. MacCallum and E. Herlt (1980), *Exact Solutions of Einsten's Field Equations*. Cambridge University Press.

R. Geroch (1971), *J. Math. Phys.*, **12** 918.

R. Geroch (1972), *J. Math. Phys.*, **13** 394.

C. M. Cosgrove (1980), *J. Math. Phys.*, **21** 2417.

C. M. Cosgrove (1981), *J. Math. Phys.*, **22** 2624.

C. M. Cosgrove (1982), *J. Math. Phys.*, **23** 615.

W. Kinnersley (1977), *J. Math. Phys.*, **18** 1529.

W. Kinnersley and D. M. Chitre (1977), *J. Math. Phys.*, **18** 1538.

W. Kinnersley and D. M. Chitre (1978a), *J. Math. Phys.*, **19** 1926.

W. Kinnersley and D. M. Chitre (1978b), *J. Math. Phys.*, **19** 2037.

J. Hauser and F. J. Ernst (1979), *Phys. Rev. D*, **20** 362.

J. Hauser and F. J. Ernst (1980), *J. Math. Phys.*, **21** 1126.

J. Hauser and F. J. Ernst (1981), *J. Math. Phys.*, **22** 1051.

B. K. Harrison (1978), *Phys. Rev. Lett.*, **41** 1197.

B. K. Harrison (1980), *Phys. Rev. D*, **21** 1695.

V. A. Belinskiĭ and V. E. Zakharov (1978), *Sov. Phys. JETP*, **48** 985.

V. E. Zakharov and A. B. Shabat (1974), *Funct. Anal. Appl.*, **8** 266.

V. E. Zakharov and A. B. Shabat (1979), *Funct. Anal. Appl.*, **13** 166.

V. E. Zakharov (1978), *The Method of Inverse Scattering Problem.* Lecture Notes in Mathematics, Springer Verlag.

E. Verdaguer (1993), *Phys. Rep.*, **229**.

R. T. Jantzen (1979), *Commun. Math. Phys.*, **64** 211.

N. Manojlović and G. Mena Marugán (1993), *Phys. Rev. D*, **48** 3704.

R. T. Jantzen (1980), *Il Nuovo Cimento*, **59B** 287.

V. A. Belinskiĭ and M. Francaviglia (1982), *Gen. Rel. Grav.*, **14** 213.

R. H. Gowdy (1974), *Ann. Phys. (NY)*, **83** 203.

N. Manojlović, G. Stephens and T. Vukašinac, work in progress.

Hamiltonian reduction and the R–matrix of the Calogero model

M. Talon

1 Introduction

Let us consider dynamical systems with a finite number of degrees of freedom, whose equations of motion can be encoded in Lax form [1]:

$$\dot{L} = [L, M] \tag{1.1}$$

This happens frequently for systems which can be naturally interpreted as motions on Lie algebras, such as top–like systems.

It is then obvious that the quantities:

$$I_n = \operatorname{Tr} L^n \tag{1.2}$$

are conserved quantities under the evolution given by the equation (1.1). Moreover assuming that the Lax matrix is globally defined on the phase space of the dynamical system, the invariants I_n are also globally defined. Then the Liouville theorem, as refined by Arnold [2] allows to assert that the dynamical system is integrable if and only if the conserved quantities I_n are in involution under the Poisson bracket, i.e.

$$\{I_n, I_m\} = 0 \tag{1.3}$$

Note that such a condition is only significant when the I_n's are globally defined since one can always complete the hamiltonian locally to a Poisson commuting set of quantities, due to the Darboux theorem.

The involutory character of the conserved quantities is equivalent to the existence of an R–matrix as shown in [3]. Here one speaks about a classical R–matrix, a concept whose connection with that of the quantum R–matrix is quite loose [4]. To be brief this means that one can write:

$$\{L_1 \overset{\otimes}{,} L_2\} = [R_{12}, L_1] - [R_{21}, L_2] \tag{1.4}$$

with $L_1 = L \otimes \mathbf{1}$, $L_2 = \mathbf{1} \otimes L$ and so on. Hence the R–matrix encodes all the information on the Poisson brackets of the system, but may be any sort of object with values in the tensor product of two spaces in which L lives, in particular R may very well depend on dynamical variables. For the usual simple dynamical systems in which R is introduced e.g. the Toda chain, this happens not to be the case.

We shall now see that for an other famous integrable dynamical system, the Calogero model [5], the associated R–matrix suffers from all the diseases that are potentially present in the general case, that is, it is dynamical, non antisymmetric, non invertible. This means that the connection of such a classical R–matrix to a quantum counterpart is highly hypothetical, notwithstanding the fact that the Calogero model is also a perfectly respectable integrable quantum system.

This model consists of N points sitting on a line with a non–"local" interaction given by:

$$H = \frac{1}{2}\sum_i p_i^2 + \frac{1}{2}\sum_{i\neq j} \frac{1}{\sinh^2(q_i - q_j)} \tag{1.5}$$

The hyperbolic sine may be replaced by an ordinary sine or even by $(q_i - q_j)^2$ without any changes in the discussion below. It may even be replaced by some elliptic function but the present methods are not adequate to discuss this integrable situation. The associated equations of motion can be recast in Lax form with an L–matrix given by

$$L_{ii} = p_i \quad , \quad L_{ij} = \frac{i}{\sinh(q_i - q_j)} \text{ if } i \neq j \tag{1.6}$$

so that $L^+ = L$ and $H = 1/2\,\mathrm{Tr}\,L^2$ and some appropriate M. There even exists a Lax pair with spectral parameter found by Krichever [6], and $L(\alpha)$ is obtained by adding to L $1/\alpha$ times the matrix with ones everywhere, which commutes to M.

This model is classically integrable because one can exhibit the R–matrix, obtained in [7] by brute force computation:

$$R_{12} = \sum_{k\neq l} \coth(q_k - q_l) E_{kl} \otimes E_{lk} + \frac{1}{2}\sum_{k\neq l} \frac{1}{\sinh(q_k - q_l)} E_{kk} \otimes (E_{kl} - E_{lk}) \tag{1.7}$$

The very unusual features alluded to above are easily observed:

- The first half of R is antisymmetric under exchange of k and l which is equivalent to the transposition in the tensor product but not the second half, hence R is non antisymmetric.

- R is non invertible means that the dualized form is not an invertible operator since diagonal terms E_{kk} appear on the left but not on the right.

- Finally R obviously depends on the dynamical variables q_j.

In the following we shall present a simplified version of the discussion of Olshanetsky and Perelomov [8] showing why the above seemingly complicated dynamical system may be obtained by Hamiltonian reduction from a very simple and obviously integrable one. We shall then benefit from this viewpoint to explain how easily one can obtain the R–matrix, following the analysis in [9].

2 The Calogero model as Hamiltonian reduction of the geodesic flow on a Lie group

We work in the phase space $M = T^*G$, that is the cotangent bundle of a semi–simple Lie group G with its canonical symplectic form, and consider the geodesic flow whose solutions are simply:

$$g(t) = g_0 \, e^{tX} \quad X \in \mathcal{G} = \mathrm{Lie}(G)$$

As a matter of fact any point in M can be parametrized as:

$$\omega = (g, \xi) \quad g \in G, \; \xi \in \mathcal{G}^* \tag{2.1}$$

so that on any tangent vector $v \in T_g(G)$ the one-form ω takes the value $\omega(v) = \xi(g^{-1}v)$. Then the Hamiltonian of the geodesic flow is $H = 1/2||\xi||^2$ and the equations of motion are solved by ξ constant and g as above. This dynamical system has a very big symmetry group allowing to find interesting systems under Hamiltonian reduction.

More precisely let us consider a non compact Lie group G and a maximal compact subgroup H such that G/H is a non compact *symmetric space.* In the following we shall restrict ourselves to the case $G = SL(N, \mathbf{C})$ and $H = SU(N)$ although the general case may be treated along essentially the same lines, see [9]. Then $H_L \times H_R$ acts on the left and on the right on M and in the above coordinates this action reads:

$$(h_L, h_R).(g, \xi) = (h_L g h_R^{-1}, h_R \xi h_R^{-1}) \tag{2.2}$$

Note that we have written everything as a left action and that h_L does not move ξ due to its definition while h_R acts by the coadjoint action. This is obviously a symmetry of the geodesic flow since $||\xi||^2$ is invariant and of the symplectic structure by naturality.

From Noether's theorem there are associated conserved momenta ($\mathcal{H}$ is the Lie algebra of H):

$$\begin{aligned} H_{(X_L,X_R)}(\omega) &= \alpha((X_L, X_R).\omega) \leftarrow \text{infinitesimal action} \\ X_L, X_R \in \mathcal{H} \quad & \quad \uparrow \text{canonical } 1-\text{form} \\ &= < g\xi g^{-1}, X_L > - < \xi, X_R > \end{aligned} \tag{2.3}$$

so that we can write the momentum $\mathcal{P}$ with values in the Lie algebra of $H_L \times H_R$ identified with its dual under the Killing form:

$$\mathcal{P}(\omega) = (P_{\mathcal{H}}\, g\xi g^{-1}, -P_{\mathcal{H}}\, \xi) \tag{2.4}$$

where $P_{\mathcal{H}}$ is the orthogonal projector on $\mathcal{H}$. Note that the symmetry group $H_L \times H_R$ acts on $\mathcal{P}$ by the adjoint action.

The Hamiltonian reduction consists in:

- Constraining ourselves to a surface of fixed momentum:

$$N_\mu = \{\omega \in M |\ \mathcal{P}(\omega) = \mu\} \tag{2.5}$$

  for a given momentum μ.

- Eliminating the extra degrees of freedom due to the fact that the stabilizer of μ in $H_L \times H_R$ is a residual symmetry on N_μ.

In order to fix this reduction we must now recall some facts about symmetric spaces: the Lie algebra $\mathcal{G}$ decomposes into the Lie algebra $\mathcal{H}$ of H and its orthogonal complement $\mathcal{K}$. Here $\mathcal{H}$ is the algebra of traceless anti–hermitean matrices, and $\mathcal{K}$ the algebra of traceless hermitean matrices. Under exponentiation we have the Cartan decomposition $G \simeq HK$ and $K = \exp(\mathcal{K}) \simeq$ symmetric space (G/H). Finally any element of K can be brought to an element of a so-called Cartan algebra of G/H, here the diagonal matrices of K, by conjugation with an element of H (Cartan conjugation theorem). The net result is that one can write:

$$\forall g \in G, \quad g = h_1 Q h_2, \quad h_1, h_2 \in H, \quad Q \in K \text{ diagonal} \tag{2.6}$$

(Of course this is directly obvious in the present case.) Moreover this decomposition is *unique* up to the redefinitions:

$$h_1 \to h_1 d \quad h_2 \to d^{-1} h_2 \quad d \in H \text{ diagonal} \tag{2.7}$$

i.e. d has phases on the diagonal. This diagonal ambiguity is unfortunately a major source of troubles in the following.

To get the Calogero model one *chooses* a peculiar value of the momentum:

$$\mu_L = i \begin{pmatrix} 0 & 1 & \cdots & 1 \\ 1 & 0 & \cdots & 1 \\ 1 & 1 & \ddots & 1 \\ 1 & 1 & \cdots & 0 \end{pmatrix} = i(vv^+ - \mathbf{1}) \quad \mu_R = 0 \tag{2.8}$$

The rationale behind these choices is the following:

- $\mu_R = 0$ implies that the whole H_R stabilises μ, hence one works on the symmetric space G/H.

- A matrix U in H stabilises μ_L i.e. $U\mu_L U^{-1} = \mu_L$ if and only if $Uv = e^{i\theta}v$, where v entering the definition of μ_L is a vector with all components equal to 1. It is then obvious that the corresponding stabilizer is:

$$H_\mu \simeq SU(N-1) \times U(1)$$

and is a *maximal subgroup* of H. Hence the reduced dynamical system is the minimal non trivial one obtained by such a process.

Finally any element $\omega = (g, \xi)$ in N_μ can be brought to the form (Q, L) with Q in K diagonal, and L some element in $\mathcal{G}$ through the action of the residual symmetry group $H_{\mu L} \times H_R$ and the decomposition $g = h_1 Q h_2$ on N_μ is *unique* since no diagonal element of H can fix v.

We can now take $\omega = (Q, L) \in N_\mu$ and write $\mathcal{P}(\omega) = \mu$. Since Q is the exponential of a diagonal traceless real matrix we write $Q = \text{Diag}(e^{q_i})$, with $\sum q_i = 0$. Then the condition on the right momentum says $P_{\mathcal{H}} L = 0$ i.e. $L \in \mathcal{K}$, and the condition on the left momentum reads $P_{\mathcal{H}} QLQ^{-1} = \mu_L$, i.e.

$$\frac{1}{2}(e^{q_i - q_j} L_{ij} - e^{q_j - q_i} \bar{L}_{ji}) = \begin{cases} 0 & i = j \\ i & i \neq j \end{cases}$$

Since $\bar{L}_{ji} = L_{ij}$ one immediately finds that L has exactly the form of the Lax matrix (1.6) with unfixed real diagonal terms p_i's.

It is easy to solve the equations of motion. In fact starting from the initial point (Q_0, L_0) the evolution according to the geodesic flow is obviously $g(t) = Q_0\, e^{L_0 t}$, $\xi = L_0$, and it remains to decompose $g(t) = h_1(t) Q(t) h_2(t)$. But we have:

$$g g^+ = h_1 Q^2 h_1^{-1} = Q_0\, e^{2L_0 t}\, Q_0$$

so one only needs to diagonalize this matrix. In particular for t small, the eigenvalues will be $e^{2q_i(t)}$, and one expands up to order t^2, which leads to (dropping the subscript 0):

$$e^{2q_i}(1 + 2t\dot{q}_i + t^2(\ddot{q}_i + 2\dot{q}_i^2) + O(t^3).$$

On the other hand we have to compute the eigenvalues of:

$$Q^2 + 2QLQt + 2QL^2Qt^2 + O(t^3)$$

which is readily done using the canonical orthonormal basis and standard quantum mechanical perturbation theory at first and second order. One gets exactly the equations of motion of the Calogero model at the initial point, but this is arbitrary. Finally this decomposition allows to express $L(t) = h_2 L_0 h_2^{-1}$ since ξ is constant, hence $L(t)$ evolves as in the Lax equation (1.1) with $M = \dot{h}_2(t) h_2^{-1}(t)$ and one can make choices so that $h_2 \in H_\mu$ instead of h_1 leading to $Mv = 0$, that is the sum of the elements of a line in M is 0. This allows to find easily M which happens to be equal to that introduced in [8].

3 Poisson brackets on the reduced phase space

We first sketch the geometric situation to which we arrived:

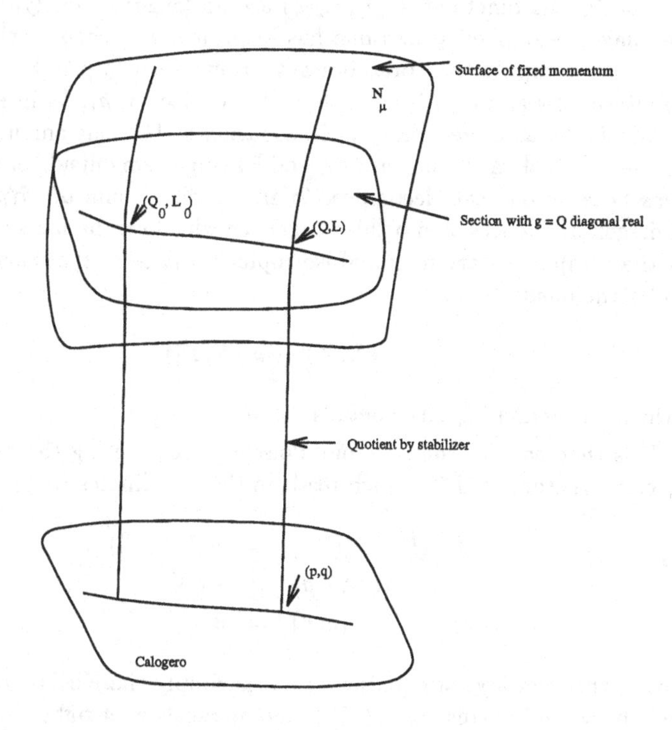

Notice that a function F on Calogero phase space is equivalent to a function on the section of N_μ characterized by $g = Q$, itself equivalent to a function F on N_μ *invariant* under the action of the stabilizer $H_\mu \times H_R$. One then extends F *arbitrarily* to a function on the whole $M = T^*G$. Then the Hamiltonian reduction of the phase space is equivalent to using Dirac brackets on M of the form:

$$\{F, G\}_{\text{red}} = \{F, G\} + \frac{1}{2}((v_G.\omega).F - (v_F.\omega).G) \tag{3.1}$$

Here the element v_F of the Lie algebra $\mathcal{H}$ of the symmetry group is determined by the equation

$$\forall X \in \mathcal{H} \quad \mu([X, v_F]) = (X.\omega).F$$

It generates a vector field $v_F.\omega$ at the point $\omega \in M$ which itself acts on the function G.

We want to compute for any two $X, Y \in \mathcal{K}$ the reduced Poisson bracket $\{L(X), L(Y)\}$ which is equivalent to $\{L_1 \overset{\otimes}{,} L_2\}$ on the Calogero phase space under the well-known dualization operation. At the generic point of N_μ: $(g, \xi) = (h_1(g)Qh_2(g), h_2^{-1}Lh_2)$ the invariant extension of $L(x)$ is obviously:

$$F_X(g, \xi) = \xi(h_2^{-1}Xh_2)$$

Out of N_μ the functions $h_1(g)$, $h_2(g)$ are no longer uniquely defined due to the diagonal ambiguity, and one has in principle to choose arbitrary extensions and compute the Dirac bracket. However it is possible to show that infinitesimally around N_μ one can fix the choices of h_1, h_2 in such a way as to kill the Dirac correction term in equation (3.1). This amounts to putting all the diagonal variations into h_2, and leaving them outside of h_1. This happens because one can decompose $\mathcal{H}$ into a direct sum of: $\mathcal{H}_\mu$, the algebra of diagonal matrices, and a third subspace which can be nicely characterized as the complementary maximal isotropic subspace of the diagonal matrices under the quadratic form:

$$(X, Y) \to \mu([X, Y])$$

(which has kernel $\mathcal{H}_\mu$). For details we refer to [9].

It is then easy to compute our Poisson bracket using the canonical symplectic structure on T^*G which reads in the coordinates (g, ξ):

$$\begin{aligned} \{\xi(X), \xi(Y)\} &= -\xi([X, Y]) \\ \{\xi(X), g\} &= -gX \\ \{g, g\} &= 0 \end{aligned} \tag{3.2}$$

Notice that the first of equations (3.2) is simply the Kirillov bracket on $\mathcal{G}^*$, and the second means that $\xi(X)$ is the momentum of right translations.

In the vicinity of the above defined section, for $X, Y \in \mathcal{K}$, we have using equations (3.2):

$$\begin{aligned} \{L(X), L(Y)\} &= \{F_X, F_Y\}_{\text{red}} \\ &= \{\xi(h_2^{-1}Xh_2), \xi(h_2^{-1}Yh_2)\}_{|h_2=1\ \xi=L} \\ &= -L([X, Y]) + L([X, \nabla_Q h_2(Y)] + [\nabla_Q h_2(X), Y]) \end{aligned}$$

where $\nabla_g h(X) = \frac{d}{dt}h(ge^{tX})_{|t=0}$. But since G/H is a symmetric space we have $[X, Y] \in \mathcal{H}$ hence $L([X, Y]) = 0$, and we recognize that the Poisson bracket is given by an R-matrix structure (the dualization of equation (1.4):

$$\{L(X), L(Y)\} = L([X, RY] + [RX, Y]) \tag{3.3}$$

with $R(X) = (\nabla_Q h_2)(X)$.

It only remains to compute this variation taking into account the above mentioned choices on the variations of h_1 and h_2 in the diagonal directions.

As a matter of fact since any g can be written $g = h_1 Q h_2$, given the tangent vector v at Q one can decompose it as:

$$v = X_L.Q + Q.T + Q.X_R$$

where X_L, $X_R \in \mathcal{H}$, $T \in \mathcal{K}$ diagonal and X_R is the sought-after $\nabla_Q h_2(Q^{-1}v)$. This decomposition becomes unique when X_L does not contain diagonal elements. Here we have to compute X_R when $v = Q.X$ with $X \in \mathcal{K}$, i.e. we solve:

$$X = Q^{-1} X_L Q + T + X_R$$

In the present case X_L, X_R are antihermitean while X, T are hermitean so this equation immediately determines the non–diagonal elements of X_L, X_R, while to fix the phases on the diagonal it is necessary to use the above hypothesis. Finally after dualization one exactly recovers the R–matrix given in equation (1.7).

The benefits of this approach to the computation of the R–matrix are:

- The existence of the R–matrix is almost obvious, stemming from a similar formula in T^*G.
- The above analysis can be extended directly to other symmetric spaces and allows to find the R–matrix for some fancy–looking Calogero models which are untractable by the direct computational method.
- A similar method has already been introduced in [10] to study the phase space of the Toda model. In this case the Iwasawa decomposition is used, a unique decomposition which greatly simplifies the discussion [9].

References

[1] P.D. Lax (1968), 'Integrals of non linear equations of evolution and solitary waves', *Comm. Pure Appl. Math.*, **21** 467.

[2] V. Arnold (1976), *Méthodes mathématiques de la mécanique classique'*, MIR, Moscou.

[3] O. Babelon and C.M. Viallet (1989), 'Hamiltonian Structures and Lax Equations', *Phys. Lett.*, **237B** 411.

[4] M. Semenov Tian Shansky (1983), 'What is a classical r-matrix', *Funct. Anal. and its Appl.*, **17, 4** 17.

[5] F. Calogero (1975), 'Exactly Solvable One-Dimensional Many-Body Problems', *Lett.Nuovo Cim.*, **13** 411-416.

[6] I.M. Krichever (1980), 'Elliptic solutions of the Kadomtsev-Petviashvili equation and many body problems', *Func. Anal. Appl.*, **14** 45.

[7] J. Avan and M. Talon (1993), 'Classical R-matrix Structure for the Calogero Model', *Phys. Lett.*, **B303** 33-37.

[8] M.A. Olshanetsky and A.M. Perelomov (1981), 'Classical Integrable Finite Dimensional Systems Related to Lie Algebras', *Physics Reports*, **71** 313-400.

[9] J. Avan, O. Babelon and M. Talon (1994), 'Construction of the classical R–matrices for the Toda and Calogero models', *Algebra and Analysis*, in press.

[10] L.A. Ferreira and D.I. Olive (1985), 'Non-Compact Symmetric Spaces and the Toda Molecule Equations', *Commun. Math. Phys.*, **99** 365-384.

Intrinsic approach to the Legendre transformation in supermechanics

José F. Cariñena and Héctor Figueroa

In recent years, there has been an increasing interest to provide supermechanics with a solid geometrical base. Some of the difficulties are right at the beginning: there is a general consent about the configuration space, which is taken to be a supermanifold, but there is no general agreement on what the velocity phase space should be. Naturally, the candidate for that should be a generalization of the tangent bundle in the context of supergeometry; unfortunately there are several different (and reasonable) notions that could serve that purpose. The tangent supermanifold introduced by Ibort and Marín–Solano in [8] seems to be the right candidate from the point of view of a physicist since a good deal of supermechanics can rigorously be developed; nevertheless, the main disadvantage of the tangent supermanifold is that it is not a bundle in any sense, which forces a local coordinate approach that does not give too much insight, or a purely algebraic approach, in practice difficult to handle; besides, in classical mechanics one usually takes advantage of the fact that vector fields, for instance, are, after all, sections of a bundle. In this sense, the tangent superbundle introduced by Sánchez–Valenzuela [10,11,2] seems to be quite appropriate, from a theoretical point of view, since there is a one–to–one correpondence between sections of this superbundle and supervector fields regarded as superderivations. Unfortunately, the tangent superbundle is too big, its dimension is $(2m+n, 2n+m)$ if the dimension of the configuration space is (m,n). Our idea is then to develop the theory in the tangent superbundle, but read off the results in the tangent supermanifold, since, as we shall see, the latter can be considered as a subsupermanifold of the former.

From a geometrical point of view, classical mechanics can be described in a very concise way, although not so well known, in terms of the concept of a vector field along a differentiable map, or more generally in terms of sections of a vector bundle along a differentiable map [5–7]. Vector fields along a map were originally introduced to deal with non–point transformations; nevertheless, they also allow one to work intrinsically without having to use points which might just be a little pedantic in the classical setting but that turns out to be essential in graded geometry. Thus, using the notion of section of a supervector bundle along a morphism of graded manifolds, together with the idea previously outlined we shall introduce the super Legendre transformation and obtain a correspondence between the Lagrangian and the Hamiltonian formulations of supermechanics, similar to the one in classical mechanics.

Given a supervector space $V = V_0 \oplus V_1$ the simplest structure of a graded manifold is the so called affine supermanifold: $S(V) := \big(V_0, C^\infty(V_0) \otimes \bigwedge(V_1^*)\big)$. Nevertheless, when dealing with supervector bundles, it becomes necessary [10,2] to replace the affine supermanifold by $V_S = S(V \oplus \Pi V)$, where Π is the change of parity functor [9], because only then [10], one has a one–to–one correspondence between equivalence classes of locally free sheaves of $\mathcal{A}_M$–modules over $\mathcal{M}$ of rank (r,s) of a graded manifold $\mathcal{M} = (M, \mathcal{A}_M)$, of dimension (m,n), and equivalence classes of supervector bundles over $\mathcal{M}$ of rank (r,s); in particular, the tangent superbundle $ST\mathcal{M} := (STM, ST\mathcal{A})$, is defined as the supervector bundle of rank (m,n) that corresponds to the sheaf of $\mathcal{A}_M$–modules $\operatorname{Der}\mathcal{A}$.

The definition of a supervector bundle is a generalization of the standard definition of a vector bundle: it is a quadruplet $\{(E, \mathcal{A}_E), \Pi, (M, \mathcal{A}_M), V_S\}$ such that $\Pi\colon (E, \mathcal{A}_E) \to (M, \mathcal{A}_M)$ is a submersion of graded manifolds, V is a real (r,s)–dimensional supervector space and every $q \in M$ lies in a coordinate neighbourhood $\mathcal{U} \subseteq M$ for which there exista an isomorphism $\Psi_{\mathcal{U}}\colon \big(\pi^{-1}(\mathcal{U}), \mathcal{A}_E(\pi^{-1}(\mathcal{U}))\big) \to (\mathcal{U}, \mathcal{A}_M(\mathcal{U})) \times V_S$ such that $P_1 \circ \Psi_{\mathcal{U}} = \Pi$, where P_1 is the projection onto the first factor. Thus, $(E, \mathcal{A}_E)$ is locally isomorphic to a graded manifold of the form $(\mathcal{U}, \mathcal{A}(\mathcal{U})) \times V_S$, so if $\{q^i, \theta^\alpha\}$, are local supercoordinates on $\mathcal{U} \subseteq M$, abusing of the notation, the local supercoodinates of $(E, \mathcal{A}_E)$ will be denoted by $\{q^i, v^j, \pi\zeta^\beta, \theta^\alpha, \zeta^\beta, \pi v^j\}$, where $\{v^j, \pi\zeta^\beta, \zeta^\beta, \pi v^j\}$ are the supercoordinates of $(E, \mathcal{A}_E)$ corresponding to the local supercoordinates of V_S under the isomorphism $\Psi_{\mathcal{U}}$. In particular, the set of local conditions $\pi v^j = 0$ and $\pi\zeta^\beta = 0$, consistently defines a subsupermanifold of $(E, \mathcal{A}_E)$ of dimension $(m+r, n+s)$, which in the case of $ST\mathcal{M} := (STM, ST\mathcal{A})$, corresponds to the tangent supermanifold $T\mathcal{M} = (TM, T\mathcal{A})$ introduced by Ibort and Marín–Solano in [8]. The one–to–one correspondence just mentioned is explicit and allow us to give a Batchelor description of the tangent superbundle [4]. In particular, if one adopt the point of view used in [8,3], that is, if the graded manifold $\mathcal{M}$ has been constructed out of a family of superdomains $\{(\mathcal{U}_\alpha, \mathcal{A}_\alpha)\}$, then $ST\mathcal{M}$ turns out to be the graded manifold built out of the family of superdomains $\{(ST\mathcal{U}_\alpha, ST\mathcal{A}_\alpha)\}$ where

$$ST\mathcal{U}_\alpha := \mathcal{U}_\alpha \times \mathbb{R}^{m+n}, \quad ST\mathcal{A}_\alpha := C^\infty(\mathcal{U}_\alpha \times \mathbb{R}^{m+n}) \otimes \bigwedge(\mathbb{R}^{2n+m}), \tag{1}$$

together with the transition functions:

$$q'^i = \Phi^1_{\mathcal{U}',\mathcal{U}}(q'^i) = \phi^i_0(q) + \phi^i_{\alpha\beta}(q)\,\theta^\alpha\theta^\beta + \cdots, \tag{2a}$$

$$\theta'^\alpha = \Phi^1_{\mathcal{U}',\mathcal{U}}(\theta'^\alpha) = \psi^\alpha_\beta(q)\,\theta^\beta + \psi^\alpha_{\beta\gamma\delta}(q)\,\theta^\beta\theta^\gamma\theta^\delta + \cdots, \tag{2b}$$

$$v'^i = \Phi^1_{\mathcal{U}',\mathcal{U}}(v'^i) = \sum_{j=0}^{m} \frac{\partial q'^i}{\partial q^j} v^j - \sum_{\beta=0}^{n} \frac{\partial q'^i}{\partial \theta^\beta}\zeta^\beta, \tag{2c}$$

$$\pi\zeta'^{\alpha} = \Phi^1_{\mathcal{U}',\mathcal{U}}(\pi\zeta'^{\alpha}) = -\sum_{j=0}^{m}\frac{\partial\theta'^{\alpha}}{\partial q^j}\pi v^j + \sum_{\beta=0}^{n}\frac{\partial\theta'^{\alpha}}{\partial\theta^{\beta}}\pi\zeta^{\beta}, \tag{2d}$$

$$\zeta'^{\alpha} = \Phi^1_{\mathcal{U}',\mathcal{U}}(\zeta'^{\alpha}) = \sum_{j=0}^{m}\frac{\partial\theta'^{\alpha}}{\partial q^j}v^j + \sum_{\beta=0}^{n}\frac{\partial\theta'^{\alpha}}{\partial\theta^{\beta}}\zeta^{\beta}, \tag{2e}$$

$$\pi v'^{i} = \Phi^1_{\mathcal{U}',\mathcal{U}}(\pi v'^{i}) = \sum_{j=0}^{m}\frac{\partial q'^{i}}{\partial q^j}\pi v^j + \sum_{\beta=0}^{n}\frac{\partial q'^{i}}{\partial\theta^{\beta}}\pi\zeta^{\beta}. \tag{2f}$$

Using the augmentation map it is clear that the underlying manifold is $STM = TM \oplus E$, where the vector bundle $E \to M$ is the Batchelor structural bundle of the graded manifold $\mathcal{M}$; see [8] for a description of E in terms of the transition functions. Furthermore, from these equations we also notice [8], that the sheaf $ST\mathcal{A}$ can be identified with the sections of the Grassmannian of a vector bundle $E' \to STM$ whose transition functions $\Psi_{\alpha',\alpha}: (\mathcal{U}' \times \mathbb{R}^{m+n}) \cap (\mathcal{U} \times \mathbb{R}^{m+n}) \to \mathrm{GL}(2n+m, \mathbb{R})$ are given by

$$\Psi_{\alpha',\alpha}(q, v, \pi\zeta) = \begin{pmatrix} \psi^{\alpha}_{\beta}(q) & 0 & 0 \\ \frac{\partial\psi^{\alpha}_{\beta}}{\partial q^i}v^i & \psi^{\alpha}_{\beta}(q) & 0 \\ -2\phi^{i}_{\alpha\beta}(q)\pi\zeta^{\beta} & 0 & \frac{\partial\phi^i_0}{\partial q^j} \end{pmatrix}; \tag{3}$$

but since we may choose the transition functions so that $\phi^{i}_{\alpha\beta}(q) = 0$ [8]; we notice that the vector bundle E' is isomorphic to the Whitney sum $TE \oplus TM \to TM \oplus E$. Moreover, not only one has a one–to–one correspondence between the supervector bundles and the locally free sheaves of graded $\mathcal{A}$–modules, but their geometrical sections are also in an explicit one–to–one correspondence with the sections of the corresponding locally free sheaf. This correspondence is done in two steps: first, one notices that

$$\mathrm{Der}\,\mathcal{A}(\mathcal{U}) \cong \mathcal{A}(\mathcal{U})^m \oplus \mathcal{A}(\mathcal{U})^n \cong \mathrm{Maps}\Big((\mathcal{U}, \mathcal{A}(\mathcal{U})), V_S\Big); \tag{4}$$

on the other hand, if $ST\mathcal{A}(\mathcal{U})$ is a notation for $ST\mathcal{A}(\tau^{-1}(\mathcal{U}))$, there is a mapping from $\mathrm{Maps}\Big((\mathcal{U}, \mathcal{A}(\mathcal{U})), V_S\Big)$ to $\Gamma\Big((\mathcal{U}, \mathcal{A}(\mathcal{U})), (\tau^{-1}(\mathcal{U}), ST\mathcal{A}(\mathcal{U}))\Big)$, the set of sections of $(E, \mathcal{A}_E)$ on $\mathcal{U}$ [4], which is an isomorphism. If $X \in \mathrm{Der}\,\mathcal{A}(\mathcal{U})$ is written in local coordinates as $X = \sum_{i=1}^{m} X^i\partial_{q^i} + \sum_{\alpha=1}^{n}\chi^{\alpha}\partial_{\theta^{\alpha}}$, then the correspondence of $\mathrm{Der}\,\mathcal{A}(\mathcal{U})$ with $\Gamma\Big((\mathcal{U}, \mathcal{A}(\mathcal{U})), (\tau^{-1}(\mathcal{U}), ST\mathcal{A}(\mathcal{U}))\Big)$, is implemented by the morphism $X \to \Sigma_X$, where $\Sigma_X = (\sigma_X, \sigma^*_X)$ is the morphism of graded manifolds described by the morphism of superalgebras $\sigma^*_X: ST\mathcal{A}(\mathcal{U}) \to \mathcal{A}(\mathcal{U})$ given by the assignments

$$\begin{array}{llllll} q^i \mapsto q^i & v^i \mapsto X^i_0 & \pi\zeta^{\alpha} \mapsto \chi^{\alpha}_0, \\ \theta^{\alpha} \mapsto \theta^{\alpha}, & \zeta^{\alpha} \mapsto \chi^{\alpha}_1 & \pi v^i \mapsto X^i_1; \end{array} \tag{5}$$

the subindices 0 and 1 stand respectively for the even and odd part of the superfunction.

A basic tool in supermechanics is the notion of a supervector field along a morphism $\Phi = (\phi, \phi^*): (N, \mathcal{B}_N) \to (M, \mathcal{A}_M)$ of graded manifolds, which were introduced in [3] as those morphisms of sheaves over M, $X: \mathcal{A} \to \Phi_*\mathcal{B}$, such that for each open subset $\mathcal{U}$ of M, $X(fg) = X(f)\,\phi^*_{\mathcal{U}}(g) + (-1)^{|X||f|}\phi^*_{\mathcal{U}}(f)\,X(g)$, whenever $f \in \mathcal{A}(\mathcal{U})$ is homogeneous. $\mathfrak{X}(\Phi)$, the sheaf of supervector fields along Φ, is a locally free sheaf of $\Phi_*\mathcal{B}$–modules over $\mathcal{M}$. If $\{\partial_{q^i}, \partial_{\theta^\alpha}\}$ is the standard local basis on $\operatorname{Der}\mathcal{A}(\mathcal{U})$ attached to the supercoordinates (q^i, θ^α), then $\partial_{\hat{q}^i} := \phi^*_{\mathcal{U}} \circ \partial_{q^i}$, $\partial_{\hat{\theta}^\alpha} := \phi^*_{\mathcal{U}} \circ \partial_{\theta^\alpha}$ is a local basis of $\mathfrak{X}(\Phi)(\mathcal{U})$. In particular, any $X \in \mathfrak{X}(\Phi)(\mathcal{U})$ can be written as $X = \sum_{i=1}^m X^i \partial_{\hat{q}^i} + \sum_{\alpha=1}^n \chi^\alpha \partial_{\hat{\theta}^\alpha}$, where $X^i = X(q^i)$ and $\chi^\alpha = X(\theta^\alpha)$ are superfunctions in $\mathcal{B}(\phi^{-1}(\mathcal{U}))$. As supervector fields, supervector fields along a morphism have a geometric counterpart: given a supervector bundle $\{(E, \mathcal{A}_E), \Pi, (M, \mathcal{A}_M), V_S\}$ over $\mathcal{M}$, a local section $\Sigma = (\sigma, \sigma^*)$ of $\mathcal{E} := (E, \mathcal{A}_E)$ along Φ over an open subset $\mathcal{U}$ of M is a morphism

$$\Sigma: \left(\phi^{-1}(\mathcal{U}), \mathcal{B}(\phi^{-1}(\mathcal{U}))\right) \to \left(\pi^{-1}(\mathcal{U}), \mathcal{A}_E(\mathcal{U})\right), \tag{6}$$

where again $\mathcal{A}_E(\mathcal{U}) := \mathcal{A}_E(\pi^{-1}(\mathcal{U}))$, satisfying the condition $\Phi_{\mathcal{U}} = \Pi_{\mathcal{U}} \circ \Sigma_{\mathcal{U}}$, where the subscript $\mathcal{U}$ means the restriction of the morphism to the corresponding open graded submanifold. The set of such sections will be denoted by $\Gamma_\Phi(\Pi|_{\mathcal{U}})$. In the particular case when the supervector bundle is the tangent superbundle $ST\mathcal{M}$, using similar ideas as before we obtain an isomorphism $\mathfrak{X}(\Phi)(\mathcal{U}) \cong \Gamma_\Phi(\mathcal{T}|_{\mathcal{U}})$, where $\mathcal{T} = (\tau, \tau^*)$ is the natural projection of $ST\mathcal{M}$. As before, the explicit correspondence between a "derivation" $X \in \mathfrak{X}(\Phi)(\mathcal{U})$ and a local section along Φ is given by $X \longmapsto \Sigma_X$, where $\sigma^*_X: ST\mathcal{A}(\tau^{-1}(\mathcal{U})) \to \mathcal{B}(\phi^{-1}(\mathcal{U}))$ is now defined by the assignments

$$\begin{array}{lllllllll} q^i & \mapsto & \phi^*(q^i), & v^i & \mapsto & X^i_0, & \pi\zeta^\alpha & \mapsto & \chi^\alpha_0, \\ \theta^\alpha & \mapsto & \phi^*(\theta^\alpha), & \zeta^\alpha & \mapsto & \chi^\alpha_1, & \pi v^i & \mapsto & X^i_1, \end{array} \tag{7}$$

and the superfunctions X^i and χ^α are now as in (5). The sheaf of graded 1–forms is the dual sheaf of $\operatorname{Der}\mathcal{A}$ and the corresponding supervector bundle $(ST^*\mathcal{M}, \Pi, \mathcal{M}, V_S)$ will be called the cotangent superbundle of $\mathcal{M}$. The explicit correspondence between sections of the supercotangent bundle $ST^*\mathcal{M} = (ST^*M, ST^*\mathcal{A})$ and graded 1–forms is accomplished using the same procedure as above: if $\omega \in \Omega^1\mathcal{A}(\mathcal{U})$, the corresponding section Σ_ω in $\Gamma\Big(\big(\mathcal{U}, \mathcal{A}(\mathcal{U})\big), \big(\pi^{-1}(\mathcal{U}), ST^*\mathcal{A}(\pi^{-1}(\mathcal{U}))\big)\Big)$ is the morphism of graded manifolds corresponding to the morphism of superalgebras $\sigma^*_\omega: ST^*\mathcal{A}(\pi^{-1}(\mathcal{U})) \to \mathcal{A}(\mathcal{U})$ defined by the assignments

$$\begin{array}{lllllllll} q^i & \mapsto & q^i, & p^i & \mapsto & w^i_0, & \pi\eta^\alpha & \mapsto & \omega^\alpha_0, \\ \theta^\alpha & \mapsto & \theta^\alpha, & \eta^\alpha & \mapsto & \omega^\alpha_1, & \pi p^i & \mapsto & w^i_1, \end{array} \tag{8}$$

where $w^i = \omega(\partial_{q^i})$ and $\omega^\alpha = \omega(\partial_{\theta^\alpha})$ are the coordinates of ω relative to the basis $\{ dq^i, d\theta^\alpha \}$ and the subindices 0 or 1 stand for the even or odd components. In analogy with the tangent superbundle, the natural subsupermanifold $T^*\mathcal{M} = (T^*M, T^*\mathcal{A})$ of $ST^*\mathcal{M}$, obtained from the set of local conditions $\pi p^i = 0$ and $\pi\eta^\alpha = 0$, will be called the cotangent supermanifold. Notice that it is a graded manifold of dimension $(2m, 2n)$. The sheaf of graded 1-forms along Φ is the sheaf of $\phi_*\mathcal{B}$–modules dual to the sheaf $\mathfrak{X}(\Phi)$. Thus, $\Omega^1(\Phi) = \mathfrak{X}(\Phi)^* = \mathrm{Hom}(\mathfrak{X}(\Phi), \phi_*\mathcal{B})$.

In general, k-superforms are defined as $\Omega^k(\Phi) := \bigwedge^k(\Omega^1(\Phi))$, where the wedge product is to be understood in the sense of graded algebras. It is easy to check that $d\hat{q}^i := \phi^* \circ dq^i$ and $-d\hat{\theta}^\alpha := -\phi^* \circ d\theta^\alpha$ is the dual basis of $\{ \partial_{\hat{q}^i}, \partial_{\hat{\theta}^\alpha} \}$. Thus, any graded 1–form ω along Φ can, locally, be written as

$$\omega = \sum_{i=1}^{m} w^i \, d\hat{q}^i + \sum_{\alpha=1}^{n} \omega^\alpha \, d\hat{\theta}^\alpha, \tag{9}$$

where the superfunctions w^i and ω^α of $\mathcal{B}(\phi^{-1}(\mathcal{U}))$ are given by $w^i = \omega(\partial_{\hat{q}^i})$ and $\omega^\alpha = \omega(\partial_{\hat{\theta}^\alpha})$. The correspondence $\Omega^1(\Phi)(\mathcal{U}) \cong \Gamma_\Phi(\Pi|_\mathcal{U})$, is carried out along the same lines and is explicitly given by $\omega \longmapsto \Sigma_\omega$, where σ^* is the morphism of superalgebras $\sigma^*_\omega \colon ST^*\mathcal{A}(\pi^{-1}(\mathcal{U})) \to \mathcal{B}(\phi^{-1}(\mathcal{U}))$ defined by the assignments

$$\begin{array}{llllllllll} q^i & \mapsto & \phi^*(q^i), & p^i & \mapsto & w^i_0, & \pi\eta^\alpha & \mapsto & \omega^\alpha_0 \\ \theta^\alpha & \mapsto & \phi^*(\theta^\alpha), & \eta^\alpha & \mapsto & \omega^\alpha_1, & \pi p^i & \mapsto & w^i_1. \end{array} \tag{10}$$

On the cotangent bundle $ST^*\mathcal{M}$, there exists a unique graded 1–form Θ_0 [10], such that $\Sigma_\omega\Theta_0 = \omega$, for every graded 1–form $\omega \in \Omega^1\mathcal{A}(\mathcal{U})$, which in local supercoordinates is given by

$$\Theta_0 = \sum_{i=1}^{m} (p^i + \pi p^i) \, dq^i + \sum_{\alpha=1}^{n} (\eta^\alpha + \pi\eta^\alpha) \, d\theta^\alpha. \tag{11}$$

Although Θ_0 is formally equal to the canonical 1–form of the cotangent bundle in non–graded geometry, it happens that the graded 2–form $-d\Theta_0$ is degenerate; nevertheless, if we restrict Θ_0 to the supercotangent manifold $T^*\mathcal{M}$, then $-d\Theta_0$ is a non–degenerate graded 2–form that will be called the canonical graded 2–form and will be denoted by Ω_0. As in non–graded geometry [6], another way to define the canonical forms is obtained when one regards graded 1–forms as geometric sections: if $\Phi = \Pi = (\pi, \pi^*)$ is the canonical projection of $ST^*\mathcal{M}$ onto $\mathcal{M}$, then $\mathrm{id} \colon ST^*\mathcal{M} \to ST^*\mathcal{M}$ belongs to $\Gamma_\Pi(\Pi) \cong \Omega^1(\Phi)$ and therefore corresponds [3] to a Π–semibasic graded 1–form Θ_0 on $ST^*\mathcal{M}$, which is the canonical graded form introduced previously. In analogy with ordinary Lagrangian mechanics, the Cartan graded

1–form associated to a given Lagrangian superfunction L in $ST\mathcal{A}$ is defined by

$$\Theta_L := dL \circ S, \tag{12}$$

where S is the vertical superendomorphism [4,8]. In local supercoordinates

$$\Theta_L = \Big(\frac{\partial L}{\partial v^i} - (-1)^{|L|}\frac{\partial L}{\partial \pi v^i}\Big)dq^i + \Big(\frac{\partial L}{\partial \pi\zeta^\alpha} - (-1)^{|L|}\frac{\partial L}{\partial \zeta^\alpha}\Big)d\theta^\alpha. \tag{13}$$

The Cartan graded 2–form is defined as the closed graded form $\Omega_L = -d\Theta_L$. On the other hand, if Y is a vertical supervector field with respect to $\mathcal{T}$ (i.e. $Y \circ \tau^* = 0$), then $\Theta_L(Y) = 0$, hence Θ_L is a $\mathcal{T}$–semibasic graded 1–form, and since $\mathcal{T}$ is a submersion, it has associated a unique graded 1–form $\widehat{\Theta}_L$ along $\mathcal{T}$ [3]. In terms of the basis $\{ d\hat{q}^i, d\hat{\theta}^\alpha \}$, $\widehat{\Theta}_L$ has the same coordinates as Θ_L corresponding to the elements $\{ dq^i, d\theta^\alpha \}$ (which is not a full basis of $\Omega^1\mathcal{A}(\mathcal{U})$), hence

$$\widehat{\Theta}_L = \Big(\frac{\partial L}{\partial v^i} - (-1)^{|L|}\frac{\partial L}{\partial \pi v^i}\Big)d\hat{q}^i + \Big(\frac{\partial L}{\partial \pi\zeta^\alpha} - (-1)^{|L|}\frac{\partial L}{\partial \zeta^\alpha}\Big)d\hat{\theta}^\alpha. \tag{14}$$

In analogy with non–graded geometry, see [6], the section $\mathcal{FL}: ST\mathcal{M} \to ST^*\mathcal{M}$ along $\mathcal{T}$ that corresponds to the graded 1–form $\widehat{\Theta}_L$ could be considered as the Legendre transformation; nevertheless, the matrix associated to Ω_L is of the form

$$\Omega_L = \begin{pmatrix} A_1 & A_2 & A_3 & B_1 & B_2 & B_3 \\ -A_2^t & 0 & 0 & B_4 & 0 & 0 \\ -A_3^t & 0 & 0 & B_5 & 0 & 0 \\ C_1 & C_4 & C_5 & D_1 & D_2 & D_3 \\ C_2 & 0 & 0 & D_2^t & 0 & 0 \\ C_3 & 0 & 0 & D_3^t & 0 & 0 \end{pmatrix}, \tag{15}$$

where $C_i = -(-1)^{|L|}B_i^t$, and therefore Ω_L will be degenerate for every superfunction $L \in ST\mathcal{M}$. Hence, we shall restrict our attention to the case when the super Lagrangian $L \in T\mathcal{M} \subset ST\mathcal{M}$, (i.e., when L does not depend on the variables $\pi v^i$and $\pi\theta^\alpha$) and consider the restriction of $\mathcal{FL}$ to $T\mathcal{M}$. In other words, given a super Lagrangian $L \in T\mathcal{M}$, the super Legendre transformation FL associated to L is the restriction of the map $\mathcal{FL}$ to $T\mathcal{M}$. In particular, $FL: T\mathcal{M} \longrightarrow ST^*\mathcal{M}$. When $L \in T\mathcal{M}$ the matrix of Ω_L reduces to

$$\Omega_L = \begin{pmatrix} A_1 & A_2 & B_1 & B_2 \\ -A_2^t & 0 & B_4 & 0 \\ C_1 & C_4 & D_1 & D_2 \\ C_2 & 0 & D_2^t & 0 \end{pmatrix}, \tag{16}$$

and to analyse its degeneracy it is necessary to consider the parity of L. If L is even then Ω_L is non–degenerate if and only if, the matrices A_2 and D_2 are invertible, in other words, exactly when

$$\frac{\partial^2 L}{\partial v^i \partial v^j} \quad \text{and} \quad \frac{\partial^2 L}{\partial \theta^\alpha \partial \theta^\beta} \quad \text{are invertible.} \tag{17}$$

We also notice that if $|L| = 0$ then FL takes values in $T^*\mathcal{M}$. In fact, locally $FL = (fl, fl^*)$ is determined by the morphism of superalgebras $fl^*: T^*\mathcal{A}(\pi^{-1}(\mathcal{U})) \to T\mathcal{A}(\tau^{-1}(\mathcal{U}))$ described by the relations:

$$q^i \mapsto q^i, \qquad \theta^\alpha \mapsto \theta^\alpha, \qquad p^i \mapsto \frac{\partial L}{\partial v^i}, \qquad \eta^\alpha \mapsto -\frac{\partial L}{\partial \zeta^\alpha}, \tag{18}$$

which, by the inverse function theorem [9], will be a local diffeomorphism when the Jacobian is invertible, and this happens exactly when (17) holds. On the other hand, if L is odd, Ω_L is non–degenerate, if and only if, the off diagonal terms are non–degenerate, this implies that $m = n$ and that B_2 is invertible. In other words, that

$$\frac{\partial^2 L}{\partial \theta^\alpha \partial v^j} \quad \text{is invertible.} \tag{19}$$

Unlike the even case, the super Legendre transformation does not take values in $T^*\mathcal{M}$, but on a subsupermanifold of $T^*\mathcal{M}$ of dimension $(m+n, n+m)$ obtained by imposing the conditions $p^i = 0 \quad 1 \le i \le m$ and $\eta^\alpha = 0 \quad 1 \le \alpha \le n$.

Moreover, locally FL is given by the assignments

$$q^i \mapsto q^i, \qquad \theta^\alpha \mapsto \theta^\alpha, \quad \pi\zeta^\alpha \mapsto \frac{\partial L}{\partial \zeta^\alpha}, \qquad \pi v^i \mapsto \frac{\partial L}{\partial v^i}; \tag{20}$$

nevertheless, when $m = n$, again by the inverse function theorem, FL is a local diffeomorphism exactly when (19) holds. We have, therefore, proved the following

Proposition 1. *The super Legendre transformation FL is a local diffeomorphism, if and only if, the graded form Ω_L is non–degenerate. In either case, we say that the super Lagrangian L is regular.*

The super Legendre transformation has the same properties as the usual Legendre transformation [1]:

Proposition 2. *Let L be a super Lagrangian in $ST\mathcal{A}$, then $\mathcal{FL}^*(\Theta_0) = \Theta_L$. Moreover, when $L \in T\mathcal{A}$ and one restricts Θ_L and Θ_0 to the appropriate subsupermanifols (for instance to $T\mathcal{M}$ and $T^*\mathcal{M}$ respectively, when $|L| = 0$) then also $FL^*(\Theta_0) = \Theta_L$. Moreover, when L is a regular super Lagrangian there exists a unique supervector field Γ_L in $\mathfrak{X}(\mathcal{M})$ such that $i_{\Gamma_L}\Omega_L = d\,E_L$, where the superenergy is defined by $E_L := \Delta L - L$ and Δ is the Liouville supervector field [8].*

Finally one has

Proposition 3. *Let L be a super Lagrangian in $T\mathcal{A}$ such that FL is a diffeomorphism (in such case we say L is hyperregular). Then $V = (FL^{-1})^* \circ \Gamma_L \circ FL^*$ is a Hamiltonian supervector field with a Hamiltonian given by $H := (FL^{-1})^* E_L$. Reciprocally, if H is the superfunction $H := (FL^{-1})^* E_L$, then the Hamiltonian supervector field V associated to L is FL–related to Γ_L.*

References

[1] R. Abraham and J.E. Marsden (1978), *Foundations of Mechanics*, Second Edition, Benjamin Cumming, Reading, Massachusetts.

[2] C.P. Boyer and O. A. Sánchez–Valenzuela (1991), 'Lie supergroup actions on supermanifolds', *Trans. Amer. Math. Soc.*, **323** 151–175.

[3] J.F. Cariñena and H. Figueroa (1994), 'A geometrical version of Noether's theorem in supermechanics', to appear in *Reports on Mathematical Physics.*

[4] J.F. Cariñena and H. Figueroa, 'Hamiltonian versus Lagrangian formulations of supermechanics', in preparation.

[5] J.F. Cariñena, C. López and E. Martínez (1989), 'A new approach to the converse of Noether's theorem', *J. Phys. A: Math. Gen.*, **22** 4777–87.

[6] J.F. Cariñena, C. López and E. Martínez (1991), 'Sections along a map applied to Higher Order Lagrangian Mechanics. Noether's theorem', *Acta Aplicandae Mathematicae*, **25** 127–151.

[7] J.F. Cariñena, J. Fernández–Núñez (1993), 'Geometric theory of time–dependent singular Lagrangians', *Fortschr. Phys.*, **41** (6) 517–552.

[8] L.A. Ibort and J. Marín-Solano (1993), 'Geometrical foundations of Lagrangian supermechanics and supersymmetry', *Reports on Mathematical Physics*, **32** (3) 385–409.

[9] D.A. Leĭtes (1980), 'Introduction to the theory of supermanifolds', *Russian Math. Surveys*, **35:1** (1980) 1–64.

[10] O.A. Sánchez–Valenzuela (1986), 'Differential–Geometric approach to supervector bundles', *Comunicaciones técnicas IIMASS–UNAM, (Serie Naranja)*, 457, México.

[11] O.A. Sánchez–Valenzuela (1988), 'On Grassmanian supermanifolds', *Trans. Amer. Math. Soc.*, **307** 597–614.

Field-antifield description of anomalous theories

Joaquim Gomis and Jordi París

1 Introduction

Anomalous gauge theories are characterized by the breakdown of its classical gauge structure due to quantum corrections. As a consequence, some classical pure gauge degrees of freedom become propagating at quantum level. According to the ideas of covariant quantization of gauge theories, it would be useful to develop an extended formalism describing in a covariant way this phenomenon by introducing extra degrees of freedom already at classical level [1]. In the field-antifield (FA) framework [2] [3] [4], this program has been implemented for irreducible theories with closed gauge algebras in [5] [6]. In what follows, after a brief account of the FA formalism, we review the basic ideas of this proposal.

2 The field-antifield formalism

The FA formalism [4] is a powerful method for the study of gauge theories (see [7] for some reviews). It encompasses previous developments for quantizing gauge theories based on BRST invariance [2] [3] and extends them to more complicated cases (open algebras,...). At the classical level it constitutes a general algorithm to construct a gauge-fixed action $S_\Sigma(\Phi)$ and its BRST transformation δ_Σ from a given classical gauge action $S_0(\phi)$ and its associated gauge structure. At the quantum level it provides the tools to study the quantum corrections to this classical BRST symmetry and its underlying structure, which are further used to analyze unitarity and renormalizability issues.

2.1 Classical aspects

Consider a classical action $S_0(\phi)$ invariant under the gauge transformations $\delta\phi^i = R^i_\alpha \varepsilon^\alpha$, $i = 1,\ldots,n$; $\alpha = 1,\ldots,m$. The FA formalism is built on a manifold $\mathcal{M}$, locally coordinated by a set of fields Φ^A, $A = 1,\ldots,N$, – including the original fields plus ghosts, antighosts, extra ghosts, etc.– and their associated antifields Φ^*_A, with opposite Grassmann parity. This set is often collectively denoted by $z^a = \{\Phi^A, \Phi^*_A\}$, $a = 1, \ldots, 2N$. Afterwards, $\mathcal{M}$ is endowed with an odd symplectic structure, the *antibracket* $(\cdot, \cdot)$, defined as

$$(X,Y) = \frac{\partial_r X}{\partial z^a}\zeta^{ab}\frac{\partial_l Y}{\partial z^b}, \qquad \text{where } \zeta^{ab} \equiv (z^a, z^b) = \begin{pmatrix} 0 & \delta^A_B \\ -\delta^A_B & 0 \end{pmatrix}.$$

At the classical level, the basic object is an action $S(z)$ verifying the *classical master equation*

$$(S,S) = 0, \tag{2.1}$$

and subject to the boundary conditions: 1) Classical limit: $S(\Phi,\Phi^*)|_{\Phi^*=0} = S_0(\phi)$, and 2) Properness condition: $\text{rank}\left(\frac{\partial_l\partial_r S}{\partial z^a \partial z^b}\right)\big|_{\text{on-shell}} = N$, where on-shell means on the surface $\sigma = \left\{\frac{\partial_r S}{\partial z^a} = 0\right\}$. Such an S is called *proper solution.*

The expansion of S in antifields of the original basis z^a

$$S(\Phi,\Phi^*) = S_0(\phi) + \Phi^*_A R^A(\Phi) + \frac{1}{2}\Phi^*_A\Phi^*_B R^{BA}(\Phi) + \ldots, \tag{2.2}$$

generates the structure functions of the classical gauge algebra [8]. Besides, fulfillment of (2.1) provides the relations defining its structure [9]. For this reason, it is sensible to call z^a *classical basis* [10] [11] [6].

The gauge-fixed theory is better analyzed in terms of the *gauge-fixed basis* [10] [11] [6], defined by means of a suitable gauge-fixing fermion $\Psi(\Phi)$ by [4]

$$\Phi^A \to \Phi^A, \qquad \Phi^*_A \to K_A + \frac{\partial\Psi(\Phi)}{\partial\Phi^A} \equiv K_A + \Psi_A. \tag{2.3}$$

In such a basis, (2.2) adopts the form

$$\hat{S}(\Phi,K) \equiv S\left(\Phi, \Phi^* = K + \frac{\partial\Psi(\Phi)}{\partial\Phi}\right) =$$

$$\left(S_0(\phi) + \Psi_A R^A + \frac{1}{2}\Psi_A\Psi_B R^{BA} + \ldots\right) + K_A\left(R^A + \Psi_B R^{BA} + \ldots\right) + \ldots$$

Its antifield independent part is the gauge-fixed action, $S_\Sigma(\Phi)$. From this, the following boundary conditions characterizing gauge-fixed basis are inferred: 1') Gauge-fixed limit: $\hat{S}(\Phi,K)\Big|_{K=0} = S_\Sigma(\Phi)$, 2') $\text{rank}\left(\frac{\partial_l\partial_r S_\Sigma(\Phi)}{\partial\Phi^A\partial\Phi^B}\right)\Big|_\sigma = N$, so that, if Ψ in (2.3) is properly chosen, propagators are well defined and the usual perturbation theory can be developed.

On the other hand, the linear part in K_A of $\hat{S}$ is the gauge-fixed BRST transformation, $\delta_\Sigma\Phi^A = (\Phi^A, \hat{S})\Big|_{K=0}$, the quadratic part contains the on-shell nilpotency structure functions and so on. We can write

$$\hat{S}(\Phi,K) = S_\Sigma(\Phi) + K_A\tilde{R}^A(\Phi) + \frac{1}{2}K_A K_B\tilde{R}^{BA}(\Phi) + \ldots, \tag{2.4}$$

so that $\hat{S}$ appears now as the generating functional of the structure functions which define the BRST symmetry, while relations derived from $(\hat{S},\hat{S}) = 0$ characterize the structure of this classical BRST symmetry.

2.2 Quantum aspects

The quantization process may spoil the classical BRST structure due to quantum corrections acquired by the BRST transformations and the higher order structure functions in (2.4). This violation indicates the presence of anomalies. The quantum aspects of the FA formalism are most suitable studied in terms of the effective action Γ associated through the usual Legendre transformation with the generating functional

$$Z(J,K) = \int \mathcal{D}\Phi \exp\left\{\frac{i}{\hbar}\left[W(\Phi,K) + J_A\Phi^A\right]\right\}, \tag{2.5}$$

with $W(\Phi,K) \equiv \hat{S}(\Phi,K) + \sum_{p=1}^{\infty} \hbar^p M_p(\Phi,K)$, and where the *local* counterterms M_p should guarantee the finiteness of the theory while preserving (as far as possible) the BRST structure at quantum level. The functional Γ is the quantum analog of $\hat{S}$ (2.4), i.e., the coefficients in its K_A expansion

$$\Gamma(\Phi,K) = \Gamma(\Phi) + K_A\Gamma^A(\Phi) + \frac{1}{2}K_AK_B\Gamma^{BA}(\Phi) + \ldots,$$

are interpreted as the quantum counterparts of the corresponding coefficients of $\hat{S}$: $\Gamma(\Phi)$ is the 1PI generating functional for the basic fields; $\Gamma^A(\Phi)$ are their quantum BRST transformations; $\Gamma^{AB}(\Phi)$ the quantum on-shell nilpotency structure functions, etc. $\Gamma(\Phi,K)$ appears thus as the generating functional of the structure functions characterizing the quantum BRST symmetry.

The quantum BRST structure and its possible breakdown are reflected in the (anomalous) BRST Ward identity

$$\frac{1}{2}(\Gamma,\Gamma) = -i\hbar(\mathcal{A}\cdot\Gamma), \qquad \mathcal{A} \equiv \left[\Delta W + \frac{i}{2\hbar}(W,W)\right](\Phi,K), \tag{2.6}$$

where the obstruction $(\mathcal{A}\cdot\Gamma)$ denotes the generating functional of the 1PI diagrams with the insertion of $\mathcal{A}$ and Δ is defined by $\Delta \equiv (-1)^A \frac{\partial_l}{\partial\Phi^A}\frac{\partial_l}{\partial\Phi^*_A}$. $\mathcal{A}$ in (2.6) parametrizes thus potential departures from the classical BRST structure due to quantum corrections.

Quantum BRST invariance thus holds if $\mathcal{A}$ in (2.6) vanishes, i.e., upon fulfillment through a local object W of the *quantum master equation* [4]

$$\frac{1}{2}(W,W) - i\hbar\Delta W = 0,$$

which encodes at once the classical master equation (2.1) plus a set of recurrent equations for the counterterms M_p

$$\begin{aligned}
(M_1,\hat{S}) &= i\Delta\hat{S}, \qquad\qquad (2.7)\\
(M_p,\hat{S}) &= i\Delta M_{p-1} - \frac{1}{2}\sum_{q=1}^{p-1}(M_q,M_{p-q}), \qquad p \geq 2.
\end{aligned}$$

2.3 Pauli-Villars regularization and anomalies

The previous computations are only meaningful once a regularization scheme is provided. In what follows, we restrict ourselves to one-loop level and use the Pauli-Villars procedure (PV) considered in [12] [13] [10] [14]. In this proposal (2.5) is substituted by the one-loop regularized expression

$$Z_{\rm reg}(J,K)=\int\mathcal{D}\Phi\mathcal{D}\chi\exp\left\{\frac{i}{\hbar}\left[\hat{S}+\hbar M_1+S_{\rm PV}|_{\chi^*=0}+J_A\Phi^A\right]\right\},\qquad(2.8)$$

where the PV fields χ^A, introduced for each field Φ^A, have the same statistics as their partners, but with the path integral formally defined so that an extra minus sign occurs in front of their loops. Each PV field χ^A comes with its antifield χ^*_A, which is put to zero at the end of the regularization.

The regulating action $S_{\rm PV}|_{\chi^*=0}$ in (2.8) is [12] [13] [10]

$$S_{\rm PV}|_{\chi^*=0}=S^{(0)}_{\rm PV}\Big|_{\chi^*=0}+S_M=\frac{1}{2}\chi^A(TR)_{AB}\chi^B-\frac{1}{2}M\chi^AT_{AB}\chi^B,\qquad(2.9)$$

with the mass matrix T_{AB} arbitrary but invertible and

$$(TR)_{AB}=\left(\frac{\partial_l}{\partial\Phi^A}\frac{\partial_r}{\partial\Phi^B}\hat{S}(\Phi,K)\right).\qquad(2.10)$$

The semiclassical approximation yields the one-loop effective action

$$\Gamma_1(\Phi,K)=M_1(\Phi,K)+\frac{i}{2}{\rm Tr}\ln\left[\frac{(TR)}{(TR)-TM}\right],$$

and its BRST variation, $\delta\Gamma_1=(\Gamma_1,\hat{S})$, by use of (2.6), the regularized expression of $\Delta\hat{S}$ [6]

$$(\Delta\hat{S})_{\rm reg}=\delta\left\{-\frac{1}{2}{\rm Tr}\ln\left[\frac{R}{R-M}\right]\right\}={\rm Tr}\left[-\frac{1}{2}(R^{-1}\delta R)\frac{1}{(1-R/M)}\right],\qquad(2.11)$$

which verifies the Wess-Zumino consistency condition [15] $\delta(\Delta\hat{S})_{\rm reg}=0$.

Assuming now for R an expansion in antifields of the type $R(\Phi,K)=R_0(\Phi_{\rm m})+K_AR^A+\ldots$, with $R_0(\Phi_{\rm m})$ invertible and $\{\Phi_{\rm m}\}$ the minimal sector of fields, (2.11) acquires the form

$$(\Delta\hat{S})_{\rm reg}=\mathcal{B}(\Phi_{\rm m})+\mathcal{O}(K)+(\mathcal{M},\hat{S}),\qquad(2.12)$$

with $\mathcal{M}=K_AP^A(\Phi)+\ldots$, $P^A(\Phi)$ local coefficients and

$$\mathcal{B}(\Phi_{\rm m})=\delta_\Sigma\left\{-\frac{1}{2}{\rm Tr}\ln\left[\frac{R_0}{R_0-M}\right]\right\}.\qquad(2.13)$$

When dealing with "closed" theories, i.e., theories for which the proper solution S (2.2) is at most linear in the antifields, $\delta_\Sigma\Phi = \delta\Phi$ and $\delta_\Sigma^2 = 0$. In this case, $\delta[\mathcal{B}(\Phi_{\rm m})] = 0$ and $\delta[\mathcal{O}(K)] = 0$ *separately.* This implies that i) the antifield independent part of (2.12) can be studied separately from $\mathcal{O}(K)$ in a consistent way and ii) anomalies with nontrivial antifield dependence as those described in [16] are not expected to appear, due to the off-shell BRST invariance of $\mathcal{B}(\Phi_{\rm m})$ in (2.12) [6].

3 Extended field-antifield formalism

3.1 Generalities

Consider now an irreducible theory with closed algebra and restrict ourselves to the study of antifield independent aspects. The regulator R can then be written as $R(\Phi, K) = \mathcal{R}(\phi) + \hat{R}(\Phi, K)$, yielding for (2.13)

$$\mathcal{B}(\Phi_{\rm m}) = \mathrm{Tr}\left[-\frac{1}{2}(\mathcal{R}^{-1}\delta\mathcal{R})\frac{1}{(1-\mathcal{R}/M)}\right] \equiv A_\alpha(\phi)\mathcal{C}^\alpha, \tag{3.1}$$

with $\mathcal{C}^\alpha$ the usual Faddeev-Popov ghosts. Assume now that no local $M_1(\phi)$ exists satisfying the antifield independent part of (2.7), i.e.,

$$(M_1, \hat{S}) = iA_\alpha(\phi)\mathcal{C}^\alpha = ia_k A_\alpha^k(\phi)\mathcal{C}^\alpha, \tag{3.2}$$

with $\{A_\alpha^k(\phi)\}$ a basis of BRST nontrivial cocycles at ghost number one and where we assume that $A_\alpha(\phi)$ stand only for the finite pieces arising in (3.1).

The rank of the functional derivatives of the anomalies $A_\alpha(\phi)$

$$\mathrm{rank}\left(\frac{\partial A_\alpha(\phi)}{\partial\phi^i}\right) = r(\leq m), \qquad \alpha = 1, \ldots, m,$$

determines the number of anomalous gauge transformations or, equivalently, the number of pure gauge degrees of freedom which become propagating at quantum level.

Let us now illustrate the general ideas of the extended formalism [5] [6] to quantize anomalous theories in the case $r = m$. The proposal consists in introducing m new fields θ^α, and demand that their gauge transformations, $\delta\theta^\alpha = -\tilde{\mu}_\beta^\alpha(\theta, \phi)\varepsilon^\beta$, are such that $\phi^a = (\phi^i, \theta^\alpha)$[1] be still a representation of the original gauge structure. Denoting the generators as $R_\alpha^a = (R_\alpha^i, -\tilde{\mu}_\alpha^\beta)$, this requirement amounts to

$$R_{\alpha,b}^a R_\beta^b - R_{\beta,b}^a R_\alpha^b = R_\gamma^a T_{\alpha\beta}^\gamma(\phi), \tag{3.3}$$

[1] For simplicity, we will consider only bosonic fields and bosonic gauge transformations, i.e., $\epsilon(\phi^i) = 0$, $\epsilon(\varepsilon^\alpha) = 0$.

where $T^{\gamma}_{\alpha\beta}(\phi)$ are the structure functions of the original gauge algebra. The commutation relations (3.3) are verified for $\tilde{\mu}^{\alpha}_{\beta}(\theta,\phi) = \left.\frac{\partial\phi^{\alpha}(\theta',\theta;\phi)}{\partial\theta'^{\beta}}\right|_{\theta'=0}$, where $\phi^{\alpha}(\theta',\theta;\phi)$ are the composition functions of the gauge (quasi)group [17].

This extension increases the classical physical content of the extended theory. Indeed, the finite gauge transformations of the classical fields with parameters θ^{α}, $F^{i}(\phi,\theta)$, are gauge invariant, $\delta F^{i}(\phi,\theta) = 0$, and can thus be considered n classical gauge invariant degrees of freedom of the extended theory. This extension procedure is thus prepared to describe already at classical level the (quantum) appearence of new degrees of freedom.

In the Field-Antifield framework all these facts are described by the (non-proper) solution of (2.1) in the extended space

$$\tilde{S} = S - \theta^{*}_{\alpha}\tilde{\mu}^{\alpha}_{\beta}C^{\beta} \equiv S + S_{\theta}, \tag{3.4}$$

with S the original proper solution and θ^{*}_{α} the antifields associated with θ^{α}.

Consider now an extension of W at one loop: $\tilde{W} = \hat{\tilde{S}} + \hbar\tilde{M}_1$. Since pure gauge degrees of freedom become propagating at quantum level, the classical part of $\tilde{W}$, $\hat{\tilde{S}}$, should describe only the propagation of the original fields Φ^{A}, while propagation of θ^{α} should be provided by $\tilde{M}_1$. In other words, using the collective notation $\Phi^{p} = \{\Phi^{A},\theta^{\alpha}\}$, $\hat{\tilde{S}}$ and $\tilde{W}$ should verify

$$\text{rank}\left(\frac{\partial_l\partial_r\hat{\tilde{S}}}{\partial\Phi^{p}\partial\Phi^{q}}\right)\Bigg|_{\text{on-shell}} = N, \qquad \text{rank}\left(\frac{\partial_l\partial_r\tilde{W}}{\partial\Phi^{p}\partial\Phi^{q}}\right)\Bigg|_{\text{on-shell}} = N + m. \tag{3.5}$$

A convenient solution $\tilde{S}$ of (2.1) in the extended space is (3.4). Then, $\Delta\hat{\tilde{S}}$ to be used in (2.7) reads (formally)[2] $\Delta\hat{\tilde{S}} = \Delta\hat{S} - \tilde{\mu}^{\beta}_{\alpha,\beta}C^{\alpha} = \Delta\hat{S} + \Delta S_{\theta}$, i.e., the new degrees of freedom modifies $\Delta\hat{S}$. The new contribution is the unregularized logarithm of the jacobian of the BRST transformation for the θ fields. The regularization should thus be adapted to the extended theory. However, this can not be done in a direct way, because (3.5) implies that a "kinetic term" for θ^{α} is lacking in $\hat{\tilde{S}}(\Phi,K;\theta,\theta^{*})$.

This drawback can by bypassed by taking $\tilde{W} = [\hat{\tilde{S}} + \hbar M_1] + \hbar M'$, where

$$M_1(\phi,\theta) = -i\int_0^1 A_{\beta}(F(\phi,\theta t))\lambda^{\beta}_{\alpha}(\theta t,\phi)\theta^{\alpha}\mathrm{d}t, \qquad \lambda^{\alpha}_{\beta} = (\mu^{\alpha}_{\beta})^{-1}, \tag{3.6}$$

is the solution of eq.(3.2) in the extended space or Wess-Zumino term and M' a suitable counterterm taking care, if necessary, of possible antifield dependences. Indeed, $\tilde{W}_{pq}$ in (3.5) contains basically the original hessian $\hat{S}_{AB}$

[2]The gauge-fixed basis we consider in the extended theory come from gauge-fixing fermions of the type $\Psi(\Phi)$. Therefore, neither θ nor θ^{*} change under (2.3).

plus a nondiagonal block, which essentially reads

$$\left(\frac{\partial^2 M_1(\phi,\theta)}{\partial\phi^i\partial\theta^\alpha}\right) = -i\left(\frac{\partial A_\alpha}{\partial\phi^i}\right)(\phi) + \mathcal{O}(\theta),$$

giving the correct rank (3.5) for $\tilde{W}_{pq}$. It seems thus plausible to consider as starting point for PV regularization of the extended theory

$$S' = [\hat{\tilde{S}} + \hbar M_1]. \tag{3.7}$$

However, this leads to new problems: i) S' (3.7) does not verify the classical master equation, and ii) the part providing for the propagation of the θ fields in (3.7) contains $\hbar$. This ruins the $\hbar$ perturbative expansion and the tool to recognize one-loop anomalies.

3.2 Extended proper solution: background terms

A sensible PV regularization of the extended theory requires to extract from S' (3.7) a proper solution W_0. To do that, expand the θ^α and θ^*_α dependent parts of S' (3.7) in powers of θ^α in the canonical parametrization ($\lambda^\alpha_\beta\theta^\beta = \theta^\alpha$) and perform the canonical transformation $\theta'^\alpha = \sqrt{\hbar}\,\theta^\alpha$, $\theta'^*_\alpha = \frac{1}{\sqrt{\hbar}}\theta^*_\alpha$ [18]. In the resulting expressions

$$\hbar M_1(\phi,\theta) \;\rightarrow\; -i\left[\sqrt{\hbar}A_\alpha(\phi)\theta^\alpha + \frac{1}{2}\theta^\alpha\left(\frac{\partial A_\alpha}{\partial\phi^i}R^i_\beta\right)\theta^\beta + \mathcal{O}(\theta^3;1/\sqrt{\hbar})\right] \tag{3.8}$$

$$-\theta^*_\alpha\tilde{\mu}^\alpha_\beta C^\beta \;\rightarrow\; -\sqrt{\hbar}\theta^*_\alpha C^\alpha + \frac{1}{2}\theta^*_\alpha T^\alpha_{\beta\gamma}(\phi)\theta^\gamma C^\beta + \mathcal{O}(\theta^2;1/\sqrt{\hbar}), \tag{3.9}$$

$\hbar$ disappears in few terms or becomes $\sqrt{\hbar}$, but in higher order terms it appears in negative powers of $\sqrt{\hbar}$. Therefore, it seems as if the quantum treatment of Wess-Zumino terms can only be done non-perturbatively.

However, this perturbative treatment can at least be applied in a sensible way to models for which only the first two terms in (3.8), (3.9) are really present. These systems are characterized by the conditions

$$\Gamma_\gamma(D_{\alpha\beta})(\phi) = 0, \quad T^\alpha_{\beta\gamma}(\phi) = 0, \quad \text{with} \quad \Gamma_\gamma \equiv R^i_\gamma\frac{\partial}{\partial\phi^i}, \quad D_{\alpha\beta} = \Gamma_\beta A_\alpha,$$

i.e., the original theory is abelian. For these systems, S' (3.7) adopts the form

$$S' \rightarrow \left[\hat{S}(\Phi,K) - \frac{i}{2}\theta^\alpha D_{\alpha\beta}(\phi)\theta^\beta\right] - \sqrt{\hbar}\,[\theta^*_\alpha C^\alpha + iA_\alpha(\phi)\theta^\alpha] \equiv W_0 + \sqrt{\hbar}M_{1/2}, \tag{3.10}$$

whereas the content of the quantum master equation $1/2(S',S') = iA_\alpha C^\alpha$ is summarized now in the relations

$$(W_0,W_0) = 0, \qquad (W_0,M_{1/2}) = 0,$$
$$\frac{1}{2}(M_{1/2},M_{1/2}) = iA_\alpha(\phi)C^\alpha.$$

Therefore, W_0 appears as the searched-for proper solution in the extended space, while the $\sqrt{\hbar}$ terms in (3.10) generalize the well-known concept of background charges and can be called in this context *background terms.*

All these results can be generalized along the same lines to gauge theories with an abelian anomalous subgroup [6] (e.g., the bosonic string).

3.3 Regularization of the extended theory

The regulated functional of the extended theory is obtained now by substituting the exponent in (2.8) by

$$\tilde{W}_0 + W_{\rm PV}|_{\tilde{\chi}^*=0} + \sqrt{\hbar}\tilde{M}_{1/2} + \hbar\tilde{M} + J_A\Phi^A + j_a\theta^a,$$

where $\tilde{W}_0$ and $\tilde{M}_{1/2}$ are the analogs of (3.10) with the coefficients a_k (3.2) of the anomalies left undetermined. $W_{\rm PV}|_{\tilde{\chi}^*=0}$ is the PV action (2.9) adapted to the extended theory and $\tilde{M}$ a suitable extension of M' in $\tilde{W}$.

For the extended theory, the effective action up to one loop reads

$$\tilde{\Gamma} = \tilde{W}_0 + \sqrt{\hbar}\tilde{M}_{1/2} + \hbar\tilde{M} + \frac{i\hbar}{2}{\rm Tr}\ln\left[\frac{(\tilde{T}\tilde{R})}{(\tilde{T}\tilde{R}) - \tilde{T}M}\right],$$

where $(\tilde{T}\tilde{R})$ and $\tilde{T}$ are the analogs of (2.10) and the mass matrix for the extended theory. Now, from (2.6) adapted to the extended theory, absence of one-loop anomalies is acquired if

$$(\tilde{M}_{1/2}, \tilde{W}_0) = 0, \qquad (\Delta\hat{\tilde{S}})_{\rm reg} + i(\tilde{M}, \tilde{W}_0) + \frac{i}{2}(\tilde{M}_{1/2}, \tilde{M}_{1/2}) = 0, \qquad (3.11)$$

with $(\Delta\hat{\tilde{S}})_{\rm reg}$ the analog of (2.11) for the extended theory. Its antifield independent part receives two contributions, the original one (3.2) plus a new piece coming from the extra variables, $B_\alpha(\phi)\mathcal{C}^\alpha = b_k A^k_\alpha(\phi)\mathcal{C}^\alpha$. Therefore, in the antifield independent sector, the second equation in (3.11) is satisfied for

$$\tilde{M}_{1/2} = -\left[\theta^*_\alpha\mathcal{C}^\alpha + i(A_\alpha + B_\alpha)(\phi)\theta^\alpha\right] = -\left[\theta^*_\alpha\mathcal{C}^\alpha + i(a_k + b_k)A^k_\alpha(\phi)\theta^\alpha\right],$$

while the first equation in (3.11) requires $\tilde{W}_0$ to be

$$\tilde{W}_0 = \hat{S}(\Phi, K) - \frac{i}{2}\theta^\alpha\left[(a_k + b_k)D^k_{\alpha\beta}(\phi)\right]\theta^\beta, \qquad D^k_{\alpha\beta}(\phi) \equiv \left(\frac{\partial A^k_\alpha}{\partial\phi^i}R^i_\beta\right).$$

Therefore, upon imposition of absence of one-loop anomalies, the new variables produce a renormalization of the coefficients a_k of the original anomalies (3.2) or, equivalently, of the coefficients of the Wess-Zumino term (3.6). In summary, up to antifield dependent issues not treated here, the extended formalism is able to describe covariantly an anomalous gauge theory in terms of a BRST invariant one with extra degrees of freedom. Application of this method to the bosonic string [6] leads to the usual Liouville action and the well-known shift $(26 - D) \rightarrow (25 - D)$.

References

[1] L. D. Faddeev and S. L. Shatashvili (1986), *Phys. Lett.* **167B** 225.

[2] C. Becchi, A. Rouet and R. Stora (1974), *Phys. Lett.* **52B** 344; (1975), *Comm. Math. Phys.* **42** 127; (1976), *Ann. Phys.* **98** 287.
I. V. Tyutin, Lebedev preprint n° 39 (1975), unpublished.

[3] J. Zinn-Justin (1975), 'Renormalization of Gauge Theories', in *Trends in Elementary Particle Theory*, H. Rollnik and K. Dietz (eds.), Lecture Notes in Physics, Vol **37**, Springer-Verlag, Berlin.

[4] I. A. Batalin and G. A. Vilkovisky (1981), *Phys. Lett.* **B102** 27; (1983),*Phys. Rev.* **D28**, 2567; (1984)(E) **D30** 508.

[5] J. Gomis and J. París (1993), *Nucl.Phys.* **B395** 288; 'Anomalous gauge theories within BV framework', to appear in Proc. of the EPS Conference on High Energy Physics HEP93, Marseille, July 22-28, 1993.

[6] J. Gomis and J. París (1993), 'Anomalies and Wess-Zumino terms in an extended, regularized Field-Antifield formalism', preprint KUL-TF-93/50, UB-ECM-PF 93/14, UTTG-16-93.

[7] M. Henneaux and C. Teitelboim (1992), *Quantization of gauge systems*, Princeton University Press, Princeton;
J. Gomis, J. París and S. Samuel (1994), 'Antibracket, Antifields and Gauge-Theory Quantization', preprint CCNY-HEP-94/03, UB-ECM-PF 94/15, UTTG-11-94;
W. Troost and A. Van Proeyen, *An introduction to Batalin-Vilkovisky Lagrangian quantization*, Leuven Notes in Math. Theor. Phys., in preparation.

[8] B. de Wit and J. W. van Holten (1979), *Phys. Lett.* **B79** 389.

[9] I. A. Batalin and G. A. Vilkovisky (1985), *J. Math. Phys.* **26** 172;
J. M. L. Fisch and M. Henneaux (1990), *Commun. Math. Phys.* **128** 627;
M. Henneaux (1990), *Nucl.Phys.* **B** (Proc.Suppl.) **18A** 47;
J. M. L. Fisch (1990), *On the Batalin–Vilkovisky antibracket–antifield BRST formalism and its applications*, Ph. D. thesis, preprint ULB TH2/90-01.

[10] W. Troost and A. Van Proeyen (1993), 'Regularization and the BV formalism', preprint KUL-TF-93/32. To appear in Proc. of the Conference *String's 93*, Berkeley, May 1993.

[11] S. Vandoren and A. Van Proeyen (1994), *Nucl.Phys.* **B411** 257.

[12] W. Troost, P. van Nieuwenhuizen and A. Van Proeyen (1990), *Nucl.Phys.* **B333** 727.

[13] A. Van Proeyen (1991), in Proc. of the Conference *Strings & Symmetries 1991*, Stony Brook, May 20-25, 1991, eds. N. Berkovits et al. World Sc. Publ. Co., Singapore, 1992, pp.338.

[14] F. de Jonghe (1994), *The Batalin-Vilkovisky Lagrangian quantisation scheme with applications to the study of anomalies in gauge theories*, Ph. D. thesis, K. U. Leuven; hep-th/9403143.

[15] J. Wess and B. Zumino (1971), *Phys. Lett.* **B37** 95.

[16] F. Brandt (1994), *Phys. Lett.* **B320** 57;
G. Barnich and M. Henneaux (1994), *Phys.Rev.Lett.* **72** 1588.

[17] I. A. Batalin (1981), *J. Math. Phys.* **22** 1837.

[18] F. De Jonghe, R. Siebelink and W. Troost (1993), *Phys.Lett.* **B306** 295.

Transfer matrix quantization of general relativity, and the problem of time

A. Carlini and J. Greensite

The "Problem of Time" in time-parametrized theories appears at the quantum level in various ways, depending on the quantization procedure. In (Dirac) canonical quantization, the usual Schrodinger evolution equation is replaced by a constraint on the physical states $H\Psi[q] = 0$; such states depend on the generalized coordinates $\{q^a\}$, but not on any additional time evolution parameter t. In path-integral formulation, one integrates over all paths, extending over all possible proper time lapses, which connect the initial q_0^a and final q_f^a configurations. The resulting Green's function $G[q_f, q_0]$, like the physical state wavefunctions $\Psi[q^a]$, has no dependence on an extra time parameter t. In certain theories, e.g. parametrized non-relativistic quantum mechanics, or the case of a relativistic particle moving in flat Minkowski space, it is possible to identify one of the generalized coordinates as the time variable, and to associate with that variable a conserved and positive-definite probability measure. In other theories, such as a relativistic particle moving in an arbitrary curved background spacetime, or in the case of quantum gravity, it has proven very difficult to identify an appropriate evolution parameter, and a unique, positive, and conserved probablility measure. This is the "Problem of Time", reviewed in ref. [1].

Our proposal for resolving this problem begins with the rather trivial observation that, at the classical level, there is no difference between the action S and the action $S' = \text{const.} \times S$; these actions obviously have the same Euler-Lagrange equations. Consider, e.g., the action for a relativistic particle, the action of the Nambu string, and the Einstein-Hilbert action

$$\begin{aligned}
S_p &= m \int d\tau \sqrt{-g_{\mu\nu} \frac{dx^\mu}{d\tau} \frac{dx^\nu}{d\tau}} \\
S_N &= T \int d^2\sigma \sqrt{-\det\left[\eta_{\mu\nu} \frac{\partial x^\mu}{\partial \sigma^i} \frac{\partial x^\nu}{\partial \sigma^j}\right]} \\
S_{EH} &= \frac{1}{\kappa^2} \int d^4x \, \sqrt{g}(-R)
\end{aligned} \tag{1}$$

Clearly, the mass m, the string tension T, and Newton's constant $G_N = \kappa^2/16\pi$, which appear in the Hamiltonian constraints of the particle, string, and gravity theories respectively, do *not* appear in the corresponding Euler-Lagrange equations. These parameters are therefore classically irrelevant, in the sense that, e.g., the mass of a particle cannot be determined from its trajectory in free fall; neither can Newton's constant be determined from solutions of the vacuum Einstein equations $R_{\mu\nu} = 0$.

We may generalize this observation to any time-parametrized theory. Consider, as a first example, a theory with D degrees of freedom

$$\begin{aligned} S &= \int dt(p_a \frac{dq^a}{dt} - N\mathcal{H}) \\ \mathcal{H} &= \frac{1}{2m} G^{ab} p_a p_b + mV(q) \end{aligned} \tag{2}$$

where the "supermetric" G_{ab} has Lorentzian signature $\{-++...+\}$. By rescaling $p_a \rightarrow p_a/\sqrt{\mathcal{E}}$, this can be rewritten

$$S = \frac{1}{\sqrt{\mathcal{E}}} S^{\mathcal{E}} \tag{3}$$

where

$$\begin{aligned} S^{\mathcal{E}} &= \int dt\, (p_a \frac{dq^a}{dt} - N\mathcal{H}^{\mathcal{E}}) \\ \mathcal{H}^{\mathcal{E}} &= \frac{1}{2m\sqrt{\mathcal{E}}} G^{ab} p_a p_b + \sqrt{\mathcal{E}} mV \end{aligned} \tag{4}$$

Since we have not transformed the coordinates q^a, it is clear that if S is stationary at $q^a(t)$, then $S^{\mathcal{E}}$ is also stationary at this trajectory. It is in this sense that the parameter $\mathcal{E}$ is indeterminate at the classical level, since the set of all possible trajectories in configuration space generated by the Hamilton equations

$$\mathcal{H}^{\mathcal{E}} = 0 \qquad \frac{\partial q^a}{\partial t} = N\frac{\partial \mathcal{H}^{\mathcal{E}}}{\partial p_a} \qquad \frac{\partial p_a}{\partial t} = -N\frac{\partial \mathcal{H}^{\mathcal{E}}}{\partial q^a} \tag{5}$$

will be independent of $\mathcal{E}$, and therefore $\mathcal{E}$ cannot be determined from the classical trajectories $q^a_{cl}(t)$ in configuration space.

Now let us rewrite the $\mathcal{H}^{\mathcal{E}} = 0$ constraint as

$$-\mathcal{E} = \text{Æ}[p,q] \equiv \frac{\frac{1}{2m} G^{ab} p_a p_b}{mV} \tag{6}$$

and consider the set of equations generated by treating Æ as a though it were a Hamiltonian, i.e.

$$\text{Æ} = -\mathcal{E} \qquad \frac{\partial q^a}{\partial \tau} = \frac{\partial \text{Æ}}{\partial p_a} \qquad \frac{\partial p_a}{\partial \tau} = -\frac{\partial \text{Æ}}{\partial q^a} \tag{7}$$

One immediately finds that

$$\frac{\partial q^a}{\partial \tau} = \frac{\sqrt{\mathcal{E}}}{mV}\frac{\partial}{\partial p_a}\mathcal{H}^{\mathcal{E}} \quad \text{and} \quad \frac{\partial p_a}{\partial \tau} == -\frac{\sqrt{\mathcal{E}}}{mV}\frac{\partial}{\partial q^a}\mathcal{H}^{\mathcal{E}} \tag{8}$$

Suppose $\overline{q}(\tau), \overline{p}(\tau)$ is some particular solution of (8), and $Æ = -\mathcal{E}$. Then, choosing $N(\tau) = V^{-1}[\overline{q}(\tau)]$ and rescaling τ by $m/\sqrt{\mathcal{E}}$, we recover the standard Hamilton equations

$$\mathcal{H}^{\mathcal{E}} = 0 \qquad \frac{\partial q^a}{\partial t} = N\frac{\partial \mathcal{H}^{\mathcal{E}}}{\partial p_a} \qquad \frac{\partial p_a}{\partial t} = -N\frac{\partial \mathcal{H}^{\mathcal{E}}}{\partial q^a} \tag{9}$$

This means that the Poisson bracket equation

$$\partial_\tau O = \{O, Æ\} \tag{10}$$

supplemented by $Æ = -\mathcal{E}$ generates classical trajectories which are equivalent, up to a time reparametrization, to those generated by

$$\partial_t O = \{O, N\mathcal{H}^{\mathcal{E}}\} \tag{11}$$

supplemented by $\mathcal{H}^{\mathcal{E}} = 0$.

In the case of a diffeomorphism-invariant field theory involving fields $q^a(x)$, we may write the action as

$$\begin{aligned} S &= \int d^4x[p_a\frac{\partial q^a}{\partial t} - N\mathcal{H}_x - N_i\mathcal{H}^i_x] \\ \mathcal{H}_x &= \kappa^2 G^{ab}p_a p_b + \sqrt{g}U(q) \end{aligned} \tag{12}$$

where the $q^a(x)$ are the set of all fields, gravitational and non-gravitational, G_{ab} is the supermetric, N and N_i the lapse and shift functions, and $\mathcal{H}^i_x$ are the supermomenta, linear in the canonical momenta $p_a(x)$. It is convenient to rescale all non-gravitational fields by an appropriate power of κ so that all $q^a(x)$, and all G^{ab}, are dimensionless. Again rescaling p_a by $\mathcal{E}^{-1/2}$, we find that S is equivalent at the classical level to

$$\begin{aligned} S^{\mathcal{E}} &= \int d^4x[p_a\frac{\partial q^a}{\partial t} - N\mathcal{H}^{\mathcal{E}}_x - N_i\mathcal{H}^i_x] \\ \mathcal{H}^{\mathcal{E}}_x &= \frac{\kappa^2}{\sqrt{\mathcal{E}}}G^{ab}p_a p_b + \sqrt{\mathcal{E}}\sqrt{g}U(q) \end{aligned} \tag{13}$$

Then the Hamiltonian and supermomentum constraints $\mathcal{H}^{\mathcal{E}}_x = \mathcal{H}^i_x = 0$ are readily seen to be equivalent to the requirement that $Æ = -\mathcal{E}$, for arbitrary choice of lapse/shift, where

$$\begin{aligned} Æ[p, q, N, N_i] &\equiv \int d^3x \left[\frac{N\kappa^2 G^{ab}p_a p_b}{\int d^3x' \sqrt{g}NU(q)} + \frac{1}{m_P}N_i\mathcal{H}^i_x\right] \\ &= \frac{1}{m_P}\int d^3x \left\{\tilde{N}\kappa^2 G^{ab}p_a p_b + N_i\mathcal{H}^i_x\right\} \end{aligned} \tag{14}$$

and where

$$\tilde{N}(x) \equiv \frac{m_P N(x)}{\int d^3x' \sqrt{g} N U(q)} \tag{15}$$

The parameter m_P is an arbitrary constant with dimensions of mass. It is quite easy to show that treating Æ as a Hamiltonian, i.e

$$\frac{dq^a(x)}{d\tau} = \frac{\delta Æ}{\delta p_a(x)}, \quad \frac{dp_a(x)}{d\tau} = -\frac{\delta Æ}{\delta q^a(x)}, \quad 0 = \frac{\delta Æ}{\delta N(x)}, \quad 0 = \frac{\delta Æ}{\delta N^i(x)} \tag{16}$$

imposing $Æ = -\mathcal{E}$, and rescaling τ by $m_P/\sqrt{\mathcal{E}}$, leads to

$$\begin{aligned}
\frac{dq^a(x)}{dt} &= \int d^3x' \left[\tilde{N}(x') \frac{\delta}{\delta p_a(x)} \mathcal{H}^{\mathcal{E}}_{x'} + N_i(x') \frac{\delta}{\delta p_a(x)} \mathcal{H}^i_{x'}\right] \\
\frac{dp_a(x)}{dt} &= -\int d^3x' \left[\tilde{N}(x') \frac{\delta}{\delta q^a(x)} \mathcal{H}^{\mathcal{E}}_{x'} + N_i(x') \frac{\delta}{\delta q^a(x)} \mathcal{H}^i_{x'}\right] \\
\mathcal{H}^{\mathcal{E}}_x &= \frac{\kappa^2}{\sqrt{\mathcal{E}}} G^{ab} p_a p_b + \sqrt{\mathcal{E}} \sqrt{g} U = 0 \\
\mathcal{H}^i_x &= 0
\end{aligned} \tag{17}$$

These are the same equations of motion and constraints which would be obtained from the Hamiltonian

$$H^{\mathcal{E}} = \int d^3x \, [N(x)\mathcal{H}^{\mathcal{E}}_x + N_i(x)\mathcal{H}^i_x] \tag{18}$$

and then choosing a lapse, in the equations of motion, satisfying a single global restriction

$$\int d^3x \, \sqrt{g} N U = \text{const.} \tag{19}$$

This is sufficient to show the equivalence of the classical evolution equations

$$\partial_\tau Q = \{Q, Æ\} \quad , \quad \partial_\tau Q = \{Q, H^{\mathcal{E}}\} \tag{20}$$

supplemented by the constraint $Æ = -\mathcal{E}$ (or $H^{\mathcal{E}} = 0$), required to hold for all choices of lapse/shift.

This brings us to our proposal: Since $\mathcal{E}$ is indeterminate at the classical level, we see no reason that it should be a fixed parameter at the quantum level. Instead, let $-\mathcal{E}$ just denote the eigenvalues of the operator Æ, and replace the Poisson bracket evolution of the classical theory by

$$\partial_\tau < Q > = < -\frac{i}{\hbar}[Q, Æ] > \tag{21}$$

or equivalently

$$i\hbar \partial_\tau \Psi = Æ \Psi \tag{22}$$

in the quantum theory. The space of physical states is spanned by those eigenstates $\Phi_{\mathcal{E}}$ which satisfy the eigenvalue equation

$$Æ\Phi_{\mathcal{E}} = -\mathcal{E}\Phi_{\mathcal{E}} \tag{23}$$

for every choice of lapse/shift; and this requirement is identical (up to operator ordering ambiguities) to the Dirac constraints $\mathcal{H}^{\mathcal{E}}_x\Phi_{\mathcal{E}} = 0$ and $\mathcal{H}^i_x\Phi_{\mathcal{E}} = 0$. In essence, having noted that the classical Hamiltonian contains an undetermined parameter $\mathcal{E}$, we have replaced the usual Wheeler-DeWitt equation by a family of such equations, each with a different value of $\mathcal{E}$. The space of physical states is spanned by states $\{\Phi_{\mathcal{E}}\}$, each of which satisfies a Wheeler-DeWitt equation with a particular value of the parameter $\mathcal{E}$. Thus, in our proposal, solutions of the Wheeler-DeWitt equation are in fact stationary s-tates; time-development comes about, just as in ordinary non-parametrized quantum mechanics, by superimposing states of varying $\mathcal{E}$, i.e.

$$\Psi[q,\tau] = \sum_{\mathcal{E}} a_{\mathcal{E}}\Phi_{\mathcal{E}}[q]e^{i\mathcal{E}\tau/\hbar} \tag{24}$$

Of course, this approach to formulating the quantum dynamics of time-parametrized theories leaves open the question of the operator ordering in $Æ[p,q]$, and the related issue of integration measure.

We will now approach the problem of time evolution from quite a different direction, namely, the transfer matrix formulation. In ordinary euclidean quantum mechanics, the transfer matrix $\mathcal{T}_\epsilon$ is an operator which evolves the wavefunction by a euclidean time-step $\Delta t = \epsilon$, and is given by

$$\begin{aligned}\psi(q',t+\epsilon) &= \mathcal{T}_\epsilon\psi(q',t)\\ &= \int d^Dq\ \mu(q)\exp[-S_\epsilon(q',q)]\psi(q,t)\end{aligned} \tag{25}$$

Denoting by $S[q_2,q_1;\Delta t]$ the action of a classical solution $q(t)$ running between the initial configuration q_1^a at time t and final q_2^a at time $t+\Delta t$, the expression S_ϵ in euclidean quantum mechanics is given by the continuation of S to imaginary time lapse

$$S_\epsilon(q_2,q_1) \equiv iS[q_2,q_1;i\epsilon]/\hbar \tag{26}$$

and the infinitesmal time evolution operator (i.e. the Hamiltonian) is given by the limit

$$H = \lim_{\epsilon\rightarrow 0}(-\frac{\hbar}{\epsilon})\ln[\mathcal{T}_\epsilon] \tag{27}$$

However, in the case of time-parametrized theories such as (2), this prescription breaks down at eq. (26); the problem is that in parametrized theories the classical action between fixed initial and final configurations does not depend on any additional time parameter, i.e. $\Delta S = S[q_2,q_1]$, which would result in a transfer matrix that is independent of the time-step. In order to have meaningful time-evolution in the transfer matrix approach, eq. (26) must be modified. The simplest possible modification is just to multiply $S[q_2, q_1]$ by an ϵ-dependent factor. Let us indicate the dependence of the action of the classsical trajectory on the supermetric G_{ab} by including it in the list of arguments, i.e. $\Delta S = S[q_2, q_1, G_{ab}]$. Our proposal for euclidean quantization is then to replace (26) by

$$S_\epsilon(q_2, q_1) = iS[q_2, q_1, G^E_{ab}]/\sqrt{\epsilon\hbar} \tag{28}$$

with integration measure

$$\mu^{-1}(q') = (\sqrt{\epsilon\hbar})^D \lim_{\epsilon\to 0} \int \frac{d^D q}{(\sqrt{\epsilon\hbar})^D} \exp(-S_\epsilon(q', q)) \tag{29}$$

and where G^E_{ab} denotes a rotation of the signature of the supermetric from Lorentzian to $\pm$ the unit matrix according to the following rule:

$$\begin{aligned} G_{ab}(q) &= E^i_a(q)\eta_{ij}E^j_b(q) \\ G^E_{ab}(q) &\equiv \mathrm{sign}[V(q)]E^i_a(q)\delta_{ij}E^j_b(q) \end{aligned} \tag{30}$$

Equivalently, defining $\mathcal{G}_{ab} \equiv 2VG_{ab}$, the rotation (30) takes $\mathcal{G}_{ab}$ to euclidean signature. This rotation is made to ensure that S_ϵ is real-valued. After computing $\mathcal{T}_\epsilon$ from (25), the final step is to compute the infinitesmal time-evolution operator, again denoted by Æ, undoing the signature rotation

$$\text{Æ} = \left[\lim_{\epsilon\to 0}(-\frac{\hbar}{\epsilon})\ln[\mathcal{T}_\epsilon]\right]_{\left\{\substack{\mathcal{G}^E\to\mathcal{G} \\ \det(\mathcal{G}^E)\to|\det(\mathcal{G})|}\right\}} \tag{31}$$

Following the prescription above, we find for the action (2) the evolution operator

$$\text{Æ} = -\frac{D+1}{2m_0^2}\hbar^2\frac{1}{\sqrt{|\mathcal{G}|}}\frac{\partial}{\partial q^n}\sqrt{|\mathcal{G}|}\mathcal{G}^{nm}\frac{\partial}{\partial q^m} + \hbar^2\frac{D+1}{6m_0^2}\mathcal{R} \tag{32}$$

where $\mathcal{G} \equiv \det(\mathcal{G}_{ab})$, and $\mathcal{R}$ is the curvature scalar formed from $\mathcal{G}_{ab}$. The integration measure $\mu(q)$ turns out to be proportional to $\sqrt{|\mathcal{G}|}$; it is easy to see that Æ is hermitian in this measure. Apart from an (irrelevant) factor of $D+1$, which can be absorbed into a rescaling of the τ-parameter, the classical limit of the Æ operator is simply the expression Æ$[p, q]$ defined in eq. (6), which we already know generates the correct classical motion.

The analogous steps for gravitation theory lead to some highly singular expressions, and here we are hampered by the lack of an invariant, non-perturbative regulator. Nevertheless, under some rather general assumptions about the regulator we find, up to operator-ordering terms

$$\text{Æ} = \frac{1}{m_P}\int d^3x \left[-\hbar^2\tilde{N}\kappa^2 G^{ab}\frac{\delta^2}{\delta q^a \delta q^b} + N_i\mathcal{H}^i_x\right] \tag{33}$$

with

$$\tilde{N}(x) \equiv m_P \frac{N(x)}{\int d^3x' \sqrt{g} N U(q)} \tag{34}$$

Once again, the classical limit of this operator is identical to the expression for Æ$[p, q]$ in eq. (14). The details can be found in ref. [2, 3].

We conclude with some general remarks. First, by superimposing states of different $\mathcal{E}$, it appears possible to construct physical states which are sharply peaked around particular 3-geometries at a given time τ. Roughly speaking, the minimal dispersion around a given 3-geometry is inversely proportional to the dispersion in $\mathcal{E}$. This may have important consequences for measurement theory. Secondly, a dispersion in $\mathcal{E}$ can be interpreted as a dispersion in the effective value of Planck's constant. This is because, in the constraint equation Æ$\Phi_\mathcal{E} = -\mathcal{E}\Phi_\mathcal{E}$, the constant $\mathcal{E}$ can be absorbed into a rescaling of Planck's constant $\hbar_{eff} = \hbar/\sqrt{\mathcal{E}}$. Thus, in our view, the non-stationarity of the Universe is due to an underlying uncertainty in the effective value of Planck's constant. Depending on the magnitude of this dispersion, there could conceivably be observable effects. We will not attempt, however, to discuss such effects here.

References

[1] K. Kuchar (1992), 'Time and interpretations of quantum gravity', in *Proceedings of the 4th Canadian Conference on General Relativity and Astrophysics*, ed. G. Kunstatter et. al., World Scientific, Singapore; C. J. Isham (1992), Imperial College preprint IMPERIAL-TP-91-92-25, gr-qc/9210011.

[2] J. Greensite (1994), *Phys. Rev.*, **D49** 930.

[3] A. Carlini and J. Greensite, 'Fundamental Constants and the Problem of Time,' in preparation.

The W_3-particle

Eduardo Ramos and Jaume Roca

Abstract

We show that W_3 is the algebra of symmetries of the "rigid-particle", whose action is given by the integrated extrinsic curvature of its world line. This is easily achived by showing that its equation of motion can be written in terms of the Boussinesq operator. We also show how to obtain the equations of motion of the standard relativistic particle provided it is consistent to impose the "zero-curvature gauge", and comment about its connection with the KdV operator.

1 Introduction

The geometrical interpretation of W-type symmetries has attracted the attention of many mathematical physicists in recent years. Although a plethora of interesting results are now at our disposal it is commonly agreed that we have not yet a complete understanding of the underlying geometry. It is clear that simple mechanical systems enjoying W symmetry could be an unvaluable tool in this difficult task. On one hand they could provide us with some geometrical and/or physical interpretation for W-transformations (W-morphisms), while on the other hand they could give us some hints about which are the relevant structures associated with W-gravity – the paradigmatic example being provided by the standard relativistic particle and diffeomorphism invariance (W_2).

It is well known by now the connection between W-morphisms and the extrinsic geometry of curves and surfaces [1]. Therefore, it seems natural to look for a W-particle candidate among the geometrical actions depending on the extrinsic curvature. The canonical analysis of those models has been carried out by several authors [2]. In particular M.S. Plyushchay studied in [3] the action given by

$$S = \alpha \int \sqrt{|\kappa^2|}\, ds, \tag{1.1}$$

where α is a dimensionless coupling constant[1] and the extrinsic curvature κ is given by

$$\kappa^2 = g_{\mu\nu} \frac{d^2 x^\mu}{ds^2} \frac{d^2 x^\nu}{ds^2}, \tag{1.2}$$

[1] In our conventions the coordinates are dimensionless.

where $g_{\mu\nu}$ stands for minkowskian, euclidean, or any x-independent metric. He showed that this dynamical system posseses two gauge invariances, and that one of them is the expected invariance under diffeomorphisms. It is the main purpose of this note to show that these gauge invariances are nothing but W_3.

2 Gauge structure analysis

We now briefly review the main results concerning the gauge structure of the action (1.1).

In an arbitrary parametrization $x^\mu(t)$ the lagrangian is given by

$$L = \alpha\sqrt{\left|\frac{\ddot{x}_\perp^2}{\dot{x}^2}\right|}, \tag{2.1}$$

where $\dot{x}^\mu = dx^\mu/dt$ and $\ddot{x}_\perp^\mu = \ddot{x}^\mu - \dot{x}^\mu(\ddot{x}\dot{x})/\dot{x}^2$.

For the study of the system in the hamiltonian framework it is convenient to introduce the velocity $q^\mu = \dot{x}^\mu$ as an auxiliary variable in order to avoid the presence of second-order derivative terms in the lagrangian. The lagrangian in these new variables reads

$$L = \alpha\sqrt{\left|\frac{\dot{q}_\perp^2}{q^2}\right|} + \gamma(q - \dot{x}). \tag{2.2}$$

We introduce the canonical momenta (P_μ, p_μ, π_μ) associated with the coordinates $(x^\mu, q^\mu, \gamma^\mu)$. Their Poisson bracket algebra is given by

$$\{\mathcal{Q}^\mu, \mathcal{P}_\nu\} = \delta^\mu_\nu. \tag{2.3}$$

The definition of the momenta implies a number of primary hamiltonian contraints. Being the lagrangian linear in $\dot{x}$ we obtain the trivial second-class constraints $\gamma^\mu = -P^\mu$ and $\pi^\mu = 0$, from which we can eliminate the canonical pair (γ, π) by means of the Dirac bracket. The primary first-class contraints are

$$\phi_1 = pq \approx 0, \qquad \phi_2 = \frac{1}{2}\left(|p^2| - \frac{\alpha^2}{|q^2|}\right) \approx 0. \tag{2.4}$$

Stabilization of these constraints through the standard Dirac procedure leads to the secondary

$$\phi_3 = Pq \approx 0, \qquad \phi_4 = Pp \approx 0, \tag{2.5}$$

and tertiary

$$\phi_5 = P^2 \approx 0, \tag{2.6}$$

first-class hamiltonian constraints, which form a closed Poisson bracket algebra.

In the constrained submanifold time evolution is generated by the hamiltonian

$$H = \phi_3 + v_1\phi_1 + v_2\phi_2. \tag{2.7}$$

The functions v_1 and v_2 have a definite expression in terms of lagrangian quantities,

$$v_1 = \frac{q\dot{q}}{q^2}, \qquad v_2 = \frac{1}{\alpha}\sqrt{|q^2\dot{q}_\perp^2|}, \tag{2.8}$$

but they should be regarded as arbitrary functions of time in the canonical formalism. Time evolution becomes unambiguous once we assign them definite values, which is nothing but a choice of gauge. The hamiltonian stabilization procedure guarantees the consistency of the expressions chosen for the v_i with their lagrangian expression (2.8).

The presence of two arbitrary functions reveals the existence of a second gauge symmetry in addition to the familiar reparametrization invariance. We shall show in what follows that such symmetries are precisely W_3.

3 Equations of motion, Boussinesq operator and W_3 symmetry

The hamiltonian equations of motion can be sugestively written as

$$\begin{aligned} \dot{x}^\mu &= q^\mu, \\ \begin{pmatrix} \dot{P}^\mu \\ \dot{p}^\mu \\ \dot{q}^\mu \end{pmatrix} &= \begin{pmatrix} 0 & 0 & 0 \\ -1 & -v_1 & -\frac{\alpha^2 v_2}{q^4} \\ 0 & v_2 & v_1 \end{pmatrix} \begin{pmatrix} P^\mu \\ p^\mu \\ q^\mu \end{pmatrix}. \end{aligned} \tag{3.1}$$

Notice that, neglecting the first equation, which simply states the definition of q^μ, the equations of motion can be casted in the "Drinfeld-Sokolov" form associated with a particular $SL(3)$ connection. We will come back to this point later on.

It will be convenient in what follows to introduce a different and more geometrical parametrization for v_1 and v_2 Let us define e and λ as

$$e^2 = |q^2|, \qquad \lambda^2 = |\kappa^2|. \tag{3.2}$$

Notice that e^2 is nothing but the modulus of the induced metric, and λ is basically the extrinsic curvature. In terms of e and λ the gauge degrees of freedom are given by $v_1 = \dot{e}/e$ and $v_2 = \lambda e^3/\alpha$.

From (3.1) we can obtain a single equation for the velocity vector q^μ,

$$\dddot{q}^\mu + u_1\ddot{q}^\mu + u_2\dot{q}^\mu + u_3 q^\mu = 0,$$

with

$$\begin{aligned}
u_1 &= -2\frac{d\ln(e^3\lambda)}{dt}, \\
u_2 &= e^4\lambda^2 + 15\frac{\dot{e}^2}{e^2} + 7\frac{\dot{e}\dot{\lambda}}{e\lambda} + 2\frac{\dot{\lambda}^2}{\lambda^2} - 4\frac{\ddot{e}}{e} - \frac{\ddot{\lambda}}{\lambda}, \\
u_3 &= e^4\dot{\lambda}\lambda + e^3\dot{e}\lambda^2 - 15\frac{\dot{e}^3}{e^3} - 7\frac{\dot{e}^2\dot{\lambda}}{e^2\lambda} - 2\frac{\dot{e}\dot{\lambda}^2}{e\lambda^2} + 10\frac{\dot{e}\ddot{e}}{e^2} + 2\frac{\ddot{e}\dot{\lambda}}{e\lambda} + \frac{\dot{e}\ddot{\lambda}}{e\lambda} - \frac{\dddot{e}}{e}.
\end{aligned} \tag{3.3}$$

Notice that u_1 is a total derivative. This allows us to remove the term with the second derivative by a simple local rescaling of q^μ. Indeed, if we define

$$y^\mu = \frac{1}{e^2\lambda^{\frac{2}{3}}} q^\mu, \tag{3.4}$$

the equation for the new velocity vector y^μ can be written in terms of the Boussinesq Lax operator

$$\dddot{y}^\mu + T\dot{y}^\mu + (W + \frac{\dot{T}}{2})y^\mu = 0, \tag{3.5}$$

with T and W given by

$$\begin{aligned}
T &= e^2\lambda^2 - \frac{1}{3}\frac{\dot{e}^2}{e^2} - \frac{\dot{e}\dot{\lambda}}{e\lambda} - \frac{4}{3}\frac{\dot{\lambda}^2}{\lambda^2} + 2\frac{\ddot{e}}{e} + \frac{\ddot{\lambda}}{\lambda}, \\
W &= e\dot{e}\lambda^2 + 3\frac{\dot{e}^3}{e^3} + \frac{5}{3}e^2\lambda\dot{\lambda} + \frac{\dot{e}^2\dot{\lambda}}{e^2\lambda} + \frac{4}{3}\frac{\dot{e}\dot{\lambda}^2}{e\lambda^2} + \frac{56}{27}\frac{\dot{\lambda}^3}{\lambda^3} \\
&\quad -4\frac{\dot{e}\ddot{e}}{e^2} - \frac{2}{3}\frac{\ddot{e}\dot{\lambda}}{e\lambda} - \frac{\dot{e}\ddot{\lambda}}{e\lambda} - \frac{8}{3}\frac{\dot{\lambda}\ddot{\lambda}}{\lambda^2} + \frac{\dddot{e}}{e} + \frac{2}{3}\frac{\dddot{\lambda}}{\lambda}.
\end{aligned} \tag{3.6}$$

The algebra of symmetries of equations of the type $\mathsf{L}\Psi = 0$ with L a differential operator of the form $\partial^n + ...$ has been studied by Radul in [4]. From his general construction it is easily deduced that for the particular case of $\mathsf{L} = \partial^3 + T\partial + W + \dot{T}/2$, which is our case of interest, the symmetry algebra is W_3. Rather than reproducing here the general arguments leading to this result, we will work out explicitly, for convenience of the less mathematically oriented reader, the case at hand.

The most general (local) variation of Ψ preserving the structure of L , up to equations of motion, is given by

$$\delta_{\epsilon,\rho}\Psi = \rho\ddot{\Psi} + (\epsilon - \frac{1}{2}\dot{\rho})\dot{\Psi} - (\dot{\epsilon} + \frac{1}{6}\ddot{\rho} + \frac{2}{3}\rho T)\Psi, \tag{3.7}$$

where the parametrization has been chosen so that ϵ and ρ denote the parameters associated with diffeomorphisms and pure W_3-morphisms respectively.

The corresponding transformations for T and W which leave the equations of motion invariant are given by

$$\begin{aligned}
\delta_\epsilon^{(T)}T &= 2\dddot{\epsilon} + 2\dot{\epsilon}T + \epsilon\dot{T}, \\
\delta_\epsilon^{(T)}W &= 3\dot{\epsilon}W + \epsilon\dot{W}, \\
\delta_\rho^{(W)}T &= 3\dot{\rho}W + 2\rho\dot{W}, \\
\delta_\rho^{(W)}W &= -\frac{1}{6}\rho^{(v)} - \frac{5}{6}\dddot{\rho}T - \frac{5}{4}\ddot{\rho}\dot{T} - \dot{\rho}(\frac{3}{4}\ddot{T} + \frac{2}{3}T^2) - \rho(\frac{1}{6}\dddot{T} + \frac{2}{3}T\dot{T}).
\end{aligned} \tag{3.8}$$

It is now a long but straightforward exercise to check that these transformations obey the algebra of W_3 transformations, *i.e.*

$$\begin{aligned}
[\delta_{\epsilon_1}^{(T)}, \delta_{\epsilon_2}^{(T)}] &= \delta_\epsilon^{(T)}, \quad \text{where} \quad \epsilon = \epsilon_1\dot{\epsilon}_2 - \dot{\epsilon}_1\epsilon_2, \\
[\delta_\epsilon^{(T)}, \delta_\rho^{(W)}] &= \delta_{\tilde{\rho}}^{(W)}, \quad \text{where} \quad \tilde{\rho} = \epsilon\dot{\rho} - 2\dot{\epsilon}\rho, \\
[\delta_{\rho_1}^{(W)}, \delta_{\rho_2}^{(W)}] &= \delta_\epsilon^{(T)}, \quad \text{where} \quad \epsilon = \frac{2}{3}\rho_1\dot{\rho}_2T - \frac{1}{4}\dot{\rho}_1\ddot{\rho}_2 + \frac{1}{6}\rho_1\dddot{\rho}_2 - (\rho_1 \leftrightarrow \rho_2),
\end{aligned} \tag{3.9}$$

where we have assumed for simplicity that the parameters are field independent. The most general case can be equally worked out from the results of [4].

Notice that the consistency of the procedure in our case is based on the fact that T and W, through their dependence in λ and e, are gauge degrees of freedom themselves, therefore before any solution to the equations is found they should be given definite values. In this language the W-morphisms given by (3.8) are to be understood as gauge transformations.

Because the relationship between y^μ and q^μ is not easily invertible the previous analysis does not yield directly the transformation rules for q^μ. These transformations are interesting by themselves because the action is naturally written in terms of q^μ. Here is where standard techniques in constrained dynamical systems come to our rescue. The gauge transformations for x^μ, q^μ, and γ^μ can be computed using the general method of [5] to give

$$\begin{aligned}
\delta_{\beta,\eta}x^\mu &= \beta q^\mu + \dot{\eta}\frac{\alpha}{e^3\lambda}\dot{q}_\perp^\mu - 2\eta\gamma^\mu, \\
\delta_{\beta,\eta}q^\mu &= \dot{\beta}q^\mu + \beta\dot{q}^\mu - \dot{\eta}\left(\gamma^\mu + \alpha\frac{\lambda}{e}q^\mu + \alpha\frac{\dot{e}}{e^4\lambda}\dot{q}_\perp^\mu\right) + \ddot{\eta}\frac{\alpha}{e^3\lambda}\dot{q}_\perp^\mu, \\
\delta_{\beta,\eta}\gamma^\mu &= 0.
\end{aligned} \tag{3.10}$$

A direct computation now shows that the tranformations given by (3.10) are not only a symmetry of the action, but also that, as a soft algebra, reproduce the algebra of W_3 transformations if we do the following identifications:

$$\epsilon = \beta + \frac{\alpha}{2}\frac{\ddot{\eta}}{e^3\lambda} + \dot{\eta}\frac{\alpha}{e^3\lambda}\left(\frac{1}{6}\frac{\dot{\lambda}}{\lambda} - \frac{1}{2}\frac{\dot{e}}{e}\right),$$

$$\rho = -\frac{\alpha}{e^3\lambda}\dot{\eta}. \tag{3.11}$$

Notice that in this parametrization the variation of x^μ has a nonlocal expression. This is not, to some extent, surprising due to the fact that the natural variable for the system seem to be supplied by q^μ rather than the coordinates themselves[2]. Moreover, as expected, the variations induced on y^μ through (3.10) coincide with the ones obtained via (3.7), tightening up the formalism, and showing the nice interplay between the hamiltonian and the W-algebraic methods of tackling the problem.

There is also a more indirect, but interesting in its own, way to show the invariance of the action (1.1) under W_3-morphisms with the help of the Miura transformation. In order to do so we should return to the expression of the hamiltonian equations of motion in terms of the $SL(3)$ connection. If we write (3.1) as

$$\left(\frac{d}{dt}-\Lambda\right)\begin{pmatrix} P^\mu \\ p^\mu \\ q^\mu \end{pmatrix} = 0, \tag{3.12}$$

it is clear that this equations have local gauge invariance under

$$\begin{pmatrix} P^\mu \\ p^\mu \\ q^\mu \end{pmatrix} \rightarrow M \begin{pmatrix} P^\mu \\ p^\mu \\ q^\mu \end{pmatrix} \quad \text{and} \quad \Lambda \rightarrow M\Lambda M^{-1} - M\frac{dM^{-1}}{dt},$$

with $M \in SL(3)$. It will be convenient for our purposes to work in the Miura gauge, *i.e.*

$$\Lambda = \begin{pmatrix} -\varphi_1-\varphi_2 & 0 & 0 \\ 1 & \varphi_2 & 0 \\ 0 & 1 & \varphi_1 \end{pmatrix}. \tag{3.13}$$

The matrix $M \in SL(3)$ which brings Λ to this form is

$$M = \frac{\lambda^{\frac{1}{3}}}{\alpha^{\frac{1}{3}}}\begin{pmatrix} e & 0 & 0 \\ 0 & -e & -i\alpha/e \\ 0 & 0 & -\alpha/e^2\lambda \end{pmatrix}.$$

Some straightforward algebra now yields

$$\begin{aligned} \varphi_1 &= -\frac{\dot{e}}{e} - \frac{2}{3}\frac{\dot{\lambda}}{\lambda} - i\lambda e \\ \varphi_2 &= \frac{1}{3}\frac{\dot{\lambda}}{\lambda} + i\lambda e. \end{aligned} \tag{3.14}$$

The key point, for our present interest, is given by the fact that the lagrangian (2.1) can be written as

$$L \sim \mathrm{Im}(\varphi_1) \sim \mathrm{Im}(\varphi_2). \tag{3.15}$$

[2]This is somehow reminiscent of what happens for the case of the two dimensional massless boson, where the natural variable is supplied by its associated $U(1)$ current.

The transformation under W_3-morphisms of the Miura fields can be computed using the Kupershmidt-Wilson theorem [6]. If we denote by ϵ the parameter associated with diffeomorphisms, and by ρ the one associated with pure W_3-morphisms, these variations are given by

$$\begin{aligned}
\delta_{\epsilon,\rho}\varphi_1 &= \frac{d}{dt}\left(\dot{\epsilon} + \epsilon\varphi_1 + \frac{1}{6}\ddot{\rho} + \frac{1}{2}(\dot{\varphi}_1\rho + \varphi_1\dot{\rho})\right.\\
&\qquad \left. -\frac{1}{3}(\frac{1}{2}\dot{\varphi}_1 - 2\dot{\varphi}_2 - \varphi_1^2 + 2\varphi_2^2 + 2\varphi_1\varphi_2)\rho\right),\\
\delta_{\epsilon,\rho}\varphi_2 &= \frac{d}{dt}\left(\epsilon\varphi_2 - \frac{1}{3}\ddot{\rho} - \frac{1}{2}(\dot{\varphi}_2\rho + \varphi_2\dot{\rho}) - \dot{\varphi}_1\rho + \varphi_1\dot{\rho}\right.\\
&\qquad \left. +\frac{1}{3}(\dot{\varphi}_1 + \frac{1}{2}\dot{\varphi}_2 - 2\varphi_1^2 + \varphi_2^2 - 2\varphi_1\varphi_2)\rho\right), \qquad (3.16)
\end{aligned}$$

which is a total derivative! And from this follows directly the invariance of the action.

4 The relativistic particle

It is easy to recover the equations of motion for the standard relativistic particle from (3.1) whenever it is consistent to impose the gauge condition $v_2 = 0$. Notice that since v_2 is proportional to the extrinsic curvature, this gauge can only be consistently imposed when it can be taken to be zero (initial conditions may be incompatible with this gauge choice). In that case the hamiltonian equations of motion collapse to

$$\ddot{x}^\mu - \frac{\dot{e}}{e}\dot{x}^\mu = 0, \qquad (4.1)$$

which is the equation of motion for the relativistic particle. Notice that (4.1) can be sugestively written as $\ddot{x}^\mu_\perp = 0$.

It is natural to ask now, in the light of our previous discussion, which role, if any, is played by the KdV Lax operator in this case. The answer to this question is easily obtained by realising that via the redefinition

$$x^\mu \to e^{\frac{1}{2}}x^\mu \qquad (4.2)$$

the equations of motion for the new variable can be written as

$$\ddot{x}^\mu + Tx^\mu = 0, \qquad (4.3)$$

with

$$T = \frac{1}{2}\frac{\ddot{e}}{e} - \frac{3}{4}\frac{\dot{e}^2}{e^2}, \qquad (4.4)$$

and being e a gauge degree of freedom for the relativistic particle we can apply all the same arguments as above.

5 Final comments and digressions

We hope to have convinced the reader that W_3-symmetry does play an important role in the symmetry structure of the rigid particle. And that standard techniques in constrained hamiltonian systems and W-algebras can be intertwined as powerful tools for the better understanding of both disciplines. But, of course, much is still to be done. For example, we do not yet completely understand the appearance of the $SL(3)$ structure in the problem, which should be the key to generalize these procedures to other W-algebras. Moreover, under quantization [3] the rigid particle is associated with massless representations of the Poincare group with integer helicity, therefore being a potential candidate for a particle description of photons, gravitons and higher spin fields. It is a tantalizing possibility that W_3 symmetry can play a role in such physical systems.

We would not like to finish without a few sentences about the possible relevance of our results to W_3-gravity. It was shown in [7] that a naive coupling to gravity of the action (1.1), *i.e.* to consider arbitrary x-dependent metrics, yields inconsistencies. This is due to the fact that the constraint algebra no longer closes in curved space-time. We believe this to be a signal indicating that the W_3-particle is only consistently coupled to W_3-gravity. Of course this obscure sentence needs some explanation. In the standard particle case the relevant bundle is the tangent bundle of the manifold, TM. In this case it is well known how to equip the bundle with a metric structure and all the powerful and well understood machinery of Riemmanian geometry is at our disposal. In the language of jet bundles TM is nothing but $J^1(\mathbf{R}, M)$, and it is our understanding that the relevant bundle for the W_3-particle is provided by $J^2(\mathbf{R}, M)$, which is not itself a vector bundle. It is well known that $J^2(\mathbf{R}, M)$ is an associated bundle to F^2M, the frame bundle of second order, and it is our belief that the required structure should be a "natural" structure in F^2M, as it is the metric in FM. It is our hope that the action of the W_3-particle will provide us with a "W_3-line element", thus offering some valuable insight about which generalized structures in F^2M should be considered. Work on this is in progress.

Acknowledgements

We would like to thank J.M. Figueroa-O'Farrill for many useful discussions about the W-orld. One of us, J.R. (no pun intended), is also thankful to the Spanish Ministry of Education for financial support.

References

[1] G. Sotkov and M. Stanishkov (1991), *Nucl. Phys.* **B356** 439;
G. Sotkov, M. Stanishkov and C.J. Zhu (1991),*Nucl. Phys.* **B356** 245;
J.L. Gervais and Y. Matsuo (1992), *Phys. Lett.* **274B** 309 (hep-th/9110028);
J.L. Gervais and Y. Matsuo (1993), *Comm. Math. Phys.* **152** 317 (hep-th/9201026);
J.M. Figueroa-O'Farrill, E. Ramos and S. Stanciu (1992), *Phys. Lett.* **297B** 289, (hep-th/9209002);
J. Gomis, J. Herrero, K. Kamimura and J. Roca (1994), *Prog. Theor. Phys.* **91** 413.

[2] R.D. Pisarski (1986), *Phys. Rev.* **D34** 670;
M.S. Plyushchay (1991), *Phys. Lett.* **253B** 50;
C. Batlle, J. Gomis, J.M. Pons and N. Román-Roy (1988), *J. Phys. A: Math. Gen.* **21** 2693.

[3] M.S. Plyushchay (1989), *Mod. Phys. Lett.* **A4** 837.

[4] A.O. Radul (1989), *Sov. Phys. JETP Lett.* **50** 371; (1991),*Functional Analysis and Its Application* **25** 25.

[5] X. Gràcia and J.M. Pons (1992), *J. Phys. A: Math. Gen.* **25** 6357.

[6] L.A. Dickey (1982), *Comm. Math. Phys.* **87** 127.

[7] D. Zoller (1990), *Phys. Rev. Lett.* **65** 2236.

Pure geometrical approach to singular Lagrangians with higher derivatives

V. V. Nesterenko and G. Scarpetta

1 Introduction

The Lagrangians defined on the parametrized curves (on the trajectories) and depending on higher derivatives of coordinates of the curve have been considered recently in a number of problems. These, in particular, are: the null-dimensional (particle-like) version of the rigid string [1 – 7], the model of boson-fermion transmutations in external Chern-Simons field [8 – 11], polymer theory [12]. Lagrangians of this kind also occur when applying a modified version of the space-time interval proposed in papers [13, 18] in connection with the conjuncture about existence of a limited value of acceleration. Recently these variational problems have also become interesting for mathematicians [19]. This list of references is certainly incomplete, however, it illustrates the continuing interest in the subject.

All these Lagrangians are as a rule singular. Investigation of such models in the framework of a classical variational calculus results in very complicated nonlinear differential equations of order $2p$ for coordinates of a curve to be found (p is the highest order of the derivatives in an initial Lagrangian function). These equations are not practically subject to analysis.

However, a considerable advance can be achieved here by applying the following basic result from the classical differential geometry [20, 21]. Any smooth curve $x^\mu(s)$, $\mu = 0, 1, \ldots, D-1$ in D-dimensional flat space-time is determined (up to its rotations as a whole) by specifying $D-1$ principal curvatures of this curve $k_i(s)$, $i = 1, 2, \ldots, D-1$, where s is the curve length. Therefore, one can try to derive the Euler-Lagrange equations in terms of the principal curvatures $k_i(s)$, $i = 1, 2, \ldots, D-1$ of the curve to be found rather than in terms of its coordinates $x^\mu(s)$. The order of the differential equations for $k_i(s)$ will be wittingly lower than the order of equations for $x^\mu(s)$ because the curvature $k_i(s)$ is expressed in terms of the derivatives of x up to order $i+1$ inclusive ($i = 1, 2, \ldots, D-1$).

In the present paper it will be shown that the equations of motion for $k_i(s)$ generated by an arbitrary Lagrangian function $\mathcal{L}(k_1(s))$ are always integrable in quadratures. If one assumes that the Lagrangian $\mathcal{L}(k_1)$ defines a model of a relativistic particle, then the integration constants turn out to be the mass and spin of the particle. In this case $k_1(s)$ is a proper acceleration of the particle.

2 Euler-Lagrange equations in terms of the principal curvatures

Let us consider a reparametrization-invariant action

$$S = \int \mathcal{L}(k_1)\, ds \tag{2.1}$$

with a generic Lagrangian function $\mathcal{L}$ depending only on the first curvature k_1 (on the particle acceleration) of the world trajectory $x^\mu(s)$, $\mu = 0, 1, \ldots, D-1$ in D-dimensional Minkowski space. Here $ds^2 = dx^\mu dx_\mu$, i.e., s is a natural parameter along the world curve $x^\mu(s)$ (its length)

$$\frac{dx^\mu}{ds}\frac{dx_\mu}{ds} = 1\,. \tag{2.2}$$

The Lorentz metric $g_{\mu\nu}$ with a signature $(+, -, \ldots, -)$ will be used. For shortening the notation, the differentiation with respect to a natural parameter s will be denoted by a over-dot. With all this, the first curvature (or simply curvature) is defined by

$$k_1^2(s) = -\,\ddot{x}_\mu\,\ddot{x}^\mu = -\ddot{x}^2\,. \tag{2.3}$$

Following the Hamiltonian principle we require

$$\delta S = \delta S_1 + \delta S_2 =$$

$$= \int ds\, \mathcal{L}'(k_1)\, \delta k_1(s) + \int \mathcal{L}(k_1)\, \delta ds = 0\,, \tag{2.4}$$

where the variations $\delta k_1(s)$ and δds are generated by the variation of the world trajectory $\delta x^\mu(s)$. The prime of the Lagrangian function $\mathcal{L}(k_1)$ denotes the differentiation with respect to its argument k_1.

With any point of the world curve $x^\mu(s)$ sought for the Frene basis [20, 21] can be associated. The latter is formed by a unit time-like tangent vector

$$e_0^\mu(s) = \frac{dx^\mu}{ds}, \quad e_0^2 = \dot{x}^2 = 1 \tag{2.5}$$

and by a set of $D-1$ space-like unit vectors $e_i^\mu(s)$:

$$e_i^\mu\, e_{j\mu} = -\,\delta_{ij}, \quad e_0^\mu\, e_{j\mu} = 0\,, \quad 1 \leq i, j \leq D-1\,. \tag{2.6}$$

The vectors of the Frenet basis will be denoted by the letters from the beginning of the Greek alphabet α, β, γ, δ, ... and subscripts μ, ν, λ, ρ, σ, ... will specify the coordinates of the ambient space-time. Thus all the Greek subscripts run over the values 0, 1, 2, ..., $D-1$. Raising and lowering these subscripts are made by the corresponding metric tensors

$$g_{\alpha\beta} = \text{diag}\,(1, -1, \ldots, -1), \quad g_{\mu\nu} = \text{diag}\,(1, -1, \ldots, -1). \qquad (2.7)$$

The orthonormality condition for the Frenet basis is written as

$$e^{\mu}_{\alpha}\, e_{\beta\mu} = g_{\alpha\beta}, \quad 0 \le \alpha, \beta \le D-1, \quad \mu = 0, 1, \ldots, D-1. \qquad (2.8)$$

We shall need the Frenet equations describing the alteration of the Frenet basis under motion of its origin along the curve

$$\dot{e}^{\mu}_{\alpha} = \omega_{\alpha}{}^{\beta}\, e^{\mu}_{\beta}, \quad \omega_{\alpha\beta} + \omega_{\beta\alpha} = 0. \qquad (2.9)$$

Nonzero elements of the matrix ω are determined by the principal curvatures

$$\omega_{\alpha,\alpha+1} = -\omega_{\alpha+1,\alpha} = k_{\alpha+1}(s), \quad \alpha = 0, 1, \ldots, D-2. \qquad (2.10)$$

Let us expand the variation $\delta x^{\mu}(s)$ in terms of the Frenet basis

$$\delta x^{\mu}(s) = \varepsilon^{\alpha}(s)\, e^{\mu}_{\alpha}(s),$$

$$\mu = 0, 1, \ldots, D-1, \quad \alpha = 0, 1, \ldots, D-1. \qquad (2.11)$$

The variations δds and $\delta k_1(s)$ encountered in (2.4) can be expressed now in terms of the functions $\varepsilon^{\alpha}(s)$ from (2.11). For variation δds one obtains

$$\delta\, ds = \delta \sqrt{dx_{\mu}\, dx^{\mu}} = \frac{dx_{\mu}\, \delta\, dx^{\mu}}{ds} = \dot{x}_{\mu}\, d\,(\delta x^{\mu}). \qquad (2.12)$$

Substituting (2.12) into (2.4) and integrating by parts the second term in (2.4) acquires the form

$$\delta S_2 = -\int d\,(\mathcal{L}(k_1)\, \dot{x}_{\mu})\; \delta x^{\mu} =$$

$$= -\int \mathcal{L}'(k_1)\, \dot{k}_1\, (\dot{x}_{\mu} \delta x^{\mu})\, ds - \int \mathcal{L}(k_1)\, (\ddot{x}_{\mu} \delta x^{\mu})\, ds. \qquad (2.13)$$

By making use of eqs. (2.8) – (2.11) we obtain for the variation δS_2

$$\delta S_2 = -\int \mathcal{L}'(k_1)\, \dot{k}_1\, \varepsilon^{0}(s)\, ds - \int \mathcal{L}(k_1)\, k_1(s)\, \varepsilon^{1}(s)\, ds. \qquad (2.14)$$

Now we proceed to calculation of the variation δk_1. ¿From the definition (2.3) we deduce

$$k_1(s)\, \delta k_1(s) = -\,(\ddot{x}_{\mu}\, \delta \ddot{x}^{\mu}) = -\,(\dot{e}_{0\mu}\, \delta \ddot{x}^{\mu}). \qquad (2.15)$$

Here it should be taken into account that the operations δ and d/ds do not commute. The direct calculation shows that

$$\left[\delta, \frac{d}{ds}\right] = -\left(e_{0\mu}\frac{d}{ds}\delta x^{\mu}\right)\frac{d}{ds}\,. \tag{2.16}$$

Applying this formula one finds

$$\delta \ddot{x}^{\mu} = \frac{d^2}{ds^2}\delta x^{\mu} - 2\,\ddot{x}^{\mu}\left(e_{0\nu}\frac{d}{ds}\delta x^{\nu}\right) - e_0^{\mu}\frac{d}{ds}\left(e_{0\nu}\frac{d}{ds}\delta x^{\nu}\right). \tag{2.17}$$

Substituting (2.17) into (2.15) and taking into account the Frenet equations (2.9) we obtain

$$\delta k_1(s) = \varepsilon^0 \dot{k}_1 - \ddot{\varepsilon}^1 + \varepsilon^1(k_1^2 + k_2^2) - 2\dot{\varepsilon}^2 k_2 - \varepsilon^2 \dot{k}_2 - \varepsilon^3 k_2 k_3\,. \tag{2.18}$$

Thus, the variation of the first curvature of the curve, $k_1(s)$ depends on the variations of the curve coordinates only along the directions e_0^{μ}, e_1^{μ}, e_2^{μ}, and e_3^{μ} (on the functions $\varepsilon^{\alpha}(s)$, $\alpha = 0, 1, 2, 3$) and on the first three curvatures k_1, k_2 and k_3.

Substituting (2.18) into (2.4), integrating by parts, and taking into account (2.14) we obtain

$$\delta S = \int ds \left\{\left[(k_1^2 + k_2^2)\,\mathcal{L}'(k_1) - \frac{d^2}{ds^2}(\mathcal{L}'(k_1)) - k_1\,\mathcal{L}(k_1)\right]\varepsilon^1(s) + \right.$$
$$\left. + \left[2\frac{d}{ds}(\mathcal{L}'(k_1)\,k_2) - \dot{k}_2\,\mathcal{L}'(k_1)\right]\varepsilon^2(s) - \mathcal{L}'(k_1)\,k_2\,k_3\,\varepsilon^3(s)\right\} = 0\,. \tag{2.19}$$

Terms outside the integral are dropped here as it is done always when deriving equations of motion from the Hamiltonian principle. The functions $\varepsilon^i(s)$, $i = 1, 2, 3$ are arbitrary, therefore, we deduce from (2.19) three equations

$$\frac{d^2}{ds^2}(\mathcal{L}'(k_1)) = (k_1^2 + k_2^2)\,\mathcal{L}'(k_1) - k_1\,\mathcal{L}(k_1)\,, \tag{2.20}$$

$$2\frac{d}{ds}(\mathcal{L}'(k_1)\,k_2) = \dot{k}_2\,\mathcal{L}'(k_1)\,, \tag{2.21}$$

$$\mathcal{L}'(k_1)\,k_2\,k_3 = 0\,. \tag{2.22}$$

The set of equations equivalent to eqs. (2.20) – (2.22) has been derived in the book [19] (see also [22]) by making use of rather complicated formal mathematical methods based on the exterior Cartan forms. We consider that these complications are not justified here. In this problem the use of simple Frenet equations well known to physicists is enough.

To satisfy eq. (2.22) one has to put $k_3 = 0$. Then, it follows that all the higher curvatures $k_4, k_5, \ldots,, k_{D-1}$ also vanish [19]. Thus we have for arbitrary D

$$k_n(s) = 0, \quad n = 3, 4, \ldots, D-1\,. \tag{2.23}$$

Equation (2.21) can be explicitly integrated. This can be proved as follows. Rewriting this equation in the form

$$2\,\mathcal{L}''(k_1)\,k_2\,dk_1 \,+\, \mathcal{L}'(k_1)\,dk_2 \,=\, 0$$

we obtain

$$2\,d\,[\ln \mathcal{L}'(k_1)] \,+\, d\,(\ln k_2) \,=\, 0\,. \tag{2.24}$$

Therefore,

$$\left(\mathcal{L}'(k_1)\right)^2 k_2 \,=\, C\,, \tag{2.25}$$

where C is an integration constant.

In view of eq. (2.25), the function $k_2(s)$ is eliminated from eq. (2.20). As a result, one nonlinear equation of the second order for the world trajectory curvature, $k_1(s)$ arises

$$\frac{d^2}{ds^2}\left(\mathcal{L}'(k_1)\right) \,=\, \left(k_1^2 \,+\, \frac{C^2}{(\mathcal{L}'(k_1))^4}\right)\mathcal{L}'(k_1) \,-\, k_1\,\mathcal{L}(k_1)\,. \tag{2.26}$$

Thus, we have derived a complete set of equations (see eqs. (2.23), (2.25), and (2.26)) for principal curvatures $k_i(s)$, $i \,=\, 1, 2, \ldots, D-1$ of the world curve $x^\mu(s)$ to be found.

3 Integrability of the Euler-Lagrange equations for principal curvatures

It is remarkable that the first integral for eq. (2.26) can be find in a general form for an arbitrary Lagrangian $\mathcal{L}(k_1)$. This integral naturally arises in analysing the Euler-Lagrange equations written in terms of the curve coordinates $x^\mu(s)$ to be found. The constants of integration turn out to be the particle mass and its spin.

In this section we shall use the arbitrary parametrization of the world curve $x^\mu(\tau)$. For convenience we somewhat alter our previous notation. The dot over x will denote the differentiation with respect to the evolution parameter τ, and the prime of x means the differentiation with respect to the natural parameter s. The prime of the Lagrangian function $\mathcal{L}$ will, as before, denote the differentiation with respect to its argument k_1. The action (2.1) assumes the form

$$S \,=\, \int L(\dot{x},\,\ddot{x})\,d\tau \,=\, \int \mathcal{L}(k_1)\,\sqrt{\dot{x}^2}\,d\tau\,, \tag{3.1}$$

where

$$L(\dot{x},\,\ddot{x}) \,=\, \sqrt{\dot{x}^2}\,\mathcal{L}(k_1)\,, \tag{3.2}$$

$$k_1^2 \,=\, \frac{(\dot{x}\ddot{x})^2 \,-\, \dot{x}^2\,\ddot{x}^2}{(\dot{x}^2)^3}\,. \tag{3.3}$$

Introducing the conserved Lorentz vector of the energy-momentum

$$P^\mu = \frac{d}{d\tau}\left(\frac{\partial L}{\partial \ddot{x}_\mu}\right) - \frac{\partial L}{\partial \dot{x}_\mu}, \quad \mu = 0, 1, \ldots, D-1, \tag{3.4}$$

we write the Euler-Lagrange equations generated by (3.1) as

$$\frac{d}{d\tau} P^\mu = 0, \quad \mu = 0, 1, \ldots, D-1. \tag{3.5}$$

Now we pass, in eq. (3.4), from an arbitrary evolution variable τ to the natural parameter s employing the following relations

$$\frac{d}{d\tau} = \sqrt{\dot{x}^2}\,\frac{d}{ds}\,, \quad \dot{x}_\mu = \sqrt{\dot{x}^2}\, x'_\mu\,,$$

$$x''_\mu = \frac{\dot{x}^2 \ddot{x}_\mu - (\dot{x}\ddot{x})\,\dot{x}_\mu}{(\dot{x}^2)^2}, \quad k_1 \frac{\partial k_1}{\partial \ddot{x}_\mu} = -\frac{x''_\mu}{\dot{x}^2}, \tag{3.6}$$

$$k_1 \frac{\partial k_1}{\partial \dot{x}_\mu} = \frac{1}{(\dot{x}^2)^4}\left\{\dot{x}^2\,(\dot{x}\ddot{x})\,\ddot{x}^\mu + \left[2\,\dot{x}^2\,\ddot{x}^2 - 3\,(\dot{x}\ddot{x})^2\right]\dot{x}^\mu\right\}$$

As a result, the energy-momentum vector acquires the form

$$P^\mu = x'^\mu\,(2\,\mathcal{L}'\,k_1 - \mathcal{L}) + x''^\mu\left(\frac{\mathcal{L}'\,k_1'}{k_1^2} - \frac{\mathcal{L}''}{k_1}\,k_1'\right) - \frac{\mathcal{L}'}{k_1}\,x'''^\mu. \tag{3.7}$$

Further we shall use the direct consequences of the definitions (2.2) and (2.3)

$$x'x'' = 0, \quad x'x''' = k_1^2, \quad x''x''' = -\,k_1\,k_1'\,. \tag{3.8}$$

The second curvature, k_2, (or torsion) is defined in the following way [23]

$$k_1^4\,k_2^2 = \det_G(x', x'', x'''), \tag{3.9}$$

where $\det_G(x', x'', x''')$ is the Gramm determinant for vectors x'_μ, x''_μ, x'''_μ [24]. It enables one to express $(x''')^2$ in terms of k_1 and k_2

$$(x''')^2 = k_1^4 - (k_1')^2 - k_1^2\,k_2^2\,. \tag{3.10}$$

In view of eq. (2.25), the torsion k_2 can be eliminated from (3.10)

$$(x''')^2 = k_1^4 - (k_1')^2 - k_1^2\,\frac{C^2}{(\mathcal{L}'(k_1))^4}\,. \tag{3.11}$$

Now we square the right and the left hand sides of eq. (3.7) and take into account (3.8) and (3.11). As a result, we obtain

$$M^2 \equiv P^2 = \mathcal{L}^2 - \left(\frac{d}{ds}\mathcal{L}'\right)^2 - 2\,\mathcal{L}\,\mathcal{L}'k_1 + (\mathcal{L}')^2\,k_1^2 - \frac{C^2}{(\mathcal{L}')^2}\,. \tag{3.12}$$

It turns out that eq. (3.12) is the first integral for the Euler-Lagrange equation (2.26) derived in the preceding Section. If $\mathcal{L}'' \neq 0$, one can be convinced of this by direct differentiation of (3.12) with respect to s. For Lagrangians linear in k_1 equations (2.26) and (3.12) should be treated as independent ones because the differentiation of (3.12) identically gives zero.

Therefore, analyzing equations of motion for $x^\mu(s)$ (eqs. (3.4) and (3.5)) we derive the first integral for the Euler-Lagrange equations in terms of the principal curvatures (eq. (2.26)). Hence, the order of eq. (2.26) can be reduced by one. From (3.12) we obtain

$$\frac{dk_1}{ds} = \pm\sqrt{f(k_1)}\,, \tag{3.13}$$

where

$$f(k_1) = \frac{1}{(\mathcal{L}'')^2}\left\{\mathcal{L}^2 - 2\,\mathcal{L}\,\mathcal{L}'\,k_1 + (\mathcal{L}')^2\,k_1^2 - \frac{C^2}{(\mathcal{L}')^2} - M^2\right\}. \tag{3.14}$$

Integration of (3.13) gives

$$\int\limits_{k_{10}}^{k_1}\frac{dx}{\sqrt{f(x)}} = \pm(s - s_0)\,, \tag{3.15}$$

where $k_{10} = k_1(s_0)$.

Thus, formula (3.15) defines the curvature of the world curve to be found as a function of s. Equations (2.23) and (2.25) determine the remaining curvatures. As a result, for an arbitrary Lagrangian $\mathcal{L}(k_1)$ the problem of finding the principal curvatures of the world trajectory sought for is reduced to quadratures.

It is interesting that the analysis of eqs. (3.4), (3.5) for $x^\mu(s)$ enables one to derive the relation between the curvature and torsion (2.25) in a new way. The integration constant C turns out to be expressed in terms of the particle spin and its mass.[1] Let us show this explicitly.

The invariance of the action (2.1) under Lorentz transformations entails the conservation of the angular-momentum tensor

$$M_{\mu\nu} = \sum_{a=1}^{2}(q_{a\mu}p_{a\nu} - q_{a\nu}p_{a\mu})\,, \tag{3.16}$$

where the canonical variables q_a^μ and p_a^μ are defined as follows

$$q_{1\mu} = x_\mu, \quad q_{2\mu} = \dot{x}_\mu,$$

[1] In the models defined by action (2.1) the spin of the particle proves to be nonzero already at the classical level. Its value is ultimately determined by the initial conditions for the corresponding equations of motion.

$$p_{1\mu} = P_\mu = -\frac{\partial L}{\partial \dot{x}^\mu} - \frac{dp_2}{d\tau}, \quad p_{2\mu} = -\frac{\partial L}{\partial \ddot{x}^\mu}. \tag{3.17}$$

The spin S of the particle will be also a conserved quantity. In the case of D-dimensional space-time the spin S is defined by [25]

$$S^2 = \frac{W}{M^2}, \tag{3.18}$$

where

$$W = \frac{1}{2} M_{\mu\nu} M^{\mu\nu} p_1^2 - (M_{\mu\sigma} p_1^\mu)^2, \quad M^2 = p_1^2.$$

For $D = 4$ the invariant W is the squared Pauli-Lubanski vector with sign minus

$$W = -w_\mu w^\mu, \quad w_\mu = \frac{1}{2} \varepsilon_{\mu\nu\rho\sigma} M^{\nu\rho} p_1^\sigma. \tag{3.19}$$

In view of these equations, one finds

$$M^2 S^2 = k_2^2 (\mathcal{L}')^4. \tag{3.20}$$

Hence, the integration constant C in (2.25) is given by

$$C^2 = M^2 S^2. \tag{3.21}$$

After quantization, S^2 in (3.21) should be replaced in the following way

$$S^2 \to S(S + D - 1), \quad S = 0, 1, \ldots, \tag{3.22}$$

where D is the dimension of space-time.

Therefore, dealing with the Euler-Lagrange equations written in terms of the curve coordinates $x^\mu(s)$ we have derived the basic equations (3.12) and (3.20) for the principal curvature k_1 and k_2 and have specified integration constants.[2] Probably, the remaining equations (2.23) can be also derived in this way. However, checking this possibility is beyond the scope of our consideration.

In conclusion of this section a general note concerning the mass spectrum in the models in question should be made. From eqs. (3.12) and (3.21) it follows that the tachyonic states with $M^2 < 0$ are unavoidably present in these theories already at the classical level [26]. This takes place at any values of the particle spin S.

4 Conclusion

The formulae (2.20) – (2.26), (3.13) – (3.15), and (3.17) provide a complete solution to the problem of obtaining equations of motion for arbitrary Lagrangians $\mathcal{L}(k_1)$ in terms of the principal curvatures of the world curve to be

[2]In paper [4] the Euler-Lagrange equations in terms of $x^\mu(s)$ have been used for obtaining some auxiliary conditions that should be satisfied by $k_1(s)$.

found and integrating these equations in quadratures. It is important that the integration constants are expressed in terms of the particle mass M and spin S. For comparison we note that in the book [19] and in paper [22] the complete integrability of the equations of motion has been proved only for the simplest Lagrangians $\mathcal{L}(k_1)$ linear and quadratic in k_1. The physical meaning of the integration constant was not elucidated there.

In the present paper we do not touch upon the problem of recovering the world trajectory by making use of its principal curvatures found here. If all the curvatures k_i, $i = 1, 2, \ldots, D-1$ are constants, then this problem is solved easily (see, for example, [23]). Obviously, in the general case one encounters here certain difficulties. However, it seems to us that the key properties of the dynamics in the models under consideration are determined by us fully enough. We have related the physical characteristics of the particle (its mass and spin) with geometrical invariants (principal curvatures) of its world trajectory.

References

[1] A. M. Polyakov (1986), *Nucl. Phys.*, **B286** 406.

[2] H. Kleinert (1986), *Phys. Lett.*, **B174** 335.

[3] M. S. Plyushchay (1988), *Mod. Phys. Lett.*, **A3** 1299; (1991), *Phys. Lett.*, **B253** 50.

[4] H. Arodź, A. Sitarz and P. Węgrzyn (1989), *Acta Phys. Polonica*, **B20** 921.

[5] M. Pavsic (1988), *Phys. Lett.*, **B205** 231.

[6] J. Grundberg, J. Isberg, U. Lindström and H. Nordström (1989), *Phys. Lett.*, **B231** 61.

[7] J. Isberg, U. Lindström and H. Nordström (1990), *Mod. Phys. Lett.*, **A5** 2491.

[8] A. M. Polyakov (1988), *Mod. Phys. Lett.*, **3A** 325.

[9] V. V. Nesterenko (1992), *Class. Quantum Grav.*, **9** 1101.

[10] M. S. Plyushchay (1990), *Phys. Lett.*, **B235** 47; (1991), *ibid.*, **B262**, 71; (1991), *Nucl. Phys.*, **B362** 54.

[11] S. Iso, C. Itoi and H. Mukaida (1990), *Phys. Lett.*, **B236** 287; (1990), *Nucl. Phys.*, **B346** 293.

[12] A. L. Kholodenko (1990), *Ann. Phys.*, **202** 186.

[13] E. R. Caianiello (1980), *Il Nuovo Cimento*, **B59** 350.

[14] E. R. Caianiello (1981), *Lett. Nuovo Cimento*, **32** 65.

[15] G. Scarpetta (1984), *Nuovo Cimento*, **41** 51.

[16] E. R. Caianiello, A. Feoli, M. Gasperini, and G. Scarpetta (1990), *Int. J. Theor. Phys.*, **29** 131.

[17] E. R. Caianiello, M. Gasperini, and G. Scarpetta (1990), *Nuovo Cimento*, **105B** 259.

[18] G. Fiorentini, M. Gasperini, and G. Scarpetta (1991), *Mod. Phys. Lett.*, **A6** 2033.

[19] P. A. Griffiths (1983), *Exterior differential systems and the calculus of variations*, Birkhäuser, Boston.

[20] L. P. Eisenhart (1964), *Riemannian Geometry*, Princeton University Press, Princeton.

[21] M. P. Do Carmo (1976), *Differential Geometry of Curves and Surfaces*, Prentice-Hall, London.

[22] T. Dereli, D. H. Hartley, M. Önder, and R. W. Tucker (1990), *Phys. Lett.*, **B252** 601.

[23] Yu. A. Aminov (1987), *Differential Geometry and Topology of Curves*, Nauka, Moscow.

[24] G. A. Korn and T. M. Korn (1968), *Mathematical Handbook*, McGraw-Hill, New York.

[25] S. S. Schweber (1961), *An Introduction to Relativistic Quantum Field Theory*, Row Peterson, New York.

[26] V. V. Nesterenko (1989), *J. Phys. A: Math. Gen.*, **22** 1673.

Dirac versus reduced phase space quantization

M. S. Plyushchay and A. V. Razumov

Abstract

The relationship between the Dirac and reduced phase space quantizations is investigated for spin models belonging to the class of Hamiltonian systems having no gauge conditions. It is traced out that the two quantization methods may give similar, or essentially different physical results, and, moreover, it is shown that there is a class of constrained systems, which can be quantized only by the Dirac method. A possible interpretation of the gauge degrees of freedom is given.

1 Introduction

There are two main methods to quantize the Hamiltonian systems with first class constraints: the Dirac quantization [1] and the reduced phase space quantization [2], whereas two other methods, the path integral method [3, 2] and the BRST quantization [4] being the most popular method for the covariant quantization of gauge-invariant systems, are based on and proceed from them [2, 5]. The basic idea of the Dirac method consists in imposing quantum mechanically the first class constraints as operator conditions on the states for singling out the physical ones [1]. The reduced phase space quantization first identifies the physical degrees of freedom at the classical level by the factorization of the constraint surface with respect to the action of the gauge group, generated by the constraints. Then the resulting Hamiltonian system is quantized as a usual unconstrained system [2]. Naturally, the problem of the relationship of these two methods arises. It was discussed in different contexts in literature [6], and there is an opinion that the differences between the two quantization methods can be traced out to a choice of factor ordering in the construction of various physical operators.

We investigate the relationship of the two methods of quantization for the special class of Hamiltonian systems with first class constraints corresponding to different physical models of spinning particles. The specific general property of the examples of constrained systems considered here is the following: their constraints generate SO(2) transformations and, hence, corresponding gauge orbits topologically are one-spheres S^1. This fact implies that these systems *do not admit gauge conditions*, and, therefore, for the construction of their reduced phase spaces we shall use a general geometrical approach to the Dirac–Bergmann theory of the constrained systems [7, 8].

2 Plane spin model

The first model we are going to consider is the plane spin model, which is a subsystem of the (3+1)–dimensional models of massless particles with arbitrary helicity [9], and of the (2+1)–dimensional relativistic models of fractional spin particles [10]. The initial phase space of the model is a cotangent bundle T^*S^1 of the one–dimensional sphere S^1, that is a cylinder $S^1 \times \mathbf{R}$. It can be described *locally* by an angular variable $0 \leq \varphi < 2\pi$ and the conjugate momentum $S \in \mathbf{R}$. The symplectic two–form ω in terms of the local variables φ, S has the form $\omega = dS \wedge d\varphi$, and, thus, we have locally $\{\varphi, S\} = 1$. Actually, any 2π–periodical function of the variable φ that is considered as a variable, taking values in $\mathbf{R}$, can be considered as a function on the phase space, i.e., as an observable, and any observable is connected with the corresponding 2π–periodical function. Therefore, we can introduce the functions $q_1 = \cos\varphi$, $q_2 = \sin\varphi$, $q_1^2 + q_2^2 = 1$, as the dependent functions on the phase space of the system. For these functions we have $\{q_1, q_2\} = 0$, $\{q_1, S\} = -q_2$, $\{q_2, S\} = q_1$. Any function on the phase space can be considered as a function of dependent coordinates q_1, q_2 and S, which will be taken below as the quantities, forming a restricted set of observables whose quantum analogs have the commutators which are in the direct correspondence with their Poisson brackets.

We come to the plane spin model by introducing the 'spin' constraint

$$\psi = S - \theta = 0, \tag{2.1}$$

where θ is an arbitrary real constant. Let us consider the Dirac quantization of the system. To this end we take as the Hilbert space the space of complex 2π–periodical functions of the variable φ with the scalar product $(\Phi_1, \Phi_2) = \frac{1}{2\pi}\int_0^{2\pi} \overline{\Phi_1(\varphi)}\Phi_2(\varphi)\, d\varphi$. The operators $\hat{q}_1$ and $\hat{q}_2$, corresponding to the functions q_1 and q_2, are the operators of multiplication by the functions $\cos\varphi$ and $\sin\varphi$, respectively, whereas the operator $\hat{S}$ is defined by $\hat{S}\Phi = (-id/d\varphi + c)\Phi$, where c is an arbitrary real constant. The operators $\hat{q}_1$, $\hat{q}_2$ and $\hat{S}$ are Hermitian operators with respect to the introduced scalar product, and they satisfy the relation $[\hat{A}, \hat{B}] = i\{A, B\}$, $A, B = q_1, q_2, S$. The quantum analog of the constraint (2.1) gives the equation for the physical state wave functions: $(\hat{S} - \theta)\Phi_{phys} = 0$. Decomposing the function $\Phi_{phys}(\varphi)$ over the orthonormal basis, formed by the functions $e^{ik\varphi}$, $k \in \mathbf{Z}$, we find this equation has a nontrivial solution only when $c = \theta + n$, where n is some fixed integer, $n \in \mathbf{Z}$. In this case the corresponding physical normalized wave function is $\Phi_{phys}(\varphi) = e^{in\varphi}$. The only physical operator [5], i.e., an operator commuting with the quantum constraint $\hat{\psi}$ here is $\hat{S}$, which is reduced to the constant θ on the physical subspace.

Now we come back to the classical theory in order to construct the reduced phase space of the model. Let us show that for the surface, defined by Eq. (2.1), there is no 'good' gauge condition, but, nevertheless, the reduced phase space of the system can be constructed. Indeed, it is clear that the one–parameter group of transformations, generated by the constraint ψ, consists of the rotations of the phase space. This group acts transitively on the constraint surface, and we have only one gauge orbit, which is the constraint surface itself. The gauge conditions must single out one point of an orbit. In our case we have to define only one gauge condition, let us denote it by χ. The function χ must be such that the pair of equations $\psi = 0$, $\chi = 0$ would determine a set, consisting of only one point, and in this point we should have $\{\psi, \chi\} \neq 0$. Recall that any function on the phase space of the system under consideration can be considered as a function of the variables φ and S, which is 2π–periodical with respect to φ. Thus, we require the 2π–periodical function $\chi(\varphi, S)$ turn into zero at only one point $\varphi = \varphi_0$ from the interval $0 \leq \varphi < 2\pi$ when $S = \theta$. Moreover, we should have $\{\psi, \chi\}(\varphi_0, \theta) = -\left.\partial\chi(\varphi, \theta)/\partial\varphi\right|_{\varphi=\varphi_0} \neq 0$. It is clear that such a function does not exist. Nevertheless, here we have the reduced phase space that consists of only one point. Therefore, the reduced space quantization is trivial: physical operator $\hat{S}$ takes here constant value θ in correspondence with the results obtained by the Dirac quantization method. When the described plane spin model is a subsystem of some other system, the reduction means simply that the cylinder T^*S^1 is factorized into a point, where $S = \theta$, and that wave functions do not depend on the variable φ.

Let us point out one interesting analogy in interpretation of the situation with nonexistence of a global gauge condition. Here the condition of 2π–periodicity can be considered as a 'boundary' condition. If for a moment we forget about it, we can take as a gauge function any monotonic function $\chi(\varphi, S)$, $\chi \in \mathbf{R}$, such that $\chi(\varphi_0, \theta) = 0$ at some point $\varphi = \varphi_0$, and, in particular, we can choose the function $\chi(\varphi, S) = \varphi$. The 'boundary' condition excludes all such global gauge conditions. In this sense the situation is similar to the situation in the non–Abelian gauge theories where without taking into account the boundary conditions for the fields it is also possible to find global gauge conditions, whereas the account of those leads, in the end, to the nonexistence of global gauge conditions [11].

3 Rotator spin model

Let us consider now the rotator spin model [12]. The initial phase space of the system is described by a spin three–vector $\boldsymbol{S}$ and a unit vector $\boldsymbol{q}$, $\boldsymbol{q}^2 = 1$, being orthogonal one to the other, $\boldsymbol{qS} = 0$. The variables q_i and S_i, $i = 1, 2, 3$, can be considered as dependent coordinates in the phase space of the system. The Poisson brackets for these coordinates are $\{q_i, q_j\} = 0$, $\{S_i, S_j\} = \epsilon_{ijk}S_k$, $\{S_i, q_j\} = \epsilon_{ijk}q_k$. Using these Poisson brackets, we find the following expression for the symplectic two–form: $\omega = dp_i \wedge dq_i = d(\epsilon_{ijk}S_jq_k) \wedge dq_i$. Introducing the spherical angles φ, ϑ $(0 \leq \varphi < 2\pi, 0 \leq \vartheta \leq \pi)$ and the corresponding momenta $p_\varphi, p_\vartheta \in \mathbf{R}$, we can write the parameterization for the vector $\boldsymbol{q}$, $\boldsymbol{q} = (\cos\varphi\sin\vartheta, \sin\varphi\sin\vartheta, \cos\vartheta)$ and corresponding parameterization for the vector $\boldsymbol{S} = \boldsymbol{S}(\vartheta, \varphi, p_\varphi, p_\vartheta)$, whose explicit form we do not write down here (see Ref. [8]). Then for the symplectic two–form we get the expression $\omega = dp_\vartheta \wedge d\vartheta + dp_\varphi \wedge d\varphi$. From this relation we conclude that the initial phase space of the system is symplectomorphic to the cotangent bundle T^*S^2 of the two–dimensional sphere S^2, furnished with the canonical symplectic structure.

The rotator spin model is obtained from the initial phase space by imposing the constraint

$$\psi = \frac{1}{2}(\boldsymbol{S}^2 - \rho^2) = 0, \qquad \rho > 0, \tag{3.1}$$

fixing the spin of the system. Using the Dirac method, we quantize the model in the following way. The state space is a space of the square integrable functions on the two–dimensional sphere. The scalar product is $(\Phi_1, \Phi_2) = \int_{S^2} \overline{\Phi_1(\varphi,\vartheta)}\Phi_2(\varphi,\vartheta)\sin\vartheta d\vartheta d\varphi$. The above mentioned parameterization allows us to use as the operator $\hat{\boldsymbol{S}}$ the usual orbital angular momentum operator expressed via spherical angles. The wave functions as the functions on a sphere are decomposable over the complete set of the spherical harmonics: $\Phi(\varphi,\vartheta) = \sum_{l=0}^{\infty}\sum_{m=-l}^{l}\Phi_{lm}Y_m^l(\varphi,\vartheta)$, and, therefore, the quantum analog of the first class constraint (3.1),

$$(\hat{\boldsymbol{S}}^2 - \rho^2)\Phi_{phys} = 0, \tag{3.2}$$

leads to the quantization condition for the constant ρ:

$$\rho^2 = n(n+1), \tag{3.3}$$

where $n > 0$ is an integer. Only in this case equation (3.2) has a nontrivial solution of the form $\Phi^n_{phys}(\vartheta,\varphi) = \sum_{m=-n}^{n}\Phi_{nm}Y_m^n(\varphi,\vartheta)$, i.e., with the choice of (3.3) we get the states with spin equal to n: $\hat{\boldsymbol{S}}^2\Phi^n_{phys} = n(n+1)\Phi^n_{phys}$. Thus, we conclude that the Dirac quantization leads to the quantization (3.3) of the parameter ρ and, as a result, the quantum system describes the states with integer spin n.

Let us turn now to the construction of the reduced phase space of the system. The constraint surface of the model can be considered as a set composed of the points specified by two orthonormal three–vectors. Each pair of such vectors can be supplemented by a unique third three–vector, defined in such a way that we get an oriented orthonormal basis in three dimensional vector space. It is well known that the set of all oriented orthonormal bases in three dimensional space can be smoothly parameterized by the elements of the Lie group SO(3). Thus, the constraint surface in our case is diffeomorphic to the group manifold of the Lie group SO(3).

The one–parameter group of canonical transformations, generated by the constraint ψ, acts in the following way: $\boldsymbol{q}(\tau) = \boldsymbol{q}\cos(S\tau) + (\boldsymbol{S}\times\boldsymbol{q})S^{-1}\sin(S\tau)$,

$\boldsymbol{S}(\tau) = \boldsymbol{S}$, where $S = \sqrt{\boldsymbol{S}^2}$. Hence, we see that the gauge transformations are the rotations about the direction, given by the spin vector. Thus, in the case of a general position the orbits of the one–parameter group of transformations under consideration are one dimensional spheres. Note, that only the orbits, belonging to the constraint surface where $S = \rho \neq 0$, are interesting to us. It is clear that an orbit is uniquely specified by the direction of the spin three–vector $\boldsymbol{S}$ whose length is fixed by the constraint ψ. As a result of our consideration, we conclude that the reduced phase space of the rotator spin model is the coset space SO(3)/SO(2), which is diffeomorphic to the two–dimensional sphere S^2. Due to the reasons discussed for the preceding model there is no gauge condition in this case either. In fact, since SO(3) is a nontrivial fiber bundle over S^2, we can neither find a mapping from S^2 to SO(3) whose image would be diffeomorphic to the reduced phase space. In other words, in this case the reduced phase space cannot be considered as a submanifold of the constraint surface.

Our next goal is to write an expression for the symplectic two–form on the reduced phase space. We can consider the variables S_i as dependent coordinates in the reduced phase space, and the symplectic two–form on it may be expressed in terms of them. With the help of an orthonormal basis formed by the vectors $\boldsymbol{q}$, $\boldsymbol{s} = \boldsymbol{S}/S$ and $\boldsymbol{q} \times \boldsymbol{s}$, we get for the symplectic two-form on the reduced phase space the following expression [8]:

$$\omega = -\frac{1}{2\rho^2}(\boldsymbol{S} \times d\boldsymbol{S}) \wedge d\boldsymbol{S}. \tag{3.4}$$

Thus, we see that the dependent coordinates S^i in the reduced phase space of the system provide a realization of the basis of the Lie algebra so(3):

$$\{S_i, S_j\} = \epsilon_{ijk} S_k. \tag{3.5}$$

The quantization on the reduced phase space can be performed with the help of the geometric quantization method proceeding from the classical relations (3.4), (3.5) and $\boldsymbol{S}^2 = \rho^2$. This was done in detail, e.g., in Ref. [17], and we write here the final results of this procedure. The constant ρ is quantized:

$$\rho = j, \qquad 0 < 2j \in \mathbf{Z}, \tag{3.6}$$

i.e., it can take only integer or half-integer value, and the Hermitian operators, corresponding to the components of the spin vector, are realized in the form: $\hat{S}_1 = \frac{1}{2}(1 - z^2)d/dz + jz$, $\hat{S}_2 = \frac{i}{2}(1 + z^2)d/dz - ijz$, $\hat{S}_3 = zd/dz - j$, where $z = e^{-i\varphi} \tan \vartheta/2$, or, in terms of the dependent coordinates, $z = (S_1 - iS_2)/(\rho + S_3)$. Operators $\hat{S}_i$ act in the space of holomorphic functions $f(z)$ with the scalar product $(f_1, f_2) = \frac{2j+1}{\pi} \int\int \overline{f_1(z)} f_2(z)(1 + |z|^2)^{-(2j+2)} d^2 z$, in which the functions $\psi_j^m \propto z^{j+m}$, $m = -j, -j+1, ..., j$, form the set of eigenfunctions of the operator $\hat{S}_3$ with the eigenvalues $s_3 = m$. These operators satisfy the relation $\hat{\boldsymbol{S}}^2 = j(j+1)$, and, therefore, we have the $(2j+1)$–dimensional irreducible representation D_j of the Lie group SU(2).

Thus, we see that for the rotator spin model the reduced phase space quantization method leads to the states with integer or half–integer spin, depending on the choice of the quantized parameter ρ, and gives in general the results physically different from the results obtained with the help of the Dirac quantization method. Let us stress once again here that within the Dirac quantization method in this model the spin operator $\hat{\boldsymbol{S}}$ has a nature of the orbital angular momentum operator, and it is this nature that does not allow spin to take half-integer values [18].

4 Top spin model

Let us consider now the top spin model [13]. The initial phase space of the model is described by the spin three–vector $\boldsymbol{S}$, and by three vectors $\boldsymbol{e}_i$ such that $\boldsymbol{e}_i\boldsymbol{e}_j = \delta_{ij}$, $\boldsymbol{e}_i \times \boldsymbol{e}_j = \epsilon_{ijk}\boldsymbol{e}_k$. Denote the components of the vectors $\boldsymbol{e}_i$ by E_{ij}. The components S_i of the vector $\boldsymbol{S}$ and the quantities E_{ij} form a set of dependent coordinates in the phase space of the system. The corresponding Poisson brackets are

$$\{E_{ij}, E_{kl}\} = 0, \quad \{S_i, E_{jk}\} = \epsilon_{ikl}E_{jl}, \quad \{S_i, S_j\} = \epsilon_{ijk}S_k. \tag{4.1}$$

The vectors $\boldsymbol{e}_i$ form a right orthonormal basis in $\mathbf{R}^3$. The set of all such bases can be identified with the three–dimensional rotation group. Taking into account Eqs. (4.1) we conclude that the initial phase space is actually the cotangent bundle $T^*\mathrm{SO}(3)$, represented as the manifold $\mathbf{R}^3 \times \mathrm{SO}(3)$. Using Eqs. (4.1), one can get the following expression for the symplectic two–form ω on the initial phase space: $\omega = \frac{1}{2}d(\boldsymbol{S} \times \boldsymbol{e}_l) \wedge d\boldsymbol{e}_l = \frac{1}{2}d(\epsilon_{ijk}S_jE_{lk}) \wedge dE_{li}$.

It is useful to introduce the variables $J_i = \boldsymbol{e}_i\boldsymbol{S} = E_{ij}S_j$. For these variables we have the following Poisson brackets: $\{J_i, E_{jk}\} = -\epsilon_{ijl}E_{lk}$, $\{J_i, J_j\} = -\epsilon_{ijk}J_k$. Note, that we have the equality $S_iS_i = J_iJ_i$.

The phase space of the top spin model is obtained from the phase space, described above, by introducing two first class constraints

$$\psi = \frac{1}{2}(\boldsymbol{S}^2 - \rho^2) = 0, \qquad \chi = \boldsymbol{S}\boldsymbol{e}_3 - \kappa = 0, \tag{4.2}$$

where $\rho > 0$, $|\kappa| < \rho$. Consider now the Dirac quantization of the model. Let us parameterize the matrix E, which can be identified with the corresponding rotation matrix, by the Euler angles, $E = E(\alpha, \beta, \gamma)$, and use the representation where the operators, corresponding to these angles are diagonal. In this representation state vectors are functions of the Euler angles, and the operators $\hat{S}_i$ and $\hat{J}_i$ are realized as linear differential operators, acting on such functions [19]. The quantum analogs of the constraints ψ and χ turn into the equations for the physical states of the system:

$$(\hat{\boldsymbol{S}}^2 - \rho^2)\Phi_{phys} = 0, \qquad (\hat{J}_3 - \kappa)\Phi_{phys} = 0. \tag{4.3}$$

An arbitrary state vector can be decomposed over the set of the Wigner functions, corresponding to either integer or half–integer spins [19]: $\Phi(\alpha,\beta,\gamma) = \phi_{jmk}D^j_{mk}(\alpha,\beta,\gamma)$, where $j = 0, 1, \ldots$, or $j = 1/2, 3/2, \ldots$, and $k, m = -j, -j+1, \ldots, j$. The Wigner functions D^j_{mk} have the properties: $\hat{\boldsymbol{S}}^2 D^j_{mk} = j(j+1)D^j_{mk}$, $\hat{S}_3 D^j_{mk} = mD^j_{mk}$, $\hat{J}_3 D^j_{mk} = kD^j_{mk}$. Using the decomposition of the state vector, we see that Eqs. (4.3) have nontrivial solutions only when $\rho^2 = j(j+1)$, and $\kappa = k$, for some integer or half–integer numbers j and k, such that $-j \leq k \leq j$. In other words we get the following quantization condition for the parameters of the model:

$$\rho^2 = j(j+1), \quad \kappa = k, \qquad -j \leq k \leq j, \quad 0 < 2j \in \mathbf{Z}.$$

The corresponding physical state vectors have the form $\Phi_{phys}(\alpha,\beta,\gamma) = \sum_{m=-j}^{j} \varphi_m D^j_{mk}(\alpha,\beta,\gamma)$. Thus, we see that the Dirac quantization of the top spin model leads to an integer or half–integer spin system.

Proceed now to the construction of the reduced phase space of the system. As the constraints ψ and χ have zero Poisson bracket, we can consider them consecutively. Let us start with the constraint ψ. From the expressions for the Poisson brackets (4.1) it follows that the group of gauge transformations, generated by the constraint ψ, acts in the initial phase space variables as follows: $\boldsymbol{e}_i(\tau) = \boldsymbol{e}_i \cos(S\tau) + (\boldsymbol{S} \times \boldsymbol{e}_i)S^{-1}\sin(S\tau) + \boldsymbol{S}(\boldsymbol{S}\boldsymbol{e}_i)S^{-2}(1-\cos(S\tau))$, $\boldsymbol{S}(\tau) = \boldsymbol{S}$, where $S = \sqrt{\boldsymbol{S}^2}$. We see that the transformation under consideration have the sense of the rotation by the angle $S\tau$ about the direction of the spin vector. Let us consider the initial phase space of the system being diffeomorphic to $\mathbf{R}^3 \times \mathrm{SO}(3)$ as a trivial fibre bundle over $\mathbf{R}^3$ with the fibre SO(3). The gauge transformations act in fibres of this bundle. It is clear that the constraint surface, defined by the constraint ψ, is a trivial fibre subbundle $S^2 \times \mathrm{SO}(3)$. As $\mathrm{SO}(3)/\mathrm{SO}(2) = S^2$, then after the reduction over the action of the gauge group we come to the fibre bundle over S^2 with the fibre S^2. As it follows from general theory of fibre bundles [20], this fibre bundle is again trivial. Thus the reduced phase space, obtained using only the constraint ψ, is the direct product $S^2 \times S^2$. The symplectic two–form on this reduced space can be written in the form [8]: $\omega = -(2\rho^2)^{-1}(\epsilon_{ijk}S_i dS_j \wedge dS_k - \epsilon_{ijk}J_i dJ_j \wedge dJ_k)$. Here the quantities S_i and J_i form a set of dependent coordinates in the reduced phase space under consideration: $S_iS_i = J_iJ_i = \rho^2$.

Let us turn our attention to the constraint χ. It is easy to get convinced that the transformations of the gauge group, generated by this constraint act in the initial phase space in the following way: $\boldsymbol{e}_i(\tau) = \boldsymbol{e}_i \cos\tau + (\boldsymbol{e}_3 \times \boldsymbol{e}_i)\sin\tau$, $i = 1, 2$, $\boldsymbol{e}_3(\tau) = \boldsymbol{e}_3$, $\boldsymbol{S}(\tau) = \boldsymbol{S}$. So, we see that the gauge group, generated by the constraint χ, acts only in one factor of the product $S^2 \times S^2$, which is a reduced phase space obtained by us after reduction with the help of the constraint ψ. Thus we can consider only that factor, which is evidently described by the quantities J_i. From such point of view, the constraint surface, defined by the constraint χ, is a one dimensional sphere S^1, where the group of gauge transformations acts transitively. Hence, after reduction we get only one point. Thus, the final reduced phase space is a two–dimensional sphere S^2, and the symplectic two–form on the reduced phase space has the form given by Eq. (3.4). Therefore, the reduced phase space we have obtained, coincides with the reduced phase space for the rotator spin model. Hence the geometric quantization method gives again the quantization condition (3.6) for the parameter ρ, while the parameter κ remains unquantized here. Therefore, while for this model unlike the previous one, two methods of quantization lead to the quantum system, describing either integer or half-integer spin states, nevertheless, the corresponding quantum systems are different: the Dirac method gives discrete values for the observable $\hat{J}_3$, whereas the reduced phase space quantization allows it to take any value κ, such that $\kappa^2 < j^2$ for a system with spin j.

Let us note here one interesting property of the system. We can use a combination of the Dirac and reduced phase space quantization methods. After the first reduction with the help of the constraint ψ, the system, described by the spin vector and the 'isospin' vector [13] with the components $I_i = -J_i$, $S_i S_i = I_i I_i$, can be quantized according to Dirac by imposing the quantum analog of the constraint χ on the state vectors for singling out the physical states. In this case we have again the quantization of the parameter κ as in the pure Dirac quantization method, and, therefore, here the observable $\hat{J}_3$ can take only integer or half–integer value. Hence, in this sense, such a combined method gives the results coinciding with the results of the Dirac quantization method.

5 Discussion and conclusions

The first considered model gives an example of the classical constrained system with finite number of the degrees of freedom for which there is no gauge condition, but nevertheless, the reduced phase space can be represented as a submanifold of the constraint surface. As we have seen, Dirac and reduced phase space quantization methods lead to the coinciding physical results for this plane spin model. Moreover, we have revealed an interesting analogy in interpretation of the situation with nonexistence of a global gauge condition for this simple constrained system with the situation taking place for the non-Abelian gauge theories [11].

The rotator and top spin models give examples of the classical systems, in which there is no global section of the space of gauge orbits. In spite of impossibility to impose gauge conditions such systems admit the construction of the reduced phase space. These two models demonstrate that the reduced phase space and the Dirac quantization methods can give essentially different physical results.

Thus, for Hamiltonian systems with first class constraints we encounter two related problems.

The first problem consists in the choice of a 'correct' quantization method for such systems. From the mathematical point of view any quantization leading to a quantum system, which has the initial system as its classical limit, should be considered as a correct one, but physical reasonings may distinguish different quantization methods. Consider, for example, the above mentioned systems. The rotator spin model, quantized according to the Dirac method, represents by itself the orbital angular momentum system with additional condition (3.2) singling out the states with a definite eigenvalue of angular momentum operator $\hat{\boldsymbol{S}}^2$. This eigenvalue, in turn, is defined by the concrete value of the quantized parameter of the model: $\rho^2 = n(n+1) > 0$. On the other hand, the reduced phase space quantization of the model gives either integer or half–integer values for the spin of the system. If we suppose that the system under consideration is to describe orbital angular momentum, we must take only integer values for the parameter ρ in the reduced phase space quantization method. But in this case we must, nevertheless, conclude, that the reduced phase space quantization method of the rotator spin model describes a more general system than the quantum system obtained as a result of the Dirac quantization of that classical system.

The Dirac quantization of the top spin model, or its combination with the reduced phase space quantization gives us a possibility to interpret this system as a system having spin and isospin degrees of freedom (with equal spin and isospin: $\hat{\boldsymbol{S}}^2 = \hat{I}_i\hat{I}_i = j(j+1)$), but in which the isospin degrees of freedom are 'frozen' by means of the condition $\hat{I}_3\Phi_{phys} = -k\Phi_{phys}$. On the other hand, as we have seen, the reduced space quantization method does not allow one to have such interpretation of the system since it allows the variable I_3 to take any (continuous) value $-\kappa$ restricted only by the condition $\kappa^2 < j^2$, i.e., the operator $\hat{I}_3$ (taking here only one value) cannot be interpreted as a component of the isospin vector operator. From this point of view a 'more correct' method of quantization is the Dirac quantization method.

In this respect it is worth to point out that there is a class of physical models, for which it is impossible to get the reduced phase space description, and which, therefore, can be quantized only by the Dirac method.

Indeed, there are various pseudoclassical models containing first class nilpotent constraints of the form [14]–[16]:

$$\psi = \xi_{i_1}...\xi_{i_n}G^{i_1...i_n} = 0, \tag{5.1}$$

where ξ_{i_k}, are real Grassmann variables with the Poisson brackets $\{\xi_k, \xi_l\} = -ig_{kl}$, g_{kl} being a real nondegenerate symmetric constant matrix. Here it is supposed that $G^{i_1 \dots i_n}$, $n \geq 2$, are some functions of other variables, antisymmetric in their indices, and all the terms in a sum have simultaneously either even or odd Grassmann parity. For our considerations it is important that constraints (5.1) are the constraints, nonlinear in Grassmann variables, and that they have zero projection on the unit of Grassmann algebra. In the simplest example of relativistic massless vector particle in (3+1)–dimensional space–time [14] the odd part of the phase space is described by two Grassmann vectors ξ^a_μ, $a = 1, 2$, with brackets $\{\xi^a_\mu, \xi^b_\nu\} = -i\delta^{ab} g_{\mu\nu}$, and the corresponding nilpotent first class constraint has the form:

$$\psi = i\xi^1_\mu \xi^2_\nu g^{\mu\nu} = 0, \tag{5.2}$$

where $g_{\mu\nu} = \mathrm{diag}(-1, 1, 1, 1)$. This constraint is the generator of the SO(2)–rotations in the 'internal isospin' space: $\xi^1_\mu(\tau) = \xi^1_\mu \cos\tau + \xi^2_\mu \sin\tau$, $\xi^2_\mu(\tau) = \xi^2_\mu \cos\tau - \xi^1_\mu \sin\tau$. The specific property of this transformation is that having $\xi^a_\mu(\tau)$ and ξ^a_μ, we cannot determine the rotation angle τ because there is no notion of the inverse element for an odd Grassmann variable. Another specific feature of the nilpotent constraint (5.2) is the impossibility to introduce any, even local, gauge constraint for it. In fact, we cannot find a gauge constraint χ such that the Poisson bracket $\{\psi, \chi\}$ would be invertible. Actually, it is impossible *in principle* to construct the corresponding reduced phase space for such a system. Obviously, the same situation arises for the constraint of general form (5.1). It is necessary to note here that in the case when the constraint ψ depends on even variables of the total phase space (see, e.g., ref. [16]), and, therefore, generates also transformations of some of them, we cannot fix the transformation parameter (choose a point in the orbit) from the transformation law of those even variables, because the corresponding parameter is present in them with a noninvertible factor, nonlinear in Grassmann variables. Therefore, the pseudoclassical systems containing the constraints of form (5.1) can be quantized only by the Dirac method, that was done in original papers [14]–[16].

Let us come back to the discussion of the revealed difference between two methods of quantization, and point out that the second related problem is clearing up the sense of gauge degrees of freedom. The difference appearing under the Dirac and reduced phase space quantization methods can be understood as the one proceeding from the quantum 'vacuum' fluctuations corresponding to the 'frozen' (gauge) degrees of freedom. Though these degrees of freedom are 'frozen' by the first class constraints, they reveal themselves through quantum fluctuations, and in the Dirac quantization method they cannot be completely 'turned off' due to the quantum uncertainty principle. Thus, we can suppose that the gauge degrees of freedom serve not simply for 'covariant' description of the system but have 'hidden' physical meaning, in some sense similar to the compactified degrees of freedom in the Kaluza–Klein theories. If we adopt such a point of view, we have to use only the Dirac quantization method. Further, the gauge principle cannot be considered then as a pure technical principle. From here we arrive also at the conclusion that the Dirac separation of the constraints into first and second class constraints is not technical, and nature 'distinguish' these two cases as essentially different, since gauge degrees of freedom, corresponding to the first class constraints, may reveal themselves at the quantum level (compare with the point of view advocated in Ref. [21]).

References

[1] P. A. M. Dirac (1964), *Lectures on Quantum Mechanics*, Yeshiva University, New York.

[2] L. D. Faddeev (1970), *Theor. Math. Phys.*, **1** 1.

[3] L. D. Faddeev and V. N. Popov (1967), *Phys. Lett.*, **25B** 29.

[4] C. Becchi, A. Rouet and R. Stora (1974), *Phys. Lett.*, **52B** 344;
I. V. Tyutin (1975), 'Gauge Invariance in Field Theory and Statistical Physics in Operatorial Formalism', FIAN preprint 39, Moscow, (in Russian);
E. S. Fradkin and G. A. Vilkovisky (1975), *Phys. Lett.*, **55B** 224;
I. A. Batalin and G. A. Vilkovisky (1985), *J. Math. Phys.*, **26** 172;
M. Henneaux (1985), *Phys. Rep.*, **126** 1.

[5] K. Sundermeyer (1982), *Constrained Dynamics*, Lecture Notes in Physics **169**, Springer, Berlin.

[6] K. Kuchař (1986), *Phys. Rev.*, **D34** 3044;
J. Romano and R. Tate (1989), *Class. Quant. Grav.*, **6** 1487;
K. Schleich (1990), *Class. Quant. Grav.*, **7** 1529;
R. Loll (1990), *Phys. Rev.*, **D41** 3785;
A. Ashtekar (1991), *Lectures on Nonpertubative Canonical Gravity*, World Scientific, Singapore.

[7] R. Abraham and J. E. Marsden (1978), *Foundation of Mechanics*, Benjamin, New York.

[8] M.S. Plyushchay and A.V. Razumov (1993), 'Dirac versus Reduced Phase Space Quantization: Relationship of Two Methods', preprint IHEP-93-81 (Protvino), hep-th/9306017.

[9] M. S. Plyushchay (1989), *Mod. Phys. Lett.*, **A4** 837; (1990), *Phys. Lett.*, **B243** 383.

[10] M. S. Plyushchay (1990), *Phys. Lett.*, **B248** 107; (1992), *Int. J. Mod. Phys.*, **A7** 7045.

[11] V. N. Gribov (1978), *Nucl. Phys.*, **B139** 1;
I. M. Singer (1978), *Commun. Math. Phys.*, **60** 7.

[12] M. S. Plyushchay (1990), *Phys. Lett.*, **B236** 291.

[13] M. S. Plyushchay (1990), *Phys. Lett.*, **B248** 299.

[14] V.D. Gershun and V.I. Tkach (1979), *JETP Lett.*, **29** 320;
A. Barducci and L. Lusanna (1983), *J. Phys.: Math. Gen.*, **A16** 1993;
P. S. Howe, S. Penati, M. Pernici and P. K. Townsend (1989), *Class. Quantum Gravity*, **6** 1125.

[15] M.S. Plyushchay (1993), *Mod. Phys. Lett.*, **A8** 937.

[16] J. L. Cortés, M. S. Plyushchay and L. Velázquez (1993), *Phys. Lett.*, **B306** 34.

[17] M. S. Plyushchay (1991), *Nucl. Phys.*, **B362** 54.

[18] L.C. Biedenharn and J.D. Louck (1981), *Angular Momentum in Quantum Physics: Theory and Application*, Addison-Wessley, Reading, MA.

[19] D. A. Varshalovich, A. N. Moskalev and V. K. Khersonsky (1989), *Quantum Theory of Angular Momentum*, World Scientific, Singapore.

[20] D. Husemoller (1966), *Fibre Bundles*, Mc–Graw–Hill, New York.

[21] L. Faddeev and R. Jackiw (1988), *Phys. Rev. Lett.*, **60** 1692.

Classical and quantum aspects of degenerate metric fields

Robin W. Tucker

Abstract

Degenerate space-time metrics are discussed in the framework of gravitation in interaction with scalar fields. Minisuperspace quantisations are effected leading to coherent states that exhibit enhanced fluctuations in the vicinity of classical cosmological loci. An alternative approach for implementing the Hamiltonian constraint is offered in terms of a spinor state space and an analysis of solutions to the scalar field equation on a 2-dimensional topologically non-trivial manifold with a degenerate metric is summarised.

1 Introduction

In the absence of a viable quantum description of the gravitational field attention is sometimes directed to the so called " mini-superspace" models [1] in which all but a small number of degrees of freedom of the gravitational field are suppressed and the dynamics is reduced from field theory to quantum mechanics. Such a programme is fraught with both conceptual and technical difficulties. It ignores many effects that may be of relevance in determining a viable quantum description. One restricts to quantum states describing highly symmetric geometries that possess a preferred class of spacelike foliations that may be used to order temporal phenomena in a classical spacetime. Even within this restricted framework there is no preferred way to effect a quantisation of Einstein's equations of motion tensorial nature of these equations gives rise to a constrained canonical system and there is no known criterion that singles out a particular mapping from the classical constraints to a set of quantum operators on a Hilbert space.

Mini-superspace models do however have the virtue that they provide a class of models which can be investigated in the hope of obtaining specific answers to well defined questions. Einstein's equations may be reduced to ordinary differential equations and these can often be given an interpretation in terms of a constrained particle dynamical system. The classical equations of motion and constraints are often non-linear and the quantisation procedure is not unique. One may effect an infinite number of transformations that recast such equations into forms with very different particle interpretations. Any of the resulting systems will in general give rise to a different quantum cosmology. Clearly a viable quantisation should be one that has some chance of yielding a classical limit not too far removed from the original classical predictions.

2 Scalar cosmology

In [2] a particular 4-dimensional cosmology describing a Robertson-Walker metric interacting with a real scalar field ϕ has been discussed. Certain classical solutions were put into correspondence with algebraic curves in the plane. For coordinates $\{u, v\}$ on the plane these cosmologies arise as the zero energy solutions of the Euler-Lagrange equations derived from the Lagrangian:

$$L(u,v) = (\dot{u}^2 - \omega_1^2 u^2) - (\dot{v}^2 - \omega_2^2 v^2) \tag{1}$$

with ω_1 and ω_2 real. Such a simple Lagrangian gives rise to a natural quantisation and the zero energy constraint yields a simple Wheeler-DeWitt equation for a scalar wavefunction. The general solution of this equation with good asymptotics in the plane is immediate. Remarkably there exist particular solutions displaying structures that exhibit peaks that fluctuate about particular classical loci. [3] For a specific rational number r these loci admit admit the parametrisation

$$u = \frac{2}{r} \cosh\left(\frac{2r\omega_2\beta^{3/2}}{3}\right) \tag{2}$$

$$v = 2\cosh\left(\frac{2\omega_2\beta^{3/2}}{3}\right) \tag{3}$$

with $-(3\pi/2r\omega_2)^{2/3} \leq \beta < \infty$, describing a real 4 dimensional classical spacetime manifold with topology $\mathbb{R} \times \Sigma_3$ and a metric tensor :

$$g = -\beta\, d\beta \otimes d\beta + (R(\beta))^2 g_{(3)} \tag{4}$$

where $g_{(3)}$ is the standard Euclidean metric on a compact 3-manifold Σ_3. The branch $0 \leq \beta < \infty$ describes a Lorentzian spacetime domain while $-(3\pi/2r\omega_2)^{2/3} \leq \beta < 0$ corresponds to a Euclidean domain. (If we extend the range of β less than $-(3\pi/2r\omega_2)^{2/3}$ then the Euclidean domain is covered more than once by the parametrisation.)

3 Spinor cosmology

One of the many outstanding problems in trying to construct a quantum field theory of gravitation concerns the appropriate interpretation of quantum states for configurations that make no overt reference to "time". Thus it is difficult in general to endow the theory with any traditional Hilbert space structure based on a hermitian inner product and a unitary evolution. Although many alternative schemes have been suggested difficulties in interpretation remain. Some of the difficulties are intrinsic to the infinite dimensional aspect of field quantisation and in this respect one often seeks guidance by studying truncated field configurations corresponding to situations with high symmetry. Symmetric models are a useful theoretical laboratory for testing ideas that may have more general validity, and enable one to disentangle conceptual problems from technical ones. As above we focus on a particular minisuperspace analysis that gives rise to a Hamiltonian constraint, classically describing the zero energy configuration of an isotropic oscillator ghost-oscillator pair. This gives rise to a Wheeler-DeWitt equation that has occurred in a number of different contexts. It appears in certain 4-dimensional spacetime cosmologies [4], [5], [6], [7], and we have discussed it in the context of a class of 2-dimensional dilaton-gravity models. These models have arisen either from string-inspired limits or from the suppression of inhomogeneous modes in Einstein's theory of general relativity, [8], [9], [10], [11], [12] [13]. Our interest with this class of models stems from the properties of coherent state solutions to the corresponding Wheeler-DeWitt equation and their relation to classical solutions to general relativity including those that change signature [14].

4 The model

As a toy model [15]that shares features with the 4-dimensional model above consider a 2-dimensional dilaton-gravity theory based on the classical action

$$S[g,\psi] = \int_N \left\{ \frac{1}{2}\psi \star \mathcal{R} + c d\psi \wedge \star d\psi + \star(\Lambda_0 + \alpha e^{c\psi}) \right\} \tag{5}$$

where N is some domain of a two-dimensional manifold, ψ is a real scalar field and $\mathcal{R}$ is the curvature scalar of the Levi-Civita connection associated with the metric tensor g. The operator $\star$ denotes the Hodge map of g and c, Λ_0 and α are constants. The classical cosmological sector of this theory can be solved exactly and admits solutions with a degenerate metric where the signature changes from being Lorentzian to Euclidean. The standard approach for implementing the Hamiltonian constraint in the quantum version of such theories is to search for complex scalar valued functions on the appropriate manifold of matter and space geometry configurations. With $\mathbb{R}^2$ as a minisuperspace with global coordinates $\{X, Y\}$ [14]labeling these configurations one seeks Wheeler-DeWitt solutions $\Psi : \mathbb{R}^2 \mapsto \mathbb{C}$ to the equation:

$$H\Psi = 0 \tag{6}$$

where

$$H = (\omega^2 X^2 - \frac{\partial_X^2}{4}) - (\omega^2 Y^2 - \frac{\partial_Y^2}{4}) \tag{7}$$

The Wheeler-DeWitt equation (6) endows $\mathbb{R}^2$ with a natural (Lorentzian signature) metric $\mathcal{G}$. In terms of the coordinates $\{X, Y\}$:

$$\mathcal{G} = \partial_X \otimes \partial_X - \partial_Y \otimes \partial_Y . \tag{8}$$

If # denotes the associated Hodge map then (6) may be written:

$$d\# d\Psi - W\#\Psi = 0 \tag{9}$$

where $W(X,Y) = 4\omega^2(X^2 - Y^2)$. By multiplying (9) by $\overline{\Psi}$ and subtracting from the corresponding equation obtained by complex conjugation we readily verify that

$$d\mathcal{J} = 0 \tag{10}$$

where the current 1-form

$$\mathcal{J} = Im(\overline{\Psi}\# d\Psi). \tag{11}$$

Although this current is conserved there is no preferred spacelike foliation of $\mathbb{R}^2$ that defines a "density" component of $\mathcal{J}$ that does not in general change sign. Furthermore, although the minisuperspace is flat there is no natural way to restrict solutions to have a positive definite norm. The Killing vectors of the metric (8) do not generate a symmetry of the equation (9). Thus there appears no invariant way to normalise solutions of (9), construct a Hilbert subspace of normalisable solutions and endow the quantum theory with the standard probabilistic interpretation. By contrast the traditional Klein-Gordon quantisation of the relativistic free particle in Minkowski spacetime exists because the Killing isometry of the spacetime metric induces a classical symmetry of the Klein-Gordon equation. In a domain with Euclidean signature one has no natural means of defining the spacelike and timelike components of a current. Thus the definition of a probability current must transcend any definition of any preferred classical time for classical spacetimes.

5 The Clifford algebra of minisuperspace

The natural Lorentzian null-cone structure in 2D minisuperspace endows the space with a (1,1) Clifford bundle structure [16]. Thus there is a matrix basis for the Clifford Algebra Cl(1,1) in which the 1-forms dX and dY are represented as matrices satisfying

$$dX \vee dX = 1 \tag{12}$$

$$dY \vee dY = -1 \tag{13}$$

$$dX \vee dY + dY \vee dX = 0 \tag{14}$$

where $\vee$ denotes multiplication in the Clifford algebra. In conventional gamma matrix notation: ($dX \mapsto \gamma^1, dY \mapsto \gamma^0$). The Clifford bundle has minisuperspace as base and Cl(1,1) as fibre. Let Φ be a section of this bundle:

$$\Phi = \Phi_0 + \Phi_1\, dX + \Phi_2\, dY + \Phi_{12}\, dX \vee dY. \tag{15}$$

Since the bundle is trivial the components $\Phi_j \equiv \{\Phi_0, \Phi_1, \Phi_2, \Phi_{12}\}$ may be regarded as complex functions on $\mathbb{R}^2$. Introduce the Clifford potentials:

$$V_1 = 2i\omega(-Y + X\,dX \vee dY) \tag{16}$$

$$V_2 = 2i\omega(Y + X\,dX \vee dY). \tag{17}$$

We assert that if Φ satisfies first order equation:

$$\mathcal{D}\Phi + V_1 \vee \Phi = 0 \tag{18}$$

where $\mathcal{D} = (d - \delta)$ then each component of Φ will satisfy (6). In this equation d denotes exterior differentiation and δ is the coderivative:

$$\delta = \#^{-1} d \# \eta$$

where the involution η [16] is a linear operator on Φ that preserves 0-forms, reverses the sign of 1-forms and

$$\eta(dX \vee dY) = dX \vee dY. \tag{19}$$

The above result follows from

$$(\mathcal{D} + V_2) \vee (\mathcal{D} + V_1) = (\mathcal{D}^2 - W) \tag{20}$$

and the recognition that $\mathcal{D}^2 = -(d\,\delta + \delta\,d)$ is the Laplace-Beltrami operator. Since the basis forms in Φ are all holonomic it follows that if Φ satisfies (18) then its components satisfy

$$d\#\,d\Phi_j - W\#\Phi_j = 0. \tag{21}$$

6 Spinor solutions and associated currents

We observe that since our minisuperspace is flat with respect to (8) it is possible to find solutions Ψ of (18) that lie in a minimal (left) ideal of Cl(1,1) at each point. For example we may decompose

$$\Phi = \Phi \vee P_+ + \Phi \vee P_- \tag{22}$$

where $P_\pm = \frac{1}{2}(1 \pm dX)$ and take $\Psi = \Phi \vee P_+$. Minimal ideals provide irreducible modules for the sub-group SPIN of the Clifford group of Cl(1,1) [16] and their elements are spinors. Since P_+ is a parallel idempotent in minisuperspace, if Φ is a solution of (18) then so is $\Phi \vee P_+$ so Ψ may be regarded as a spinor solution of (18). We concentrate on those spinor solutions of (14) that are asymptotically well behaved as $|X|$ or $|Y|$ tends to infinity. Such solutions may be expressed in terms of a basis of Hermite functions. Thus if

$$\Psi = (u - vdY) \vee P_+ \tag{23}$$

(14) may be expressed as the coupled partial differential equations:

$$\partial_\xi F - 2i\omega\xi\, H = 0 \tag{24}$$

$$\partial_\eta H + 2i\omega\eta\, F = 0 \tag{25}$$

where $\xi = (Y-X)/\sqrt(2)$ $\eta = (Y+X)/\sqrt(2)$, $F = u - v$, $H = u + v$. These have the solutions

$$F(X,Y) = \sum_{n=0}^{\infty} c_n e^{-(z_1^2+z_2^2)/2} H_n(z_1)H_n(z_2) \tag{26}$$

$$H(X,Y) = \sum_{n=0}^{\infty} b_n e^{-(z_1^2+z_2^2)/2} H_n(z_1)H_n(z_2) \tag{27}$$

where $z_1 = \sqrt(2\omega)X$ and $z_2 = \sqrt(2\omega)Y$. The complex coefficients $\{c_n\}$ and $\{b_n\}$ are linearly correlated by the wave equation (14). A typical " coherent spinor state " solution takes the form:

$$F = Ce^{-(\alpha(X^2+Y^2)-2\beta XY)} \tag{28}$$

$$H = \frac{i}{\omega}(\alpha+\beta)F \tag{29}$$

where α, β, C are arbitrary complex constants.

We recall that in the Dirac theory of a relativistic particle described by a spinor ψ on spacetime, the Dirac vector current with components $\overline{\psi}\gamma^\mu\psi$ is conserved and possesses a positive-definite density for any non-trivial spinor. In the language of Clifford bundles this current is the form:

$$j[\psi] = *ReS_1(\psi \vee \tilde{\psi}) \tag{30}$$

where S_1 projects out the 1-form part of its argument and $\tilde{\psi} = C^{-1} \vee \psi^{\mathcal{I}}$ for some involution $\mathcal{I}$ in the (simple) Clifford algebra that is equivalent to hermitian conjugation:

$$A^\dagger = C^{-1}A^{\mathcal{I}}C \tag{31}$$

for all elements A in the Clifford algebra. With this goal in mind we find from (18) and (30)

$$dj[\Psi] = Re\,Tr(\tilde{\Psi} \vee (V_2 - V_1) \vee \Psi) \vee \#1 \tag{32}$$

for all spinor solutions Ψ of (18) where

$$j[\Psi] = \#ReS_1(\Psi \vee \tilde{\Psi}). \tag{33}$$

Here transposition is induced by the involution $\mathcal{I} \equiv \eta\xi$ where ξ is the main anti-involution [16] of the Clifford algebra and the adjoint spinor $\tilde{\Psi} = C^{-1} \vee \overline{\Psi}^{\xi\eta}$, where $\overline{\Psi}$ denotes the complex conjugation of Ψ. In the spinor basis defined by the projectors $P_{\pm}$ that we are using, the element $C = dY$. It follows from (32) that $j[\Psi]$ defines a closed current for solutions that satisfy the condition $Re\,Tr(i\omega\tilde{\Psi} \vee \Psi) = 0$. Although such solutions do exist we shall not impose this restrictive.

For a spinor solution (22) with components $\{u, v\}$ we find in the chart $\{X, Y\}$:

$$j[\Psi] = uv\,dY - (\frac{1}{2}|u|^2 + \frac{1}{2}|v|^2)\,dX \tag{34}$$

which clearly displays the non-negativity of the density

$$\rho_K(X, Y) = -j[\Psi](K) \tag{35}$$

where K denotes the Killing vector field ∂_X. If Ψ carries a representation of SPIN, then such a density, defined by a spacelike Killing vector field will remain positive for all proper Lorentz transformations that preserve the metric. Thus it may be adopted as a probability measure for the interpretation of the theory. However as the notation indicates a choice of spacelike Killing vector K is implied. Since the current $j[\Psi]$ is not conserved for all Ψ there exists no choice of spacelike foliation such that the integral of the probability density over a particular leaf of the foliation is independent of the leaf chosen. The existence of such a leaf dependent "charge" means that one cannot identify such a leaf as an "instant of time" and interpret such a charge as a normalisation factor for a state in the traditional manner. If we adopt as the Hilbert space norm of a state Ψ

$$(\Psi, \Psi)_K = \int_{\mathbb{R}^2} \rho_K \# 1 \tag{36}$$

then this also will depend on the choice of Killing vector K. Inasmuch as any convenient norm can be used to define a Hilbert space this is not necessarily a drawback. However the probabilistic interpretation of the theory must be restricted to describing relative probabilities between configurations. Thus for $K = \partial_X$ the relative probability densities for configuration Y_1 and Y_2 irrespective of X may be defined as $\mu(Y_1)/\mu(Y_2)$ where:

$$\mu(Y) = \int_{-\infty}^{\infty} \rho_K(X, Y)\,dX \tag{37}$$

and in general $\mu(Y)$ will depend on the configuration variable Y.

To give substance to the above one requires particular quantum states that can be related to classical cosmologies. We focus on those solutions that for a given choice of Killing vector K enable one to construct functions ρ_K that have maxima in the vicinity of those loci in (mini-)superspace corresponding to parametrised solutions to the classical field equations. For a classical manifold with a proscribed topology and a proscribed signature structure we concentrate on particular classical solutions with degenerate metrics. Furthermore the manifold should enable one to perform a Hamiltonian description of the field equations so that the classical and quantum degrees of freedom can be put into correspondence [14]. Such a correspondence may be given by a parametrised curve in mini-superspace:

$$\tau \mapsto \{X = X(\tau), Y = Y(\tau)\} \qquad \tau_0 < \tau < \tau_1 \tag{38}$$

As we have observed above such a parametrisation of a classical solution may describe a Euclidean signature metric for part of the manifold and a Lorentzian signature metric elsewhere. Thus it is natural to use τ as a choice of classical evolution parameter which is a classical time in the Lorentzian domain. We may now transfer the relative probability interpretation to the class of classical observers that inhabit the classical cosmology defined by the locus of the maxima of ρ_K. The density $\rho_K(X(\tau), Y(\tau))$ now offers a means of predicting the relative probabilities for finding the classical configurations $\{X, Y\}$ at "times" τ_1 and τ_2. The freedom in choosing different parametrisations to describe the same classical solutions corresponds to the freedom in choosing different coordinate systems on the classical manifolds.

Equation (32) is reminiscent of the equation that follows from the non-relativistic
Schrödinger equation in the presence of a complex potential. Indeed the lack of hermiticity of the hamiltonian there is analogous to the property $V_1 \neq \pm V_2^{\mathcal{I}}$. In the Schrödinger situation the use of a complex potential models the absorptive properties of an open system. For a closed system a non-hermitian hamiltonian is usually regarded as pathological. However in the context of gravitation such a reaction requires caution [13]. For example if the non-unitary evolution of a pure state of matter to a mixed state via the Hawking process can be maintained when gravitational back reaction is taken into account then probability conservation in a gravitational context may not be tenable. It is clear why the conservation of our current is impossible. Since the state vanishes asymptotically in all directions in the configuration space there is no way that a flux of positive density from the peaks of the state can flow smoothly to zero . In such a scenario it is tempting to conjecture that it is the existence of degenerate classical geometries that are mandatory to accommodate the absorption of probability flux in the Euclidean domains. Just as the creation (and annihilation) of a classical

cosmology may correspond to such domains where a classical spacetime description breaks down, the same may be true at the end points of localised gravitational collapse.

7 Waves on a manifold with a degenerate metric

The question of the behaviour of tensor fields in the vicinity of a signature transition has generated considerable controversy in the literature and led to alternative frameworks for discussing the problem [17], [18], [19], [20],[21]. A number of these different frameworks implicitly use different differential structures and it is of considerable interest to extricate the physics from distinct but equivalent differential structures.

To address some of these issues we have examined the "wave equation" on a manifold with a non trivial global topology and a degenerate metric that partitions the space into Lorentzian and Eucliean domains. [22] To simplify the analysis we shall construct a 2-dimensional manifold M as follows. First remove n non-intersecting caps $\{\mathrm{Cap}^{(\alpha)}\}_{\alpha=1\ldots n}$ from a Euclidean sphere S^2. Then smoothly glue a tube onto each hole made by removing each cap. For each (α), construct a coordinate chart $(U^{(\alpha)}, \Phi^{(\alpha)})$

$$U^{(\alpha)} = M - \{\text{all the tubes except tube } (\alpha)\}$$

$$\Phi^{(\alpha)}: U^{(\alpha)} \mapsto \Phi(U^{(\alpha)}) \subset \mathbb{R}^2 \;,\; (\tau^{(\alpha)}, \phi^{(\alpha)})$$

with $0 < \phi^{(\alpha)} < 2\pi$ and $-\infty < \tau^{(\alpha)} \le \pi$, For each cap let $\tau_1^{(\alpha)}$ be the angle subtended bewteen an edge of the cap, the center of the sphere, and the center of the cap. For $\tau^{(\alpha)} > \tau_1^{(\alpha)}$ $(\tau^{(\alpha)}, \phi^{(\alpha)})$ are the usual spherical coordinates for S^2. Thus $\Phi^{(\alpha)}(U^{(\alpha)})$ has holes in it where the other $n-1$ tubes are joined, and all the holes are in the region $\tau^{(\alpha)} > \tau_1^{(\alpha)}$. We endow M with the axially symmetric metric

$$g^{(\alpha)} = f(\tau^{(\alpha)})(d\tau^{(\alpha)} \otimes d\tau^{(\alpha)}) + h(\tau^{(\alpha)})(d\phi^{(\alpha)} \otimes d\phi^{(\alpha)}) \tag{39}$$

and introduce the the regions $\mathbf{S}^{(\alpha)}$, $\mathbf{E}^{(\alpha)}$, $\mathbf{L}^{(\alpha)}$, $\mathbf{M}^{(\alpha)}$ and the ring $\Sigma^{(\alpha)}$ so that:

	$\mathbf{M}^{(\alpha)}$	$\mathbf{L}^{(\alpha)}$	$\Sigma^{(\alpha)}$	$\mathbf{E}^{(\alpha)}$	$\mathbf{S}^{(\alpha)}$
$\phi^{(\alpha)}$	$0 < \phi^{(\alpha)} < 2\pi$	$0 < \phi^{(\alpha)} < 2\pi$	$0 < \phi^{(\alpha)} < 2\pi$	$0 < \phi^{(\alpha)} < 2\pi$	$0 < \phi^{(\alpha)} < 2\pi$
$\tau^{(\alpha)}$	$-\infty < \tau^{(\alpha)} \le \tau_3^{(\alpha)}$	$\tau_3^{(\alpha)} < \tau^{(\alpha)} \le \tau_2^{(\alpha)}$	$\tau^{(\alpha)} = \tau_2^{(\alpha)}$	$\tau_2^{(\alpha)} \le \tau^{(\alpha)} < \tau_1^{(\alpha)}$	$\tau_1^{(\alpha)} \le \tau^{(\alpha)} < \pi$
$f(\tau^{(\alpha)})$	$f(\tau^{(\alpha)}) = -1$	$f(\tau^{(\alpha)}) < 0$	$f(\tau^{(\alpha)}) = 0$	$f(\tau^{(\alpha)}) > 0$	$f(\tau^{(\alpha)}) = R_1^2$
$h(\tau^{(\alpha)})$	$h(\tau^{(\alpha)}) = R_2^2$	$h(\tau^{(\alpha)}) > 0$	$h(\tau^{(\alpha)}) > 0$	$h(\tau^{(\alpha)}) > 0$	$h(\tau^{(\alpha)}) = R_1^2 \sin^2(\tau^{(\alpha)})$
g	Flat Minkowskian	Lorentzian	Degenerate	Euclidean	Euclidean - Spherical

Here f, h are constructed with the aid of bump functions (infinitely differentiable functions with compact support). We wish to solve the equation

$d \star d\psi = 0$. This is the wave equation in the domain $\mathbf{M}^{(\alpha)} \cup \mathbf{L}^{(\alpha)}$ with a metric of Lorentzian signature and Laplace's Equation in the Euclidean region $\mathbf{S}^{(\alpha)} \cup \mathbf{E}^{(\alpha)}$. We shall require that ψ be continous across the signature changing rings $\{\Sigma^{(\alpha)}\}$ and satisfy the junction condition

$$[\Sigma^\star \star (d\psi)]_\Sigma = 0 \tag{40}$$

where, for any ring Σ, $[\omega]_\Sigma$ represents the discontinuity

$$[\omega]_\Sigma = \lim_{x \mapsto x_0 \in \Sigma,\, x \in \mathbf{E}^{(\alpha)}} \omega_x - \lim_{x \mapsto x_0 \in \Sigma,\, x \in \mathbf{L}^{(\alpha)}} \omega_x \tag{41}$$

Let $\mathbb{C}$ denote the complex plane with complex variable $z^{(\alpha)}$ sharing the origin with $U^{(\alpha)}$. It is convenient to introduce the (stereographic) projection $p^{(\alpha)}$

$$p^{(\alpha)}\colon \mathbf{S}^{(\alpha)} \mapsto \mathbb{C} \quad : (\tau^{(\alpha)}, \phi^{(\alpha)}) \mapsto z^{(\alpha)} = 2R_1 \tan\left(\frac{\tau^{(\alpha)}}{2}\right) e^{i\phi^{(\alpha)}} \tag{42}$$

$$p^{(\alpha)}\colon \mathbf{S}^{(\alpha)} \cap \mathbf{E}^{(\alpha)} \cap \Sigma^{(\alpha)} \mapsto \mathbb{C} \quad : (\tau^{(\alpha)}, \phi^{(\alpha)}) \mapsto z^{(\alpha)} = \frac{2R_1}{\Xi} e^{G_E(\tau^{(\alpha)}) + i\phi^{(\alpha)}} \tag{43}$$

where

$$G_E(\tau) = \int_{\tau_2}^{\tau} \left(\frac{f(\tau')}{h(\tau')}\right)^{\frac{1}{2}} d\tau' \quad \text{and} \quad \Xi = \cot\left(\tfrac{\tau_1}{2}\right) \exp\left(G_E(\tau_1)\right) \tag{44}$$

With the aid of the 3-parameter (Mobius) transformation $M_{a,b,c}$ of the complex plane

$$z^{(\beta)} = M_{a\,b,\,c}(z^{(\alpha)}) = e^{ic} \left(\frac{-2R_1 \tan\left(\frac{b}{2}\right) + z^{(\alpha)} e^{ia}}{1 + \dfrac{\tan\left(\frac{b}{2}\right) z^{(\alpha)} e^{ia}}{2R_1}} \right) \tag{45}$$

we can extend this projection p so that

$$p^{(\alpha)}\colon \mathbf{E} = \mathbf{S}^{(\alpha)} \cup \bigcup_{\alpha=1}^{n} \mathbf{E}^{(\alpha)} \mapsto p^{(\alpha)}(\mathbf{E}) \subset \mathbb{C}.$$

$p^{(\alpha)}(\mathbf{E})$ is a closed subset of $\mathbb{C}$ whose boundary is the collection of disjoint circles corresponding to the signature changing rings $\Sigma^{(\alpha)}$. In terms of the arbitrary functions $\psi^E_\pm(z)$ the general solution to Laplace's equation on the Euclidean region $\mathbf{E}$ is given by

$$\psi(z) = \psi^E_+(z) + \overline{\psi^E_-(z)} + A \log(|z|) \tag{46}$$

The presence of the log term with arbitrary complex constant A is only necessary if any pair of tubes are attached to antipodal points of the sphere.

We choose functions that are analytic on the punctured complex plane to ensure non-singular solutions on M. The solution in the Lorentzian region $\Sigma^{(\alpha)} \cup \mathbf{L}^{(\alpha)} \cup \mathbf{M}^{(\alpha)}$ that obeys the boundary conditions above is give solely in terms of the restriction $p^{(\alpha)}(\Sigma^{(\alpha)})$ to the circle with radius $\frac{2R_1}{\Xi}$:

$$\begin{aligned} &\psi_L^{(\alpha)} : \Sigma^{(\alpha)} \cup \mathbf{L}^{(\alpha)} \cup \mathbf{M}^{(\alpha)} \mapsto \mathbb{C} \\ &\psi_L^{(\alpha)}(\tau,\phi) = \psi_+^{L(\alpha)}\left(\tfrac{2R_1}{\Xi} e^{i(\phi+G_L(\tau))}\right) + \psi_-^{L(\alpha)}\left(\tfrac{2R_1}{\Xi} e^{i(\phi-G_L(\tau))}\right) \\ &\qquad - A\, G_L(\tau) - A\log\left(\tfrac{\Xi}{2R_1}\right) \end{aligned} \tag{47}$$

where

$$\begin{aligned} \psi_+^{L(\alpha)}(z) &= \left(\tfrac{1}{2}\right)(1+i)(\psi_+^{E(\alpha)}(z) - i\overline{\psi_-^{E(\alpha)}(z)}) \\ \psi_-^{L(\alpha)}(z) &= \left(\tfrac{1}{2}\right)(1-i)(\psi_+^{E(\alpha)}(z) + i\overline{\psi_-^{E(\alpha)}(z)}) \end{aligned}$$

$$G_L(\tau) = \int_{\tau_2}^{\tau} \left(\frac{-f(\tau')}{h(\tau')}\right)^{\frac{1}{2}} d\tau' \quad \text{and} \quad \epsilon = -R_2 G_L(\tau_3) + \tau_3 \tag{48}$$

On $\mathbf{M}^{(\alpha)}$ this becomes

$$\begin{aligned} \psi_M(\tau,\phi) &= \psi_+^L\left(\tfrac{2R_1}{\Xi} e^{i(\phi+(\tau-\epsilon)/R_2)}\right) + \psi_-^L\left(\tfrac{2R_1}{\Xi} e^{i(\phi-(\tau-\epsilon)/R_2)}\right) \\ &\quad - A(\tau-\epsilon)/R_2 - A\log\left(\tfrac{\Xi}{2R_1}\right) \end{aligned} \tag{49}$$

Thus we have a complete description of the general solution in terms of a pair of functions in the complex plane.

It is of interest to examine the forms

$$\mathrm{T}_X = i_X d\psi \wedge \star d\psi + d\psi \wedge i_X \star d\psi \tag{50}$$

In Lorentzian regions where X is one of the Killing vectors $\frac{\partial}{\partial\tau}$ or $\frac{\partial}{\partial\phi}$ we may identify energy and angular momentum density 1-forms respectively. Integrating these over a ring of constant τ gives an energy and angular momentum appropriate to that hypersurface.

$$\begin{aligned} \mathcal{E}^L(\tau) = 4\left(\tfrac{-f(\tau)}{h(\tau)}\right)^{\frac{1}{2}}\Bigg(\left(\tfrac{2R_1}{\Xi}\right)^2 \int_0^{2\pi} \Big(&\psi_+^{L\prime}\left(\tfrac{2R_1}{\Xi}e^{i\phi}\right)\, \overline{\psi_+^{L\prime}\left(\tfrac{2R_1}{\Xi}e^{i\phi}\right)} \\ &+ \psi_-^{L\prime}\left(\tfrac{2R_1}{\Xi}e^{i\phi}\right)\, \overline{\psi_-^{L\prime}\left(\tfrac{2R_1}{\Xi}e^{i\phi}\right)}\Big) d\phi + \pi A\overline{A}\Bigg) \end{aligned} \tag{51}$$

Note the energy is only a constant in the Minkowsian flat regions of the manifold $\mathbf{M}^{(\alpha)}$'s where the metric is independent of τ. From the symmetry of the metric the angular momentum is constant in $\mathbf{M}^{(\alpha)} \cup \mathbf{L}^{(\alpha)} \cup \Sigma^{(\alpha)} \cup \mathbf{E}^{(\alpha)}$ and is given by:

$$\mathcal{L} = 4\left(\tfrac{2R_1}{\Xi}\right)^2 \mathrm{Im}\left(\int_0^{2\pi} \psi_+^{E\prime}\left(\tfrac{2R_1}{\Xi}e^{i\phi}\right) \psi_-^{E\prime}\left(\tfrac{2R_1}{\Xi}e^{i\phi}\right) e^{2i\phi} d\phi\right) \tag{52}$$

Particular solutions can now be constructed corresponding to a single wave mode in any particular Lorentzian domain. Such a solution indicates how this mode is transmitted across the Euclidean domain into the other/break Lorentzian domains and permits an exact analytical computation of the dependence of the energy and angular momentum on the metrical characteristics of this topological manifold.

References

[1] C W Misner (1972), in *Magic without Magic*, J Klauder (ed.), Freeman, p.441.

[2] T Dereli and R W Tucker (1991), *Class. Q. Grav.*, 10 365.

[3] T Dereli, M Onder and R W Tucker (1993), *Class. Quantum Grav.*, 10 1425.

[4] L J Garay, J J Halliwell and G A Marugán (1991), *Phys. Rev.*, D43 2572.

[5] C Kiefer (1990), *Nucl. Phys.*, B341 273.

[6] T Dereli and R W Tucker (1993), *Class. Quantum Grav.*, 10 365.

[7] T Dereli, M Önder and R W Tucker (1993), *Class. Quantum Grav.*, 10 1425.

[8] G Mandal, A M Sengupta and S R Wadia (1991), *Mod.Phys. Lett.*, A6 1685.

[9] E Witten (1991), *Phys. Rev.*, D44 314.

[10] C G Callan, Jr., S B Giddings, J A Harvey and A Strominger (1992), *Phys. Rev.*, D45 R1005.

[11] J Navarro-Salas, M Navarro and V Aldaya (1993), ' Wave Functions of the Induced 2D-Gravity', Valencia Preprint FTUV/93-3.

[12] R Jackiw (1993), 'Gauge Theories for Gravity on a Line', (*In Memoriam* M C Polivanov), MIT Preprint.

[13] M Önder and R W Tucker (1993), *Phys. Letts.*, B311 47.

[14] M Önder and R W Tucker (1994), 'On the Relation between Classical and Quantum Cosmology in a 2D Dilaton-Gravity Model', *Class. Q. Grav.* (To Appear).

[15] T Dereli, M Önder and R W Tucker (1994), *Phys. Letts.*, B323 134.

[16] I M Benn and R W Tucker (1987), *An Application to Spinors and Geometry with Applications in Physics*, Adam Hilger.

[17] G F R Ellis (1991), *Gen. Rel. Grav.*, 23

[18] G F R Ellis, A Sumruk, D Coule and C Hellaby (1992), *Class. Quantum. Grav.*, 9 1535.

[19] S A Hayward (1992), *Class. Quantum. Grav.*, 9 1851.

[20] T Dray, C Manogue, R W Tucker (1993), *Phys. Rev.*, D48 2587.

[21] M Kossowski and M Kriele (1993), *Class. Quantum Grav.*, 10 1157.

[22] J Gratus and R W Tucker (1994), ' Scalar Fields on Hedge-Hog Spaces ', Lancaster Preprint (in preparation).

BRST-antibracket cohomology in 2d conformal gravity

Friedemann Brandt, Walter Troost and Antoine Van Proeyen

Abstract

We present results of a computation of the BRST-antibracket cohomology in the space of local functionals of the fields and antifields for a class of $2d$ gravitational theories which are conformally invariant at the classical level. In particular all classical local action functionals, all candidate anomalies and all BRST–invariant functionals depending nontrivially on antifields are given and discussed for these models.

1 Introduction

Conformal invariance plays a crucial role in various two dimensional physical models. Of special interest is the question whether conformal invariance of a classical theory is maintained in the quantum theory or becomes anomalous. In string theory, for instance, vanishing of the conformal anomaly determines the critical dimension and imposes "equations of motion" in target space [1]. Since the work of Wess and Zumino [2] it is well-known that anomalies have to satisfy consistency conditions following from the algebra of the symmetries of the classical theory. The general form of these conditions can be elegantly formulated in the BV-antifield formalism [3] as the vanishing of the antibracket of the proper solution S of the classical master equation and a functional $\mathcal{A}$ of the fields and antifields representing the anomaly [4]:

$$(S, \mathcal{A}) = 0. \tag{1.1}$$

(1.1) amounts to a cohomological problem since it requires BRST invariance of $\mathcal{A}$: the BRST operator[1] s is defined on arbitrary functionals $\mathcal{F}$ of fields and antifields through

$$s\,\mathcal{F} := (S, \mathcal{F}) \tag{1.2}$$

and its nilpotency is implied by the Jacobi identity for the antibracket and by the fact that S solves the classical master equation $(S, S) = 0$:

$$s^2 = 0. \tag{1.3}$$

[1]We call s the BRST operator although this terminology often is used only when it acts on the fields, and not on antifields.

Let us denote by $H^*(s)$ the BRST cohomology in the relevant space of functionals of the fields and antifields which must be specified in each particular case (usually it is the space of local functionals whose precise definition must be adapted to the problem). Since the BRST operator increases the ghost number (gh) by one unit due to $gh(S) = 0$, $H^*(s)$ can be computed in each subspace of functionals with a definite ghost number g separately where we denote it by $H^g(s)$. Anomalies are represented by cohomology classes of $H^1(s)$, at least if the ghost number is conserved in the quantum theory which holds at tree level due to $gh(S) = 0$. However it is often useful to compute $H^g(s)$ for other values of g as well. In particular $H^0(s)$ is interesting since it contains S itself. This opens the possibility to construct S by computing $H^0(s)$ after fixing the desired field content and gauge invariances of a model. This was actually our starting point for the computation of $H^*(s)$ in a class of two dimensional models which are conformally invariant at the classical level (see next section). A computation of $H^0(s)$ can also be useful for given S since it can provide information about observables or counterterms arising in a theory. Moreover antifield–dependent solutions $\mathcal{M}$ of $(S, \mathcal{M}) = 0$ with ghost number 0 can have interesting interpretations and applications. For instance they are needed for the construction of a functional $\tilde{S} = S + \mathcal{M} + \ldots$ satisfying $(\tilde{S}, \tilde{S}) = 0$. If $\tilde{S}$ exists, it provides a nontrivial extension of the theory which is consistent in the sense that it is invariant under suitable extensions of the gauge transformations of the original theory characterized by S. This was pointed out and exemplified in [5]. BRST–invariant functionals $\mathcal{M}$ with ghost number 0 which do not satisfy $(\mathcal{M}, \mathcal{M}) = 0$ may alternatively receive an interpretation as background charges, which can cancel anomalies. This can be implemented in the BV antifield formalism [6] by formally considering $\mathcal{M} = \sqrt{\hbar}\mathcal{M}_{1/2}$ as a contribution to the quantum action W of order $\sqrt{\hbar}$ since $W = S + \sqrt{\hbar}\mathcal{M}_{1/2} + \hbar\mathcal{M}_1 + \ldots$ implies $(W, W) = \hbar(\mathcal{M}_{1/2}, \mathcal{M}_{1/2}) + \ldots$, i.e. $(\mathcal{M}, \mathcal{M})$ indeed can cancel one-loop anomalies.

2 Characterization of the models

Our aim was the computation of $H^*(s)$ for a class of two dimensional models which are conformally invariant at the classical level. To this end we did not characterize these models by specific conformally invariant classical actions but we only specified the field content and the gauge invariances of the classical theory. Then we computed $H^*(s)$ in the space of antifield–independent functionals which for ghost number 0 in particular provides the most general local classical action functional S_0 and thus characterizes more precisely the models to which our results apply. Then we completed the computation of $H^*(s)$ by inclusion of the antifields. This procedure is possible due to the closure of the algebra of gauge symmetries since in this case the BRST transformations of the fields do not depend on the antifields. The BRST transformations of the antifields however involve (functional derivatives of) S_0 and therefore their inclusion requires the knowledge of S_0.

In detail, the models which we investigated are characterized by

(i) Field content: S_0 is a local functional of the $2d$ metric $g_{\alpha\beta} = g_{\beta\alpha}$ $(\alpha, \beta \in \{+,-\})$ and a set of bosonic scalar matter fields X^μ $(\mu \in \{1, 2, \ldots, D\})$.

(ii) S_0 is invariant under $2d$ diffeomorphisms and local Weyl transformations of the metric $g_{\alpha\beta}$.

(iii) S_0 does not possess any nontrivial gauge symmetries apart from those mentioned in (ii).[2]

In order to make (i) precise we have to add the definition of local functionals we used:

(iv) A functional of a set of fields Z^A is called local if its integrand is a polynomial in the derivatives of the Z^A (without restriction on the order of derivatives) but may depend nonpolynomially on the undiffentiated fields Z^A and explicitly on the coordinates x^α of the two dimensional base manifold.

(ii) requires

$$s\, S_0 = 0 \tag{2.1}$$

where s acts on $g_{\alpha\beta}$ and X^μ according to

$$s\, g_{\alpha\beta} = \xi^\gamma \partial_\gamma g_{\alpha\beta} + g_{\gamma\beta}\partial_\alpha \xi^\gamma + g_{\alpha\gamma}\partial_\beta \xi^\gamma + c\, g_{\alpha\beta}, \quad s\, X^\mu = \xi^\alpha \partial_\alpha X^\mu. \tag{2.2}$$

Of course these are just the BRST transformations of $g_{\alpha\beta}$ and X^μ where ξ^α and c are the anticommuting ghosts of diffeomorphisms and local Weyl transformations respectively. The BRST transformations of the ghosts are chosen such that (1.3) holds on all fields $g_{\alpha\beta}, X^\mu, \xi^\alpha, c$. This leads to

$$s\, \xi^\alpha = \xi^\beta \partial_\beta \xi^\alpha, \quad s\, c = \xi^\alpha \partial_\alpha c. \tag{2.3}$$

(2.1)–(2.3) and requirement (iii) guarantee that the proper solution of the classical master equation is given by

$$S = S_0 - \int d^2x\, (s\Phi^A)\Phi^*_A \tag{2.4}$$

---

[2]Trivial gauge symmetries of an action S_0 depending on a set of (bosonic) fields φ^i are by definition of the form $\delta_\epsilon \varphi^i = P^{ij}\delta S_0/\delta\varphi^j$ where $P^{ij} = -P^{ji}$ are arbitrary functions of the φ^i, arbitrary parameters $\epsilon(x)$ and the derivatives of the φ^i and ϵ.

where we used customary collective notations $\{\Phi^A\} = \{g_{\alpha\beta}, X^\mu, \xi^\alpha, c\}$ and $\{\Phi^*_A\} = \{g^{*\alpha\beta}, X^*_\mu, \xi^*_\alpha, c^*\}$ for fields and antifields. The BRST transformation of Φ^*_A is given by the functional right derivative of S w.r.t. Φ^A (in our conventions the BRST operator acts from the left everywhere)

$$s\,\Phi^*_A = \frac{\delta_r S}{\delta \Phi^A}. \tag{2.5}$$

Remark:
Although we allow the integrands of local functionals to depend explicitly on x^α according to (iv), it turns out that integrands of BRST–invariant functionals actually do not carry an explicit x-dependence (up to trivial contributions of course). Nevertheless we need this definition of local functionals in order to cancel candidate anomalies as e.g. $\int d^2x\, \xi^\alpha L$ where L is a Weyl invariant density ($sL = \partial_\alpha(\xi^\alpha L)$). Namely these functionals are BRST invariant but not BRST exact unless we admit counterterms whose integrands depend explicitly (and in fact polynomially) on the x^α. These are well-known features of all gravitational theories (cf. [7, 8, 9]). However, if one takes into account topological properties of the base manifold, the x-independence of the integrands of nontrivial BRST–invariant functionals holds strictly only if the manifold does not allow closed p-forms with $p \neq 0$ which are not exact. The results we present in the next section therefore hold in a strict sense only under this additional assumption (cf. [9] for general remarks on this point).

3 Results

3.1 Antifield–independent functionals

We found that $H^g(s)$ vanishes for $g > 4$ in the space of local antifield–independent functionals, i.e. each local BRST–invariant functional with ghost number $g > 4$ which does not depend on antifields is the BRST variation of a local functional with ghost number $g-1$ which also does not depend on antifields. We only spell out the results for $g = 0, 1$. Those for $g = 2, 3, 4$ will be given in [10].

$H^0(s)$ provides the most general classical action S_0. It is given by

$$S_0 = \int d^2x\, \left(\tfrac{1}{2}\sqrt{g}\, g^{\alpha\beta} G_{\mu\nu}(X)\partial_\alpha X^\mu \partial_\beta X^\nu + B_{\mu\nu}(X)\partial_+ X^\mu \partial_- X^\nu\right) \tag{3.1}$$

with $g = |\det(g_{\alpha\beta})|$. $G_{\mu\nu}$ and $B_{\mu\nu}$ are arbitrary functions of the X^μ satisfying

$$G_{\mu\nu} = G_{\nu\mu}, \quad B_{\mu\nu} = -B_{\nu\mu}.$$

$B_{\mu\nu}$ is defined only up to contributions $\partial_\mu B_\nu(X) - \partial_\nu B_\mu(X)$ which yield total derivatives in the integrand of (3.1). Here and henceforth

$$\partial_\mu := \frac{\partial}{\partial X^\mu}$$

denote derivatives w.r.t. matter fields. (3.1) is the most general functional satisfying requirements (i) and (ii) listed in section 2 with the restrictions imposed by (iv). (iii) represents an additional requirement which excludes e.g. functions $G_{\mu\nu}$ and $B_{\mu\nu}$ admitting a nonvanishing solution $g^\mu(X)$ of

$$G_{\mu\nu}g^\nu = \Gamma_{\mu\nu\rho}g^\rho = H_{\mu\nu\rho}g^\rho = 0 \tag{3.2}$$

where

$$\Gamma_{\mu\nu\rho} = \tfrac{1}{2}\left(\partial_\mu G_{\nu\rho} + \partial_\nu G_{\mu\rho} - \partial_\rho G_{\mu\nu}\right), \quad H_{\mu\nu\rho} = \partial_\mu B_{\nu\rho} + \partial_\nu B_{\rho\mu} + \partial_\rho B_{\mu\nu}. \tag{3.3}$$

Namely (3.2) implies the invariance of S_0 under $\delta_\epsilon X^\mu = \epsilon(x)g^\mu(X)$ for an arbitrary function $\epsilon(x)$ and thus the presence of an additional gauge invariance which violates requirement (iii).

It is also worth noting that invariance of the theory under target space reparametrizations can be elegantly formulated in the antifield formalism as well (this kind of an invariance must not be confused with invariance of the action functional in the usual sense, of course). Namely any two action functionals $S_0'[X] := S_0[X + \delta X]$ and $S_0[X]$ which are related by an arbitrary infinitesimal target space reparametrization $\delta X^\mu = f^\mu(X)$ differ by the BRST-variation of a local antifield–dependent functional (with BRST transformation of the antifields defined by means of $S_0[X]$). Both $S_0'[X]$ and $S_0[X]$ are of the form (3.1) with functions $G'_{\mu\nu}$ and $G_{\mu\nu}$ resp. $B'_{\mu\nu}$ and $B_{\mu\nu}$ related by[3]

$$G'_{\mu\nu} = G_{\mu\nu} + 2\partial_{(\mu}f_{\nu)} - \Gamma_{\mu\nu\rho}f^\rho\,; \qquad B'_{\mu\nu} = B_{\mu\nu} + 2\partial_{[\mu}B_{\nu]} + H_{\mu\nu\rho}f^\rho \tag{3.4}$$

where $f_\mu := G_{\mu\nu}f^\nu$. We note that an analogous statement holds for *any* theory characterized by a local action functional $S_0[\phi]$ (where ϕ^i are the fields with ghost number 0). Namely consider infinitesimal (local) field redefinitions $\delta\phi^i = f^i(\phi, \partial\phi, \ldots)$ which are chosen such that

$$s\,\left(S_0[\phi + \delta\phi] - S_0[\phi]\right) = 0 \tag{3.5}$$

holds with the BRST operator encoding the gauge symmetries of $S_0[\phi]$. Then there is a (local) functional Γ^{-1} with ghost number -1 such that

$$S_0[\phi + \delta\phi] - S_0[\phi] = s\,\Gamma^{-1}[\Phi, \Phi^*]. \tag{3.6}$$

This statement holds also for theories without gauge invariances.[4] In this case s reduces to the Koszul-Tate differential $\delta_{KT} = \int d^2x(\delta S_0[\phi]/\delta\phi^i)\delta/\delta\phi^*_i$ and (3.6) holds obviously for arbitrary field redefinitions $\delta\phi^i$.

[3](Anti-)Symmetrization of indices is defined by $f_{(\mu\nu)} = \frac{1}{2}(f_{\mu\nu} + f_{\nu\mu})$ etc.

[4]Notice that transformations $\delta\phi^i$ satisfying (3.5) are more general transformations than symmetry transformations of S_0 since the latter satisfy the much stronger condition $S_0[\phi + \delta\phi] = S_0[\phi]$.

$H^1(s)$ provides the antifield–independent candidate anomalies. They are given by

$$\mathcal{A} = \mathcal{H}_+ + \mathcal{H}_- + \mathcal{X}_+ + \mathcal{X}_-, \tag{3.7}$$

$$\mathcal{H}_\pm = a_\pm \int d^2x\, c^\pm (\partial_\pm)^3 h_{\mp\mp}, \tag{3.8}$$

$$\mathcal{X}_\pm = \int d^2x\, \tfrac{1}{1-y}(\partial_\pm \xi^\pm + h_{\mp\mp}\partial_\pm \xi^\mp)\nabla_+ X^\mu \nabla_- X^\nu f^\pm_{\mu\nu}(X) \tag{3.9}$$

where a_+, a_- are constants, $f^+_{\mu\nu}, f^-_{\mu\nu}$ are arbitrary functions of the X^μ and

$$\begin{array}{ll} h_{\pm\pm} = g_{\pm\pm}/(g_{+-} + \sqrt{g}), & y = h_{++}h_{--}, \\ \nabla_\pm X^\mu = (\partial_\pm - h_{\pm\pm}\partial_\mp)X^\mu, & c^\pm = \xi^\pm + h_{\mp\mp}\xi^\mp. \end{array} \tag{3.10}$$

Using the original components of the metric, $\mathcal{X}_\pm$ read

$$\mathcal{X}_\pm = \int d^2x\, (\partial_\pm \xi^\pm + h_{\mp\mp}\partial_\pm \xi^\mp)\left(\tfrac{1}{2}\sqrt{g}\, g^{\alpha\beta} f^\pm_{(\mu\nu)}(X)\partial_\alpha X^\mu \partial_\beta X^\nu \right.$$
$$\left. + f^\pm_{[\mu\nu]}(X)\partial_+ X^\mu \partial_- X^\nu\right).$$

In fact the parts of $\mathcal{X}_\pm$ containing the symmetric and antisymmetric parts of $f^\pm_{\mu\nu}$ are separately BRST invariant. They are also nontrivial and inequivalent in the space of antifield–independent functionals. We remark however that those functionals $\mathcal{X}_+, \mathcal{X}_-$ which arise from contributions

$$2\partial_{(\mu} H^\pm{}_{\nu)} \mp 2\partial_{[\mu} H^\pm{}_{\nu]} + (H_{\mu\nu\rho} - 2\Gamma_{\mu\nu\rho})\, H^{\pm\rho} \quad (\text{with } H^\pm{}_\mu := G_{\mu\nu} H^{\pm\nu}) \tag{3.11}$$

to $f^\pm_{\mu\nu}$ are trivial in the space of local functionals of the fields and antifields where $H^{\pm\mu}(X)$ are arbitrary functions of the X^μ and the upper (lower) sign refers to contributions to $f^+_{\mu\nu}$ ($f^-_{\mu\nu}$).

3.2 Antifield–dependent functionals

The existence and explicit form of antifield–dependent BRST–invariant functionals depends of course on the specific form of S_0, i.e. on the specific choice of the functions $G_{\mu\nu}$ and $B_{\mu\nu}$ in (3.1) since they enter in the BRST transformation of X^*_μ and $g^{*\mu\nu}$, see (2.5). Nevertheless one can classify all antifield–dependent BRST–invariant functionals as follows:

a) $g \notin \{-1, 0, 1\}$:
There are no antifield–dependent cohomology classes in these cases, independently of the specific form of S_0. More precisely: If W^g is an antifield–dependent local BRST–invariant functional with ghost number $g \notin \{-1, 0, 1\}$ then there is a local functional W^{g-1} such that $\tilde{W}^g := W^g - sW^{g-1}$ does not depend on antifields anymore.

b) $g = -1$:
BRST–invariant local functionals with ghost number -1 exist if and only if $G_{\mu\nu}$ and $B_{\mu\nu}$ admit a nonvanishing solution $f^\mu(X)$ of

$$\partial_\mu f_\nu + \partial_\nu f_\mu - 2\Gamma_{\mu\nu\rho} f^\rho = 0, \quad H_{\mu\nu\rho} f^\rho = \partial_\mu H_\nu - \partial_\nu H_\mu \tag{3.12}$$

for some arbitrary functions $H_\mu(X)$. (3.12) identifies $f_\mu := G_{\mu\nu} f^\nu$ as the components of a Killing vector in target space. Any solution of (3.12) generates a continuous *global* symmetry of S_0 through

$$\delta_\epsilon X^\mu = \epsilon f^\mu(X), \quad \epsilon = const. \tag{3.13}$$

In other words: BRST–invariant local functionals with ghost number -1 correspond one-to-one to these global symmetries of S_0. They are given by

$$W^{-1} = \int d^2x\, X^*_\mu f^\mu(X). \tag{3.14}$$

c) $g = 0$:
BRST–invariant local functionals with ghost number 0 depending non-trivially on the antifields exist if and only if $G_{\mu\nu}$ and $B_{\mu\nu}$ admit a nonvanishing solution $f^\mu(X)$ of

$$0 = \partial_\mu f_\nu + \partial_\nu f_\mu - 2\Gamma_{\mu\nu\rho} f^\rho, \quad 0 = \partial_\mu f_\nu - \partial_\nu f_\mu \mp H_{\mu\nu\rho} f^\rho \tag{3.15}$$

where the second condition must be satisfied either with the + or the $-$ sign for a particular solution f^μ. Comparing (3.15) and (3.12) we conclude that (3.15) requires that S_0 possesses a global symmetry (3.13) with the additional restriction imposed by the second condition (3.15). The BRST–invariant functionals with ghost number 0 arising from a solution of (3.15) with a minus sign in front of $H_{\mu\nu\rho} f^\rho$ are given by

$$\mathcal{M}_+ = \int d^2x \left[X^*_\mu (\partial_+ \xi^+ + h_{--} \partial_+ \xi^-) - \tfrac{2}{1-y} \nabla_+ X^\nu \partial_+ h_{--} G_{\mu\nu} \right] f^\mu \tag{3.16}$$

and the functionals $\mathcal{M}_-$ arising from a solution of (3.15) with a plus sign in front of $H_{\mu\nu\rho} f^\rho$ are obtained from (3.16) by exchanging all + and $-$ indices. These solutions do not satisfy $(\mathcal{M}, \mathcal{M}) = 0$ and can thus not be added to the extended action without breaking $(S, S) = 0$, however they can be used to introduce background charges as explained in the introduction. Namely, taking $h_{++} = 0$, dropping the corresponding ξ^- ghost, and specialising to $G_{\mu\nu} = \delta_{\mu\nu}$, (3.16) becomes $\int d^2x\, (X^*_\mu \partial_+ \xi^+ - 2\partial_+ X_\mu \partial_+ h_{--})\, f^\mu$ in which one recognises the so-called background charge terms (see [6] for their inclusion in the BV formalism). Therefore, (3.16) constitutes the generalization of this chiral gauge treatment.

d) $g = 1$:
In this case we obtain (3.2) as necessary and sufficient conditions for the existence of BRST–invariant local functionals with ghost number 1 depending nontrivially on the antifields. As discussed above, (3.2) implies that S_0 possesses an additional gauge symmetry which violates requirement (iii) and thus has to be excluded. Namely in presence of additional gauge symmetries, (2.4) is not a proper solution of the classical master equation anymore. To construct a proper solution one must introduce a ghost and its antifield for each additional gauge symmetry. In the extended space of functionals depending also on these additional fields, the antifield–dependent functionals arising from solutions g^μ of (3.2) indeed are trivial. Nevertheless one of course has to reexamine the whole investigation of $H^*(s)$ in the case of a higher gauge symmetry and therefore our results do not apply to this case.

4 Sketch of the computation

In the first step of the computation, the BRST cohomology in the space of local functionals $W = \int d^2x f$ is related to the BRST cohomology in the space of local functions by means of the descent equations following from $sW = 0$:

$$s\,\omega_2 + d\,\omega_1 = 0, \quad s\,\omega_1 + d\,\omega_0 = 0, \quad s\,\omega_0 = 0 \tag{4.1}$$

where $\omega_2 = d^2x f$ is the integrand of W written as a 2-form and ω_1 and ω_0 are local 1- and 0-forms. It is well-known that the descent equations terminate in gravitational theories always with a nontrivial 0-form ω_0 (contrary to the Yang–Mills case)[5] and that their "integration" is trivial:

$$\omega_1 = b\,\omega_0, \quad \omega_2 = \tfrac{1}{2}\,bb\,\omega_0, \quad b = dx^\alpha \frac{\partial}{\partial \xi^\alpha}. \tag{4.2}$$

According to these statements which were first proved and applied in [8] (for arbitrary dimensions) it is sufficient to determine the general solution of

$$s\,\omega_0 = 0 \tag{4.3}$$

in the space of local functions of the fields and their derivatives. The BRST–invariant functionals resp. their integrands are then obtained via (4.2) from the solutions of (4.3).

The investigation of (4.3) is considerably simplified by performing it in an appropriate new basis of variables substituting the fields, antifields and their derivatives. The construction of this new basis is the second and crucial step

[5] This statement holds in a strict sense only in absence of closed p-forms ($p \neq 0$) which are not exact, cf. [9].

within the computation. The new basis contains in particular the following variables $T^{\mu}_{m,n}$ substituting one-by-one the partial derivatives $(\partial_+)^m(\partial_-)^n X^\mu$ of the matter fields:

$$T^{\mu}_{m,n} = \left(\frac{\partial}{\partial c^+}\, s\right)^m \left(\frac{\partial}{\partial c^-}\, s\right)^n X^\mu \tag{4.4}$$

where c^+ and c^- are the ghost variables defined in (3.10) and it is understood that the BRST transformations occurring in (4.4) are expressed in terms of these ghosts. The first few (and most important) T's are given by

$$T^{\mu}_{0,0} = X^\mu, \quad T^{\mu}_{1,0} = \tfrac{1}{1-y}\nabla_+ X^\mu, \quad T^{\mu}_{0,1} = \tfrac{1}{1-y}\nabla_- X^\mu$$

with $\nabla_\pm X^\mu$ as in (3.10). The most important ghost variables are

$$c^n_+ = \tfrac{1}{(n+1)!}\,(\partial_+)^{n+1}\, c^+, \quad c^n_- = \tfrac{1}{(n+1)!}\,(\partial_-)^{n+1}\, c^-, \quad n \geq -1. \tag{4.5}$$

The remarkable property of the $T^{\mu}_{m,n}$ is that they span the representation space for two copies of the "Virasoro algebra" (without central extension) whose associated ghosts are just the variables (4.5). Namely the BRST transformations of $T^{\mu}_{m,n}$ and $c^n_\pm$ can be written as

$$s\, T^{\mu}_{m,n} = \sum_{k\geq -1} (c^k_+ L^+_k + c^k_- L^-_k) T^{\mu}_{m,n}, \quad s\, c^k_\pm = \tfrac{1}{2} f_{mn}{}^k c^m_\pm c^n_\pm \tag{4.6}$$

where L^+_n and L^-_n represent on the $T^{\mu}_{m,n}$ the Virasoro algebra according to

$$[L^\pm_m, L^\pm_n] = f_{mn}{}^k L^\pm_k, \quad [L^+_m, L^-_n] = 0, \quad f_{mn}{}^k = (m-n)\delta^k_{m+n} \quad (m,n,k \geq -1). \tag{4.7}$$

$L^\pm_k T^{\mu}_{m,n}$ can be evaluated using (4.7) and

$$T^{\mu}_{m,n} = (L^+_{-1})^m\, (L^-_{-1})^n\, X^\mu, \quad L^\pm_n X^\mu = 0 \quad \forall n \geq 0. \tag{4.8}$$

The equivalence of (4.4) and the first relation (4.8) can be verified using $s^2 = 0$ and the following representation of $L^\pm_n$ on $T^{\mu}_{m,n}$ which is implied by (4.6):

$$L^+_n = \left\{ s, \frac{\partial}{\partial c^n_+} \right\}, \quad L^-_n = \left\{ s, \frac{\partial}{\partial c^n_-} \right\}, \quad n \geq -1. \tag{4.9}$$

The $T^{\mu}_{m,n}$ are called tensor fields. In fact one can extend the definition of tensor fields to the antifields. Of particular importance are those tensor fields which substitute X^*_μ. They are given by

$$\hat{X}^*_\mu = \tfrac{1}{1-y}\, X^*_\mu. \tag{4.10}$$

In the third step one proves by means of standard methods that nontrivial contributions to solutions of (4.3), written in terms of the new basis, depend on the fields, antifields and their derivatives only via the $c_\pm^n$ and the tensor fields constructed of the matter fields and the antifields since all other variables group into trivial systems of the form (a, sa) and do not enter in (4.6).

In step four we take advantage of the fact that L_0^+, L_0^- are diagonal on all tensor fields and on the ghosts (4.5) on which these generators are defined by means of (4.9). Namely one has

$$L_0^+ T_{m,n}^\mu = m T_{m,n}^\mu,\ L_0^- T_{m,n}^\mu = n T_{m,n}^\mu,\ L_0^\pm c_\pm^n = n c_\pm^n,\ L_0^\pm c_\mp^n = 0 \qquad (4.11)$$

and similar relations for the tensor fields constructed of the antifields (e.g. $\hat{X}_\mu^*$ has (L_0^+, L_0^-) weights $(1,1)$). By means of standard arguments one concludes that solutions of (4.3) can be assumed to have total weight $(0,0)$ (all other contributions to ω_0 are trivial).

The fifth and final step consists in the investigation of (4.3) in the space of those local functions of the ghosts (4.5) and the tensor fields which have total weight zero under both L_0^+ and L_0^-. It turns out that $c_+^{-1} = c^+$ and $c_-^{-1} = c^-$ are the only variables having negative weights under L_0^+ or L_0^-. In fact they have weights $(-1,0)$ and $(0,-1)$ respectively. Since the ghosts anticommute, there are only few possibilities to construct local functions with total weight $(0,0)$ at all. In fact the whole computation reduces to the investigation of functions of the following quantities:

$$\begin{aligned} &c_\pm^0,\quad \tilde{c}_\pm \equiv 2c_\pm^{-1}c_\pm^1,\quad X^\mu = T_{0,0}^\mu,\quad T_+^\mu \equiv c_+^{-1}T_{1,0}^\mu,\quad T_-^\mu \equiv c_-^{-1}T_{0,1}^\mu,\\ &T_{+-}^\mu \equiv c_-^{-1}c_+^{-1}T_{1,1}^\mu,\quad T_\mu^* \equiv c_+^{-1}c_-^{-1}\hat{X}_\mu^* \end{aligned} \qquad (4.12)$$

on which s acts according to

$$\begin{aligned} &s\,c_\pm^0 = \tilde{c}_\pm,\ s\,X^\mu = T_+^\mu + T_-^\mu,\ s\,T_+^\mu = T_{+-}^\mu,\ s\,T_-^\mu = -T_{+-}^\mu,\\ &s\,T_\mu^* = 2G_{\mu\nu}T_{+-}^\nu + (H_{\nu\rho\mu} - 2\Gamma_{\nu\rho\mu})T_+^\nu T_-^\rho \end{aligned} \qquad (4.13)$$

where sT_μ^* follows from (2.5) and thus of course requires the knowledge of (3.1) which is obtained from the solution of the antifield independent problem. Taking into account the algebraic identities relating the quantities (4.12) as a consequence of the odd grading of $c^\pm$, like $\tilde{c}_+\tilde{c}_+ = T_+^\mu\tilde{c}_+ = T_+^\mu T_+^\nu = 0$ etc., one sees that the space of nonvanishing functions of these quantities is rather small apart from the occurrence of arbitrary functions of $T_{0,0}^\mu = X^\mu$. This allows ultimately to solve (4.3) completely. The solution of (4.3) which yields (3.1) is for instance given by $T_+^\mu T_-^\nu K_{\mu\nu}(X)$ where the symmetric and antisymmetric parts of $K_{\mu\nu}$ are just $G_{\mu\nu}$ resp. $B_{\mu\nu}$.

5 Summary

We have determined the complete BRST-antibracket cohomology in the space of local functionals for theories satisfying the assumptions (i)–(iv) listed in section 2. We found that nontrivial cohomology classes exist only for ghost numbers $g = -1, \ldots, 4$. The representatives of the cohomology classes with $g = 1, \ldots, 4$ can be chosen such that they do not depend on antifields at all. Due to their special importance we summarize and comment only the results for $g = -1, 0, 1$ in detail.

The cohomology classes with $g = -1$ correspond one-to-one to the independent solutions $f^\mu(X)$ of (3.12) which can be interpreted as the Killing vectors in target space. Each of them generates a global symmetry of S_0 according to (3.13). The resulting BRST–invariant functionals are given by (3.14). This result is not surprising since it has been shown in [11] that the BRST cohomology classes with ghost number -1 correspond one-to-one to the independent nontrivial continuous global symmetries of the classical action which is part of a cohomological reformulation of Noether's theorem.

For $g = 0$ there are two types of cohomology classes. Those of the first type are represented by antifield–independent functionals and provide the most general classical action for models characterized by (i)–(iv). It is given by (3.1), with the understanding that two such actions are equivalent if they are related by a target space reparametrization (3.4). Representatives of cohomology classes of the second type depend nontrivially on the antifields. They correspond one-to-one to those Killing vectors $f^\mu(X)$ which satisfy (3.15). The corresponding BRST–invariant functionals are given by (3.16) (and an analogous expression for $\mathcal{M}_-$). They correspond to so-called background charges and might provide BRST–invariant functionals $\mathcal{M} = \mathcal{M}_+ + \mathcal{M}_-$ which, as remarked in the introduction, can be used in order to look for a consistent extension of the models or investigate an anomaly cancellation through background charges (these applications would require appropriate choices of $G_{\mu\nu}$, $B_{\mu\nu}$ and f^μ).

The cohomology classes with $g = 1$ represent candidate anomalies. One can distinguish two types of them. Representatives of the first type can be chosen to be independent of the matter fields. In fact there are precisely two inequivalent cohomology classes of this type, represented by the matter field independent functionals $\mathcal{H}_+$ and $\mathcal{H}_-$ given in eq. (3.8) (contrary to slightly misleading formulations in [12] which give the impression that there is only one cohomology class represented by a special linear combination of $\mathcal{H}_+$ and $\mathcal{H}_-$). Candidate anomalies of the second type depend nontrivially on the matter fields and are represented by the functionals (3.9). A functional (3.9) is cohomologically trivial if and only if $f^\pm_{\mu\nu}$ have the form (3.11). All other functionals (3.9) are BRST invariant and cohomological nontrivial in the complete space of local functionals of fields and antifields. This result corrects a statement given in [13] where the authors claim that matter field dependent contributions to BRST–invariant functionals with ghost number 1 can be always removed by adding trivial contributions. It is worth noting in this context that in fact both types of anomalies arise in a generic model.

The requirement that the matter field dependent anomalies vanish at the one-loop level imposes the target space "equations of motion" for $G_{\mu\nu}$ and $B_{\mu\nu}$, vanishing of the matter field independent anomalies fixes the target space dimension to $D = 26$, as discussed e.g. in [1] (the quantities corresponding to the symmetric and antisymmetric parts of $f^{\pm}_{\mu\nu}$ and to $a_{\pm}$ occurring in the sum $\mathcal{H}_+ + \mathcal{H}_-$ for $a_+{=}a_-$ are in the second ref. [1] denoted by $\beta^G_{\mu\nu}$, $\beta^B_{\mu\nu}$ and β^Φ respectively).

Finally we point out that the absence of anomaly candidates depending nontrivially on the antifields is a general feature of all models characterized by (i)–(iv) and represents a remarkable difference to the situation in Yang–Mills and Einstein–Yang–Mills theories with a gauge group containing at least two abelian factors if the classical action has at least one nontrivial global symmetry [14].

References

[1] E. Fradkin and A. Tseytlin (1985), *Nucl. Phys.*, **B261** 1;
C.G. Callan, D. Friedan, E.J. Martinec and M.J. Perry (1985), *Nucl. Phys.*, **B262** 593.

[2] J. Wess and B. Zumino (1971), *Phys. Lett.*, **B37** 95.

[3] I.A. Batalin and G.A. Vilkovisky (1981), *Phys. Lett.*, **B102** 27; (1983), *Phys. Rev.*, **D28** 2567 (E: **D30** (1984) 508).

[4] W. Troost, P. van Nieuwenhuizen and A. Van Proeyen (1990), *Nucl. Phys.*, **B333** 727;
P.S. Howe, U. Lindström and P. White (1990), *Phys. Lett.*, **B246** 430.

[5] G. Barnich and M. Henneaux (1993), *Phys. Lett.*, **B311** 123.

[6] F. De Jonghe, R. Siebelink and W. Troost (1993), *Phys. Lett.*, **B306** 295;
S. Vandoren and A. Van Proeyen (1994), *Nucl. Phys.*, **B411** 257.

[7] L. Bonora, P. Pasti and M. Tonin (1986), *J. Math. Phys.*, **27** 2259.

[8] F. Brandt, N. Dragon and M. Kreuzer (1990), *Nucl. Phys.*, **B340** 187.

[9] F. Brandt (1993), 'Structure of BRS–invariant local functionals', preprint NIKHEF-H 93-21, hep-th/9310123.

[10] F. Brandt, W. Troost and A. Van Proeyen (1994), 'A complete computation of the BRST-antibracket cohomology of 2d gravity conformally coupled to matter', in preparation.

[11] G. Barnich, F. Brandt and M. Henneaux (1994), 'Local BRST cohomology in the antifield formalism: I. General theorems', preprint ULB-TH-94/06, NIKHEF-H 94-13, hep-th/9405109.

[12] M. Werneck de Oliveira, M. Schweda and S.P. Sorella (1993), *Phys. Lett.*, **B315** 93;
A. Boresch, M. Schweda and S.P. Sorella (1994), *Phys. Lett.*, **B328** 36.

[13] G. Bandelloni and S. Lazzarini (1993), 'Diffeomorphism cohomology in Beltrami parametrization', preprint PAR-LPTM-1993, GEF-TH-YY/1993.

[14] F. Brandt (1994), *Phys. Lett.*, **B320** 57;
G. Barnich and M. Henneaux (1994), *Phys. Rev. Lett.*, **72** 1588;
G. Barnich, F. Brandt and M. Henneaux (1994), 'Local BRST cohomology in the antifield formalism: II. Application to Yang–Mills theory', preprint ULB-TH-94/07, NIKHEF-H 94-15, hep-th/9405194;
G. Barnich, F. Brandt and M. Henneaux, 'Local BRST cohomology of Einstein gravity', in preparation.

Quantisation of 2+1 gravity for g = 1 and g = 2

J.E.Nelson

Introduction

In this talk I wish to outline a programme of canonical quantisation of 2+1 dimensional gravity developed in the last few years in collaboration with T.Regge and F.Zertuche [1-6] showing how, at least for g = 1, the torus, it is equivalent to the metric quantisation [7], and presenting an operator algebra for the case of g = 2 [1]. In Section 1 I shall briefly introduce this operator algebra, or holonomy quantisation, for arbitrary g. In Section 2 I shall review the classical, metric formalism developed by Moncrief and others [7]. In Section 3 I shall show how these two approaches work for g = 1, how they are classically related through a time-dependent canonical transformation and lead to equivalent quantisations [8]. Finally in Section 4 I shall briefly present an operator algebra for g = 2 [1]. Spacetime is understood to have topology $\mathbb{R} \times \Sigma$ with signature $(-,+,+)$ where Σ is a closed Riemann surface of genus g and the cosmological constant $\Lambda < 0$.

1 Operator algebra quantisation

This is the first-order, connection approach inspired by Witten [9] and developed by myself, Regge and Zertuche [1-6]. The Einstein-Hilbert action with $\Lambda = -\frac{1}{\alpha^2} < 0$

$$S = \int d^3x\sqrt{g}(R+2\Lambda) \tag{1.1}$$

decomposes as

$$\alpha \int dt \int d^2x\, \varepsilon^{ij} \varepsilon_{ABCD} (\omega_j^{CD}\, \dot{\omega}_i^{AB} - \omega_0^{AB} R_{ij}^{CD}) \tag{1.2}$$

where ω_i^{AB} is an SO(2,2) connection, $A, B = 0,1,2,3$, $i, j = 1,2$ and $\eta_{AB} = (-1,1,1,-1)$. In terms of the original triad e_i^a and ISO(2,1) spin connection ω_i^{ab}, $a, b = 0,1,2$ it reads

$$\omega^A{}_B = \begin{pmatrix} \omega^a{}_b & -\frac{e^a}{\alpha} \\ \frac{e_b}{\alpha} & 0 \end{pmatrix} \tag{1.3}$$

The constraints which follow from (1.1) are

$$R^{AB} = d\omega^{AB} - \omega^{AC}{}_\wedge \omega_C{}^B \approx 0 \tag{1.4}$$

and can be solved locally by setting

$$\omega^{AB} = (dG\, G^{-1})^{AB} \qquad G \in SO(2,2) \tag{1.5}$$

though it is more convenient to use the spinor representation SL$(2,\mathbb{R}) \otimes$ SL$(2,\mathbb{R})$, where if S $\in$ SL$(2,\mathbb{R})$

$$G^{AB}\gamma_B = S^{-1}\gamma^A S \tag{1.6}$$

The Poisson brackets following from (1.2)

$$(\omega_i^{AB}(\mathbf{x}), \omega_j^{CD}(\mathbf{y})) = \frac{1}{2\alpha}\varepsilon_{ij}\varepsilon^{ABCD}\delta^2(\mathbf{x}-\mathbf{y}) \tag{1.7}$$

can be integrated over intersecting homotopic paths on Σ and lead to corresponding brackets for S(σ), S(γ) where $\sigma, \gamma \in \pi_1(\Sigma)$. The matrices S$(\sigma)$ therefore furnish a representation of $\pi_1(\Sigma)$ in SL$(2,\mathbb{R})$, and their traces are gauge invariant since by conjugation

$$Tr(S(\sigma)) = Tr(S(\delta\sigma\delta^{-1}))$$

These classical brackets were calculated by hand for g = 2, then generalised and quantised. There are two sets of time-independent traces (corresponding to the upper and lower spinor components) and representing intersecting paths in Σ. One set satisfies the quantum, ordered, commutator [,] algebra

$$[a_{mk}, a_{jl}] = [a_{mj}, a_{kl}] = 0 \tag{1.8}$$

$$[a_{jk}, a_{km}] = \Big(\frac{1}{K} - 1\Big)(a_{jm} - a_{jk}a_{km}) \tag{1.9}$$

$$[a_{jk}, a_{kl}] = \Big(1 - \frac{1}{K}\Big)(a_{jl} - a_{kl}a_{jk}) \tag{1.10}$$

$$[a_{jk}, a_{lm}] = \Big(K - \frac{1}{K}\Big)(a_{jl}a_{km} - a_{kl}a_{jm}) \tag{1.11}$$

where $a_{ij} = a_{ji}$, $K = e^{i\theta}$, $\tan\theta = -\hbar/8\alpha$, $m,j,k,l = 1\cdots 2g+2$. The other set of variables b_{ij}, with $[a_{ij}, b_{kl}] = 0 \;\; \forall\; ijkl$, satisfies an algebra identical to (1.8-11) but with $K \to 1/K$, $\theta \to -\theta$. The classical (Poisson bracket (,)), algebra is obtained from (1.8-11) by $(A,B) \to \frac{[A,B]}{K-1}$ when $K \to 1$. The algebra (1.8-11) uses a new presentation [4] of $\pi_1(\Sigma)$ in which all paths have mutual intersections $0, \pm 1$. This presentation is invariant under the Dehn group of transformations, for g = 1 and g = 2. The braid group B(2g+2) is a symmetry group of the classical analogue of the above algebra [3]. More importantly, the quantum algebra (1.8-11) is invariant under the quantum Dehn twists [4] generated by each element, for example $a_{ij} = \frac{\cos\Psi}{\cos\frac{\theta}{2}}$ as

$$a_{kl} \to F a_{kl} F^{-1}, \quad F = \exp\frac{-i\Psi^2}{2\theta} \tag{1.12}$$

Moreover, the a_{ij} are not all independent, but due to a number of trace and rank identities there are precisely $6g-6$ independent a_{ij}'s, at least classically [2].

In this formulation, the constraints have been solved exactly, and the Hamiltonian is zero. There is no time development.

2 Classical metric formalism

In the second-order, metric approach [7] the action (1.1) decomposes as

$$\int dt \int d^2x \, (\pi^{ij}\dot{g}_{ij} - N^i\mathcal{H}_i - N\mathcal{H}) \tag{2.1}$$

where g_{ij} is the metric on Σ with momentum π^{ij}. The constraints

$$\mathcal{H}^i = -2D_j\,\pi^{ij} \approx 0 \qquad\qquad i,j=1,2. \tag{2.2}$$

$$\mathcal{H}_\perp = g^{-\frac{1}{2}}\left(\pi_{ij}\,\pi^{ij} - (\pi^{ij}\,g_{ij})^2\right) - g^{\frac{1}{2}}({}^2R + 2\Lambda) \approx 0 \tag{2.3}$$

can also be solved exactly, but this relies on the labelling of the surfaces Σ by $T = \mathrm{Tr}\,\mathrm{K} = $ constant, where K is the extrinsic (or mean) curvature. The metric g_{ij} of Σ can always be written in the form

$$g_{ij} = e^{2\lambda}\tilde{g}_{ij}$$

where $\tilde{g}_{ij}$ is a constant curvature metric on Σ with

$${}^2R(\tilde{g}) = k\,, \qquad \begin{array}{cc} k=1 & g=0 \\ k=0 & g=1 \\ k=-1 & g>1\,. \end{array}$$

The Hamiltonian constraint (2.3) reduces to a differential equation for the conformal factor λ [7]

$$\Delta_{\tilde{g}}\lambda - \frac{T^2}{4}e^{2\lambda} + \frac{1}{2}\tilde{g}^{-1}\tilde{\pi}_{ij}\tilde{\pi}^{ij}\,e^{-2\lambda} - \frac{k}{2} - \Lambda = 0 \tag{2.4}$$

where $\tilde{\pi}^{ij}$ is the traceless transverse part of the momentum conjugate to $\tilde{g}_{ij}$, and (2.2) is automatically satisfied. A solution of (2.4) for λ always exists for $g \geq 1$ [7], and the action (2.1) becomes

$$S = \int dT\left(p^\alpha \frac{dm_\alpha}{dT} - H(p,m)\right)\,. \tag{2.5}$$

where m_α are coordinates on Teichmuller space, $\alpha = 1....6g-6$ and have conjugate momenta

$$p^\alpha = \int_\Sigma d^2x\tilde{\pi}^{ij}\frac{\partial}{\partial m_\alpha}\tilde{g}_{ij}$$

In (2.5) $H(p,m)$ is an effective Hamiltonian representing the area of Σ at time T

$$H = \int_\Sigma d^2x\, g^{\frac{1}{2}} = \int_\Sigma d^2x\, \tilde{g}^{\frac{1}{2}} e^{2\lambda} \tag{2.6}$$

where $\lambda = \lambda(\tilde{g}, \tilde{\pi}, T)$ is given by (2.4). So $H(p,m)$ generates time development in the gauge $T = \mathrm{Tr}\, K$. This programme depends heavily on this gauge and there is no explicit solution of (2.4) for g>1.

3 Classical and quantum equivalence for g = 1

For $k = 0$, $g = 1$, the torus, (2.4) is easily solved and $m = m_1 + im_2$ are the two degrees of freedom (the moduli) of the flat metric $\tilde{g}_{ij}$ of Σ, with conjugate momenta $p = p^1 + ip^2$ satisfying the Poisson brackets

$$(\bar{m}, p) = (m, \bar{p}) = -2, \qquad (m, p) = (\bar{m}, \bar{p}) = 0 \tag{3.1}$$

The effective Hamiltonian (2.6) is

$$H = \frac{m_2 |p|}{\sqrt{T^2 - 4\Lambda}} \tag{3.2}$$

and generates the T development of the modulus and momentum through

$$\frac{dp}{dT} = (p, H), \qquad \frac{dm}{dT} = (m, H) \tag{3.3}$$

The standard action of the modular group on the torus modulus,

$$\begin{aligned} S &: m \to -\frac{1}{m}, \qquad p \to \bar{m}^2 p \\ T &: m \to m + 1, \qquad p \to p \end{aligned} \tag{3.4}$$

preserves the brackets (3.1) and the Hamiltonian (3.2).

The holonomy quantisation works as follows [6]. Since for the torus there are only 3 independent paths U, V, UV, all with single intersections, satisfying

$$U\, V\, U^{-1}\, V^{-1} = id.$$

there are three independent traces for each ($\pm$) spinor component

$$x^\pm = \mathrm{Tr}\big(\mathrm{S(U)}\big) \quad \mathrm{y}^\pm = \mathrm{Tr}\big(\mathrm{S(V)}\big) \quad \mathrm{z}^\pm = \mathrm{Tr}\big(\mathrm{S(U\,V)}\big) \tag{3.5}$$

which, from (1.7) satisfy

$$(x^\pm\,,\, y^\pm) = \pm\frac{1}{4\alpha}(x^\pm\, y^\pm - z^\pm), \;\; (x^+\,,\, y^-) = 0 \tag{3.6}$$

and cyclical permutations of x, y, z. The traces (3.5) are not all independent. They must satisfy

$$F = \frac{1}{2}\mathrm{Tr}\Big(\mathrm{S}(\mathrm{U}\,\mathrm{V}\,\mathrm{U}^{-1}\,\mathrm{V}^{-1}) - \mathrm{I}\Big)$$
$$= 1 - (x^{\pm})^2 - (y^{\pm})^2 - (z^{\pm})^2 + 2x^{\pm}y^{\pm}z^{\pm} = 0\,. \qquad (3.7)$$

The expression (3.7) has zero brackets with all x, y, z and is symmetric under cyclical permutations of x, y, z. Note that $F = 0$ can be solved by setting

$$x^{\pm} = \cosh\frac{r_1^{\pm}}{2}, y^{\pm} = \cosh\frac{r_2^{\pm}}{2}, z^{\pm} = \cosh\frac{(r_1^{\pm} + r_2^{\pm})}{2}$$

and then (3.6) implies, for the two remaining variables $r_1^{\pm}$, $r_2^{\pm}$

$$(r_2^{\pm}, r_1^{\pm}) = \pm\frac{1}{\alpha}, \quad (r^{+}, r_{-}) = 0 \qquad (3.8)$$

Direct quantisation of (3.8) would then imply

$$[r_2^{\pm}, r_1^{\pm}] = \pm\frac{i\hbar}{\alpha} \qquad (3.9)$$

Alternatively, the brackets (3.6) can be quantised as follows. In (3.6) replace the bracket (x, y) by $\frac{xy-yx}{i\hbar}$ and on the R.H.S. , xy by $\frac{1}{2}\,(xy{+}yx)$. The result is, for the + variables,

$$e^{i\theta}xy - e^{-i\theta}yx = 2i\sin\theta z \qquad \textit{and} \quad \textit{cyclical} \qquad (3.10)$$

that is, the cyclical representation of $SU(2)_q$ [6] with $\tan\theta = -\hbar/8\alpha$, consistent with the commutators

$$[r_1^{\pm}, r_2^{\pm}] = \pm 8i\theta\,, \quad [r^{+}, r^{-}] = 0 \qquad (3.11)$$

when the *operators* $x^{\pm}$, $y^{\pm}$, $z^{\pm}$ are represented by

$$x^{\pm} = \sec\theta\cosh\frac{r_1^{\pm}}{2}, y^{\pm} = \sec\theta\cosh\frac{r_2^{\pm}}{2}, z^{\pm} = \sec\theta\cosh\frac{(r_1^{\pm} + r_2^{\pm})}{2}$$

The commutators (3.11) differ from the direct quantisation (3.9) by terms of $O(\hbar^3)$. They are invariant under

$$\begin{aligned} S &: r_1^{\pm} \to r_2^{\pm}, \qquad r_2^{\pm} \to -r_1^{\pm} \\ T &: r_1^{\pm} \to r_1^{\pm} + r_2^{\pm}, \qquad r_2^{\pm} \to r_2^{\pm}, \end{aligned} \qquad (3.12)$$

On the traces (3.5), the quantum modular group action is

$$\begin{aligned} S &: x^{\pm} \to y^{\pm}, \quad y^{\pm} \to x^{\pm}, \quad z^{\pm} \to (1+K)x^{\pm}y^{\pm} - z^{\pm} \\ T &: x^{\pm} \to z^{\pm}, \quad y^{\pm} \to y^{\pm}, \quad z^{\pm} \to (1+K)y^{\pm}z^{\pm} - x^{\pm} \end{aligned} \qquad (3.13)$$

corresponding to the intersection number preserving exchanges

$$\begin{aligned} S: U \to V^{-1}, \qquad V \to U \\ T: U \to UV, \qquad V \to V \end{aligned} \tag{3.14}$$

The second of these corresponds to a quantum Dehn twist for the traces x, y, z.

The above two formalisms seem very different but Carlip and I have recently shown that [8] they are related by a time-dependent canonical transformation and that the unique action can be written in the equivalent forms.

$$\begin{aligned} \int dt \int d^2x\, \pi^{ij} \dot{g}_{ij} &= \alpha \int dt \int d^2x\, \varepsilon^{ij} \varepsilon_{ABCD}\, \omega_j^{CD}\, \dot{\omega}_i^{AB} \\ &= \int \frac{1}{2} (\bar{p} dm + p d\bar{m}) + H dT - d(p^1 m_1 + p^2 m_2) \\ &= \alpha \int (r_1^- dr_2^- - r_1^+ dr_2^+) \end{aligned} \tag{3.15}$$

and is therefore expressed directly in terms of the holonomy parameters $r_1^\pm, r_2^\pm$. In (3.15) $T = -\frac{2}{\alpha} \cot \frac{2t}{\alpha}$ and

$$m = \left(r_1^- e^{it/\alpha} + r_1^+ e^{-it/\alpha} \right) \left(r_2^- e^{it/\alpha} + r_2^+ e^{-it/\alpha} \right)^{-1} \tag{3.16}$$

$$p = -\frac{i\alpha}{2 \sin \frac{2t}{\alpha}} \left(r_2^+ e^{it/\alpha} + r_2^- e^{-it/\alpha} \right)^2 \tag{3.17}$$

The Hamiltonian (3.2), invariant under (3.12) now takes the form

$$H = \frac{\alpha(r_1^- r_2^+ - r_1^+ r_2^-)}{\sqrt{T^2 - 4\Lambda}} \tag{3.18}$$

It can be checked that the brackets (3.8) imply (3.1) and that the transformations (3.12) imply (3.4). Similar results hold for $\Lambda = 0$ [10].

4 Operator algebra for g = 2

For g = 2, 2g+2 = 6, so there are 15 initial a_{ij}'s, but by use of computer algebra we were able [1] to reduce to 6 independent variables as follows. A convenient choice for the 6 independent elements is given by 3 commuting angles $\varphi_{-b} = -\varphi_b$, $b = \pm 1 \cdots \pm 3$ defined by:

$$a_{12} = \frac{\cos \varphi_1}{\cos \frac{\theta}{2}}, \qquad a_{34} = \frac{\cos \varphi_2}{\cos \frac{\theta}{2}} \qquad a_{56} = \frac{\cos \varphi_3}{\cos \frac{\theta}{2}} \tag{4.1}$$

and commuting operators M_{ab} with the properties:

$$\begin{aligned} M_{ab} &= M_{ba} \qquad a,b \;=\; \pm 1 \cdots \pm 3 \\ M_{a,-a} &= 1, \; M_{a,-b} M_{b,c} \;=\; M_{ac} \end{aligned} \tag{4.2}$$

The M_{ab} act as raising and lowering operators on the φ_a:

$$M_{\pm a,b} \varphi_a = (\varphi_a \mp \theta) M_{\pm a,b} \tag{4.3}$$

The relations (4.2-3) follows from the single sector factorisation for all $a, b = \pm 1 \cdots \pm 3$:

$$M_{ab} = M_a M_b = M_{ba} = M_b M_a \tag{4.4}$$

$$M_{-a} = M_a^{-1} \tag{4.5}$$

$$M_{\pm a} \varphi_a = (\varphi_a \mp \theta) M_{\pm a} \tag{4.6}$$

and therefore (4.4-6) can be formally satisfied by setting

$$M_a = \exp\left(-\theta \frac{\partial}{\partial \varphi_a} \right) = \exp(-i\theta p_a)$$

with

$$[\varphi_a, \varphi_b] = 0, \; [\varphi_a, p_b] = i\delta_{ab}, \; [p_a, p_b] = 0, \; a, b, = 1, 2, 3 \tag{4.7}$$

It can be checked that the 12 remaining a_{ik} can be expressed in terms of the φ_a and their conjugate momenta p_a. For example the quantum, ordered operator a_{23} can be expressed as

$$\begin{aligned} a_{23} = {} & \cos(\theta p_1) \cos(\theta p_2) \\ & + \Big(\cot \varphi_1 \cot \varphi_2 - \frac{\cos \varphi_3}{\cos\left(\frac{\theta}{2}\right) \sin \varphi_1 \sin \varphi_2} \Big) \sin(\theta p_1) \sin(\theta p_2) \\ & - i \tan\left(\frac{\theta}{2}\right) (\cot \varphi_2 \cos(\theta p_1) \sin(\theta p_2) + \cot \varphi_1 \cos(\theta p_2) \sin(\theta p_1)) \end{aligned}$$

A Hilbert space of wave functions $\Psi(z_a)$, $z_a = \cos \varphi_a$ with measure $\sigma(z) d^3 z$ can be defined by

$$||\Psi||^2 = \int |\Psi(z)|^2 \sigma(z) d^3 z \tag{4.8}$$

The weight function $\sigma(z)$ can be determined from the hermiticity of the a_{ij}, where hermiticity is defined for an operator O by $\langle \Psi, O\Phi \rangle = \langle O^\dagger \Psi, \Phi \rangle$. $\sigma(z)$ can be expressed in terms of a deformed Gamma function. The positivity of

$\sigma(z)$ depends on the domain of the φ_a but it is real and positive if at least one of the φ_a is imaginary. Further details can be found in [1].

Discussion

The very general results for arbitrary genus outlined in Sections 1 and 2 seem very distinct and unconnected. It is only for g = 1 that a relationship has been established. It would be extremely interesting if a similar relationship existed in general but this seems very difficult. Hopefully the results of Section 4 for g = 2 will suggest how to represent the algebra of observables for g > 2 though surely there will be further complications. Even more interesting would be if these studies would give us some insight into 3+1 quantum gravity though at the moment this seems very distant.

References

[1] J. E. Nelson and T. Regge (1994), 'Quantisation of 2+1 gravity for genus 2', *Phys. Rev.* **D**, to appear.

[2] J. E. Nelson and T. Regge (1993), *Commun. Math. Phys.,* **155** *561.*

[3] J. E. Nelson and T. Regge (1991), *Commun. Math. Phys.,* **141** *211.*

[4] J. E. Nelson and T. Regge (1991), *Phys. Lett.*, **B272** 213.

[5] J. E. Nelson and T. Regge (1989), *Nucl. Phys.*, **B328** 190.

[6] J. E. Nelson, T. Regge and F. Zertuche (1990), *Nucl. Phys.*, **B339** 516.

[7] V. Moncrief (1989), *J. Math. Phys.*, **30** 2907;
A. Hosoya and K. Nakao (1990), *Class. Quantum Grav.*, **7** 163;
A. Hosoya and K. Nakao (1990), *Prog. Theor. Phys.*, **84** 739;
Y. Fujiwara and J. Soda (1990), *Prog. Theor. Phys.*, **83** 733.

[8] S. Carlip and J. E. Nelson (1994), *Phys.Lett.*, **B324** 299.

[9] E. Witten (1988/89), *Nucl. Phys.*, **B311** 46-78.

[10] S. Carlip (1990), *Phys. Rev.*, **D42** 2647;
S. Carlip (1992), *Phys. Rev.*, **D45** 3584;
S. Carlip (1993), *Phys. Rev.*, **D47** 4520.

Geometry and dynamics with time-dependent constraints

Jonathan M. Evans and Philip A. Tuckey

1 Introduction

A Hamiltonian dynamical system can be described geometrically by a phase space manifold Γ (of dimension $2d$ say) equipped with a symplectic form ω and a Hamiltonian function H (see Abraham and Marsden (1978) and Arnol'd (1978)). The condition that ω is symplectic means that it is a non-degenerate closed two-form, so it can be used to introduce a Poisson bracket $\{\,,\}$ on Γ. The evolution of the system in time t is given by a particular set of trajectories on Γ, parametrized by t, such that Hamilton's equation

$$\frac{\mathrm{d}f}{\mathrm{d}t} = \frac{\partial f}{\partial t} + \{f, H\} \tag{1}$$

holds for any time-dependent function f on Γ.

Since the seminal work of Dirac (1950,1958,1964) there has been intensive study of systems of this type which can be consistently constrained to some physical phase space manifold Γ^* (of dimension $2n$ say) which is embedded in Γ in a manner we now describe. In the most general case the embedding of Γ^* in Γ can depend on time and it must therefore be defined by a family of maps

$$\varphi_t : \Gamma^* \to \Gamma\,, \tag{2}$$

depending smoothly on t, each of which is a diffeomorphism onto its image $X_t \subset \Gamma$. We assume that each of the trajectories on Γ for which (1) holds has the property that it always lies in the subspaces X_t for each t, or else that it always lies in the complements of these spaces. It is clear that trajectories of the former type correspond exactly under the embedding (2) to trajectories on Γ^*, and one can attempt to reformulate the dynamics for this subclass of trajectories in a manner which is intrinsic to Γ^*.

We define ω^* on Γ^* at time t by pulling back ω using φ_t and we assume that this is also a symplectic form (albeit a time-dependent one in general). We can then use ω^* to introduce the Dirac bracket $\{\,,\}^*$ on Γ^*. The key issue which we shall address here is whether, for a given choice of embeddings (2), one can find a Hamiltonian function H^* such that Hamilton's equation

$$\frac{\mathrm{d}f}{\mathrm{d}t} = \frac{\partial f}{\partial t} + \{f, H^*\}^* \tag{3}$$

holds for any time-dependent function f on Γ^*. When the embeddings (2) are independent of time, (3) follows easily from (1) with $H^* = H$. In the general case, however, the dynamics on Γ^* is determined not just by the dynamics on Γ but also by the time-dependence of the embeddings φ_t, and under these circumstances it is non-trivial to determine whether (3) holds for some function H^*.

The most compelling reason for studying this general situation is the fact that gauge choices with explicit time dependence are essential in order to restrict systems which are invariant under time-reparametrizations, such as the relativistic particle, string or general relativity, to their physical degrees of freedom (but see Henneaux *et al* (1992) for possible modifications of the action to allow other gauge choices). Here we summarize the solution of this problem given in Evans and Tuckey (1993) and we clarify some related issues. (We have recently learned that Mukunda (1980) has previously obtained results which are locally equivalent to ours using an algebraic approach. Related work from the Lagrangian point of view appears in (Rañada 1994).) We then give some new examples, extending the treatment of the relativistic particle in a background field (Evans 1993) to the case of a string in an arbitrary antisymmetric tensor background.

2 Extended phase space and constrained dynamics

We define extended phase space to be $\bar{\Gamma} = \Gamma \times \mathbb{R}$, where the second factor is time. We can, in a natural way, regard H and ω as living on $\bar{\Gamma}$ (by pulling back using the projection map) and we define the contact form on $\bar{\Gamma}$ to be

$$\Omega = \omega + \mathrm{d}H \wedge \mathrm{d}t \,. \tag{4}$$

(In Evans and Tuckey (1993) Ω was called the Poincaré-Cartan two-form; in Abraham and Marsden (1978) Ω is introduced as an example of a contact structure.) Any trajectory on Γ parametrized by t is clearly equivalent to a trajectory on $\bar{\Gamma}$ with parameter s chosen such that $\mathrm{d}t/\mathrm{d}s$ is nowhere zero. Let V be the tangent vector to the trajectory on $\bar{\Gamma}$. Then Hamilton's equation (1) is precisely the condition

$$i(V)\Omega = 0 \tag{5}$$

(where $i(V)$ denotes interior multiplication of a form by the vector field V).

When the system is constrained we can similarly define extended physical phase space to be $\bar{\Gamma}^* = \Gamma^* \times \mathbb{R}$. The family of embeddings (2) is equivalent to the single embedding

$$\bar{\varphi} : \bar{\Gamma}^* \to \bar{\Gamma} \,, \qquad \bar{\varphi}(x,t) = (\varphi_t(x), t) \,, \tag{6}$$

which is a diffeomorphism onto its image $\bar{X} = \{(x,t) : x \in X_t, t \in \mathbb{R}\} \subset \bar{\Gamma}$ (assuming, as stated earlier, that φ_t varies smoothly with t). Define the form Ω^* on $\bar{\Gamma}^*$ to be the pull back of Ω using $\bar{\varphi}$. In general this has the structure

$$\Omega^* = \omega^* + (\mathrm{d}H + Y) \wedge \mathrm{d}t \tag{7}$$

for some one-form Y. (Here we use the fact that any time-dependent form on Γ^* can be regarded as a smooth form on $\bar{\Gamma}^*$; when the form is time-independent this reduces to pulling back using the projection map.)

Any solution of Hamilton's equation (1) which lies in $\bar{X}$ clearly corresponds to a trajectory in $\bar{\Gamma}^*$ with tangent vector V^* which satisfies

$$i(V^*)\,\Omega^* = 0\,. \tag{8}$$

By comparison with (4) and (5) we see that Hamilton's equation (3) holds on Γ^* if and only if

$$Y = \mathrm{d}K \bmod \mathrm{d}t \qquad \text{and then} \qquad H^* = H + K \tag{9}$$

for some function K on $\bar{\Gamma}^*$. One can show that this holds locally (*ie.* in any contractible region on Γ^*) if and only if ω^* is independent of time on Γ^*, a fact which will prove useful later.

To discuss specific examples it is convenient to introduce on Γ local coordinates z^M, $M = 1,\ldots,2d$. The subsets $X_t \subset \Gamma$ are defined by a set of time-dependent constraint functions $\psi^I(z^M,t)$, $I = 1,\ldots,2(d-n)$, which are, in the language of Dirac (1950,1958,1964), second-class. The fact that these constraint functions are preserved in time is equivalent to our initial assumption that there exists a subset of trajectories confined to the subspaces X_t. The condition that the constraint functions are second-class is equivalent to our assumption that the form ω^* is symplectic.

If ξ^A, $A = 1,\ldots,2n$, are local coordinates on Γ^* then the embeddings φ_t or $\bar{\varphi}$ allow us to regard the z^M as time-dependent functions of these variables on $\bar{X}$, and we have the explicit expression

$$Y = -\frac{\partial z^M}{\partial t}\frac{\partial z^N}{\partial \xi^A}\,\omega_{MN}\,\mathrm{d}\xi^A = -\,\omega_{MN}\,\frac{\partial z^M}{\partial t}\,\mathrm{d}z^N \quad \bmod \mathrm{d}t \tag{10}$$

for the one-form appearing in (9). It is convenient in practice to specify φ_t or $\bar{\varphi}$ by giving explicit expressions for a set of functions $\xi^A(z^M,t)$, which we call physical variables; on restriction to $\bar{X}$ these functions define (the inverses of) these embeddings in terms of the local coordinates.

3 Remarks

Our result (9) establishes necessary and sufficient conditions for a family of embeddings φ_t, or a choice of physical variables $\xi^A(z^M,t)$, to result in a Hamiltonian time-evolution equation (3) on Γ^*. For a given set of constraint subspaces X_t, or equivalently a set of constraint functions $\psi^I(z^M,t)$, a family of embeddings or physical variables having this property always exists locally. This follows from Darboux's Theorem, which tells us that locally we can find embeddings φ_t or choose coordinates $(\xi^A) = (q^\alpha, p_\alpha)$ on Γ^* such that $\omega^* = \mathrm{d}q^\alpha \wedge \mathrm{d}p_\alpha$. Since this expression is manifestly independent of time on Γ^*, it satisfies the criterion which we gave following (9). On the other hand, there are clearly many embeddings or choices of physical variables for which (3) will not hold, as can be seen by performing an arbitrary time-dependent coordinate transformation to make ω^* time-dependent.

Our result does not tell us how to explicitly construct a set of embeddings or physical variables with the required property, and in general this remains an open problem. A partial result in this direction is case (B) of Evans (1991). This applies to a system with time-independent gauge symmetry generators which has imposed on it a set of time-dependent gauge-fixing conditions involving some subset of canonical variables which all commute under the Poisson bracket. It is worth pointing out that if these canonical variables are regarded as configuration space coordinates in some equivalent Lagrangian description, then the result in question can also be obtained by first gauge-fixing the Lagrangian and then passing to the Hamiltonian formalism. The new examples we shall present below lie outside the scope of case (B) of Evans (1991). Thus the result (9) is still useful for finding good sets of physical variables, even though it offers no general method for doing so.

Finally we emphasize that our main motivation for the work summarized here is the reduced phase space approach to the canonical quantisation of systems which require time-dependent gauge choices. Even at the classical level, a system whose time evolution is not described by an equation of the form of (3) falls outside the realm of conventional Hamiltonian mechanics. In passing to the quantum theory, (3) becomes the Heisenberg equation of motion, which guarantees the existence of a unitary time evolution operator. In the absence of a classical evolution equation of the form of (3), Gitman and Tyutin (1990a) have given an alternative prescription for the Heisenberg quantum evolution equation, in which extra terms appearing on the right hand side of (3) are taken over. This approach is complicated by the difficulty in obtaining an explicitly unitary time evolution. The consideration of simple examples such as the relativistic particle in an arbitrary background electromagnetic field (Evans 1993) reveals that our approach can be much simpler – compare with Gitman and Tyutin (1990b), Gavrilov and Gitman (1993). Batalin and Lyakovich (1991) have also considered the quantization of systems with time-dependent Hamiltonian and constraints.

4 Examples

We shall now apply our result (9) to discuss the gauge-fixing of relativistic particles and strings moving in d-dimensional Minkowski space-time with background gauge fields. We shall take coordinates x^μ on Minkowski space-time which are either 'orthonormal' with $\mu = 0, \ldots, d{-}1$, or of 'light-cone' type with $\mu = +, -, 1, \ldots, d{-}2$, so that the flat metric has components $-g_{00} = g_{11} = \ldots = g_{d-1\,d-1} = g_{+-} = g_{-+} = 1$ and all others vanishing. (This means that $x^\pm = (x^{d-1} \pm x^0)/\sqrt{2}$ agreeing with the conventions of Evans (1993) but not Evans (1991).) It is useful to set up some conventions regarding indices which will allow us to deal with temporal and light-cone gauge conditions in a uniform way. Thus we shall let the single index n on any vector denote either 0 or $+$, and we shall label the remaining components by $a = 1, \ldots, d-1$ or $a = -, 1, \ldots, d-2$ respectively. We shall also find it useful to denote the 'transverse' components by $i = 1, \ldots, d-2$. In what follows the ranges of these indices will always be understood.

4.1 Relativistic particle in an electromagnetic field

A particle of mass m and charge e moving in an arbitrary electromagnetic field $A_\mu(x^\nu)$ can be described by the Lagrangian

$$L = -\left[m\sqrt{-\dot{x}^2} + eA_\mu(x^\nu)\,\dot{x}^\mu\right] . \tag{11}$$

Here $x^\mu(t)$ is the particle's trajectory, t is a parameter along the worldline, and $\dot{x} = dx/dt$. Introducing the canonical momentum $p_\mu = \partial L/\partial \dot{x}^\mu$ conjugate to x^μ, we have coordinates (x^μ, p_μ) on phase space, with Poisson bracket $\{x^\mu, p_\nu\} = \delta^\mu_\nu$. There is a single, first-class constraint

$$\phi = (p + eA)^2 + m^2 = 0 . \tag{12}$$

The Hamiltonian is $H = \lambda\phi$, where λ is an arbitrary (time-dependent) function on phase space.

Consider the class of gauge-fixing conditions of the form

$$x^n = f(p_\mu, t) , \tag{13}$$

where f can be any function of its arguments which defines a good gauge choice (we shall make no attempt to be more precise concerning this last point). We define a set of physical variables

$$(\xi^A) = (x^{a*}, p_a) \quad \text{where} \quad x^{a*} = x^a - \int dp_n \frac{\partial f}{\partial p_a} \tag{14}$$

(partial derivatives and integrals of f are to be understood in terms of the functional dependence given by (13)) and it is clear that in principle the equations (12), (13) and (14) allow us to express all quantities as functions of (ξ^A, t). We claim that the system can then be described by these physical variables together with a Hamiltonian

$$H^* = -\int dp_n \frac{\partial f}{\partial t} , \tag{15}$$

which is valid for any background gauge field A_μ and any function f.

The explicit restriction to physical phase space is of course very involved for a general background field. In principle we can substitute from (13) and (14) into (12) to find p_n as a function of the physical variables and time, and substitution of this result back into (13) and (14) then determines all the x^μ as functions of (ξ^A, t). Fortunately, it is not necessary to carry out this elimination explicitly in order to verify that our chosen physical variables do indeed satisfy the criterion (9) leading to the general expression for the Hamiltonian given above. This is because the explicit time dependence of x^μ enters only through p_n and f; and by using this fact it is easy to calculate from (10) that $Y = -\mathrm{d}(\int dp_n\, \partial f/\partial t)$ mod dt. Since the original Hamiltonian H vanishes when $\phi = 0$, the result follows.

Examples are:

$$x^n = t \quad \text{giving} \quad x^{a*} = x^a\,, \quad H^* = -p_n\,, \tag{16}$$

which reproduces the temporal and light-cone results of Evans (1993);

$$\begin{aligned} x^+ = p^+ t \quad &\text{giving} \quad x^{-*} = x^- - p^- t\,, \quad x^{i*} = x^i\,, \quad H^* = -p_+ p_-\,; \\ x^0 = p^0 t \quad &\text{giving} \quad x^{a*} = x^a\,, \quad H^* = \tfrac{1}{2} p_0^2\,. \end{aligned} \tag{17}$$

These expressions are deceptively simple in appearance because they represent very complicated functions of the physical variables in the case of a general background. It is interesting that the Hamiltonians have universal forms in terms of the original momenta, in the sense that the dependence on the background field enters only through these particular functions.

4.2 Relativistic closed string in an antisymmetric tensor field

A closed string moving in an arbitrary background antisymmetric tensor field $B_{\mu\nu}(x^\rho)$ can be described by the Lagrangian

$$L = -\int_0^{2\pi} d\sigma \left[\left((\dot{x}.x')^2 - (\dot{x})^2 (x')^2 \right)^{1/2} + B_{\mu\nu}(x^\rho)\, \dot{x}^\mu x'^\nu \right]\,. \tag{18}$$

Here $x^\mu(t,\sigma)$ describes the string's trajectory, t and $0 \le \sigma \le 2\pi$ parametrize the world-sheet, and $\dot{x} = \partial x/\partial t$, $x' = \partial x/\partial\sigma$. Introducing the momentum $p_\mu(\sigma) = \delta L/\delta \dot{x}^\mu(\sigma)$ conjugate to $x^\mu(\sigma)$ as usual, we have coordinates $(x^\mu(\sigma), p_\mu(\sigma))$ on phase space, with Poisson bracket $\{x^\mu(\sigma), p_\nu(\sigma')\} = \delta^\mu_\nu \delta(\sigma - \sigma')$. The only constraints are first-class and are given by

$$(p_\mu + B_{\mu\nu} x'^\nu)^2 + (x')^2 = 0\,, \tag{19}$$

$$x'.p = 0\,. \tag{20}$$

Again the Hamiltonian H is proportional to the constraints.

It is useful to introduce the position zero-mode and total momentum of the string by

$$X^\mu(t) = \frac{1}{2\pi} \int_0^{2\pi} d\sigma\, x^\mu(t,\sigma)\,, \qquad P_\mu(t) = \int_0^{2\pi} d\sigma\, p_\mu(t,\sigma)\,. \tag{21}$$

The factor of 2π ensures X^μ and P_μ are conjugate variables. One can then write a decomposition of the string fields

$$x^\mu = X^\mu + \tilde{x}^\mu \,, \qquad p_\mu = \tfrac{1}{2\pi} P_\mu + \tilde{p}_\mu \,, \tag{22}$$

where the tilded variables represent the oscillator degrees of freedom.

We consider the class of gauge-fixing conditions of the form

$$x^n = f(P_\mu, t) \,, \tag{23}$$

$$p^n = \tfrac{1}{2\pi} P^n \,, \tag{24}$$

where f is, as before, any function of its arguments which provides a good gauge-fixing condition. The gauge conditions and constraints allow for the complete elimination of one pair of string position and momentum variables corresponding to the direction in space-time labelled by n, and they allow also for the elimination of one additional set of string oscillators. We introduce a set of physical variables

$$(\xi^A) = (X^{a*}, P_a, \tilde{x}^i(\sigma), \tilde{p}_i(\sigma)) \quad \text{where} \quad X^{a*} = X^a - \int dP_n \frac{\partial f}{\partial P_a} \tag{25}$$

(comments similar to those following (14) apply) and with a little thought one can see that the equations (19), (20), (23), (24) indeed allow us in principle to express all the original variables in terms of (ξ^A, t). At this stage the analysis looks very like that for the particle, at least as far as the string zero mode and total momentum variables are concerned. For the case of a light-cone gauge ($n = +$) this comparison is accurate: for the physical variables given above one can again calculate Y from (10) and deduce that (9) holds, yielding a Hamiltonian

$$H^* = -\int dP_n \frac{\partial f}{\partial t} \,, \tag{26}$$

which is valid for any background field $B_{\mu\nu}$ and any gauge-fixing function f. For the case of a temporal gauge condition ($n = 0$), however, the physical variables written above do not satisfy (9) in general, unless the background has some special symmetry. The previous arguments break down because the solutions for the redundant oscillator variables $\tilde{x}^{d-1}$ and $\tilde{p}_{d-1}$ in terms of (ξ^A, t) can depend explicitly on time for a general background field. This difficulty is absent if, for example, the background vanishes, $B_{\mu\nu} = 0$, and then the physical variables and Hamiltonian written above hold for any function f.

Examples are:

$$x^n = t \quad \text{giving} \quad X^{a*} = X^a \,, \quad H^* = -P_n \,; \tag{27}$$

and

$$x^+ = P^+t \quad \text{giving} \quad X^{-*} = X^- - P^-t\,, \quad X^{i*} = X^i\,, \quad H^* = -P_+P_-\,;$$
$$x^0 = P^0 t \quad \text{giving} \quad X^{a*} = X^a\,, \quad H^* = \tfrac{1}{2}P_0^2\,. \tag{28}$$

These last examples generalise previous light-cone and temporal gauge-fixing constructions (Goddard *et al* 1973, Scherk 1975, Goddard *et al* 1975).

Acknowledgments

JME is supported by a fellowship from the EU Human Capital and Mobility programme and PAT by an Alexander von Humboldt Research Fellowship.

References

Abraham R and Marsden J E (1978), *Foundations of mechanics*, Benjamin/ Cummings.

Arnol'd V I (1978), *Mathematical methods of classical mechanics*, Springer.

Batalin I A and Lyakovich S L (1991), in *Group theoretical methods in physics*, Nova Science, vol. 2, p. 57.

Dirac P A M (1950), *Canad. J. Math.*, 2 129.

Dirac P A M (1958), *Proc. Roy. Soc.*, A246 326.

Dirac P A M (1964), *Lectures on quantum mechanics*, Academic.

Evans J M (1991), *Phys. Lett.*, B256 245.

Evans J M (1993), *Class. Quantum Grav.*, 10 L221.

Evans J M and Tuckey P A (1993), *Int. J. Mod. Phys.*, A8 4055.

Gavrilov S P and Gitman D M (1993), *Class. Quantum Grav.*, 10 57.

Gitman D M and Tyutin I V (1990a), *Quantization of fields with constraints*, Springer.

Gitman D M and Tyutin I V (1990b), *Class. Quantum Grav.*, 7 2131.

Goddard P, Goldstone J, Rebbi C and Thorn C B (1973), *Nucl. Phys.*, B56 109.

Goddard P, Hanson A J and Ponzano G (1975), *Nucl. Phys.*, B89 76.

Henneaux M, Teitelboim C and Vergara J (1992), *Nucl. Phys.*, B387 391.

Mukunda N (1980), *Phys. Script.*, 21 801.

Rañada M F (1994), *J. Math. Phys.*, 35 748.

Scherk J (1975), *Rev. Mod. Phys.*, 47 123.

Collective coordinates and BRST transformations *or* Gauge theories without gauge fields

H.J.W. Müller-Kirsten and J.-z. Zhang

1 Introduction

It is well known from electrodynamics that the electric field $\vec{E}$ and the vector potential $\vec{A}$ form a canonical pair, and that the Gauss law, i. e. $\vec{\nabla} \cdot \vec{E} = \rho$, therefore represents a constraint which links various canonical momenta. The converse seems also to be true in the sense that a theory with constraints is a gauge theory, i. e. a theory with a gauge symmetry. Constraints arise naturally with a large class of field transformations. In particular the quantisation of a theory with a classical finite energy field configuration like a soliton reqires a transformation to collective and fluctuation coordinates and hence to a larger number of degrees of freedom which immediately results in the appearance of at least one momentum–dependent constraint. Quantisation in the background of a soliton or other such classical configuration implies also a perturbation expansion in its neighbourhood and hence the loop–expansion if the path–integral method is employed. It is well known that this is, in general, a complicated task which involves in particular also the problem of gauge fixing. One way to achieve the effects of gauge fixing without breaking the invariance is to reformulate the invariance as a BRST [1][2] transformation.

In the following we consider the motion of a particle in the neighbourhood of a classical Newtonian path as a relatively simple analogue to a field theory with a soliton–like classical configuration, the classical time variable beeing replaced by the function $f(q)$ where q parametrises the classical path. We first show that the momentum constraint generates a gauge symmetry of the first order Lagrangian. The BRST extension is then shown to lead to an invariant Hamiltonian in which the ghosts decouple for a suitable gauge fixing choice. It is then shown that the constraints, and only these, project out precisely the BRST and anti–BRST invariant physical states of the theory.

2 Kinematics and the first–order Lagrangian

We consider the Lagrangian [3][4]

$$L = \frac{1}{2}\dot{\vec{R}}^2 - V(R) \tag{2.1}$$

where $\vec{R}$ is an N–dimensional Euclidean vector with canonical momentum $\vec{P} = \dot{\vec{R}}$. We assume that the path $\vec{R}$ of the particle of mass $m = 1$ can be approximated by a Newtonian path $\vec{r}(f(q))$. $\vec{R}$ is then written

$$\vec{R} = \vec{r}(f(q)) + \sum_{a=2}^{n} \vec{n}_a(f(q))\eta_a \tag{2.2}$$

where $\{\vec{n}_a(f)\}$ together with $\vec{r}_f \equiv \frac{d\vec{r}}{df}$ form a moving local reference frame at the point $\vec{r}(f)$, i. e. $\vec{n}_a \cdot \vec{n}_b = \delta_{ab}$, $\vec{n}_a \cdot \vec{r}_f = 0$.

It is essential to use a convenient notation. We write therefore

$$\vec{R} = R_i(f(q))\vec{e}_i = Q_\alpha \vec{n}_\alpha(f(q)) \tag{2.3}$$

where $\{\vec{e}_i\}$ are the unit vectors of the fixed frame and $\{\vec{n}_\alpha\}$ those of the moving reference frame, with $i, \alpha = 1, \ldots, N$, and $\vec{n}_1 = \frac{\vec{r}_f}{r_f}$. Then

$$R_i = M_{i\alpha}Q_\alpha, \quad Q_\alpha = W_{\alpha i}R_i \tag{2.4}$$

where

$$M_{i\alpha} = \langle \vec{e}_i \mid \vec{n}_\alpha \rangle = \frac{\partial R_i}{\partial Q_\alpha}, \quad W_{\alpha i} = \langle \vec{n}_\alpha \mid \vec{e}_i \rangle = \frac{\partial Q_\alpha}{\partial R_i} \tag{2.5}$$

and $W_{\alpha i}M_{i\beta} = \delta_{\alpha\beta}$, $M_{i\alpha}W_{\alpha j} = \delta_{ij}$. Now,

$$p_\alpha := \frac{\partial L}{\partial \dot{Q}_\alpha} = \frac{\partial L}{\partial \dot{R}_i}\frac{\partial R_i}{\partial Q_\alpha} = P_i M_{i\alpha} \tag{2.6}$$

$$p := \frac{\partial L}{\partial \dot{q}} = \frac{\partial L}{\partial \dot{R}_i}\frac{\partial R_i}{\partial q} = P_i T_i \tag{2.7}$$

where

$$T_i = \frac{\partial R_i}{\partial q} = f'\langle \vec{e}_i \mid \vec{n}_{\alpha,f}\rangle Q_\alpha. \tag{2.8}$$

From (2.6) and (2.7) we obtain the constraint

$$\phi := p - p_\alpha W_{\alpha i}T_i = 0 \tag{2.9}$$

where $W_{\alpha i}T_i = f'\Gamma_{\alpha\beta}Q_\beta$ and $\Gamma_{\alpha\beta} := \langle \vec{n}_\alpha \mid \vec{n}_{\beta,f}\rangle = -\Gamma_{\beta\alpha}$. Since the constraint involves momenta we have to start from the first–order Lagrangian expressed in terms of coordinates and momenta and not from the momentum–integrated second–order form. Then $L = P_i\dot{R}_i - H$ where $H = \frac{1}{2}P_i^2 + V(R)$ becomes

$$L = p\dot{q} + p_\alpha \dot{Q}_\alpha - H, \quad H = \frac{1}{2}p_\alpha p_\alpha + V(Q). \tag{2.10}$$

3 The gauge transformation

The theory can be shown to possess no secondary or higher order constraints, i. e. $\{\phi, H + \lambda\phi\}_P = 0$. Then

$$L(q, Q_\alpha, \lambda; p, p_\alpha, p_\lambda) = p\dot{q} + p_\alpha \dot{Q}_\alpha - H_\lambda(q, Q_\alpha, \lambda; p, p_\alpha, p_\lambda) \quad (3.1)$$

$$H_\lambda = \frac{1}{2}p_\alpha^2 + V(R(f(q))) + \lambda\phi. \quad (3.2)$$

This new first-order Lagrangian has two constraints, i. e. $\psi_1 := \phi = 0$, $\psi_2 := p_\lambda = 0$, and we see that they commute, i. e.

$$\{\psi_i, \psi_j\}_P = 0, \quad i, j = 1, 2. \quad (3.3)$$

Thus ψ_1 and ψ_2 are first class, and (3.3) expresses the fact that the gauge transformation can be constructed from ψ_1 and ψ_2. We define the generator of this time dependent gauge transformation by

$$Q_g := -ig(t)\phi - i\dot{g}(t)p_\lambda \quad (3.4)$$

where $g(t)$ is an arbitrary real function of t.

Then for an operator Ω, and $U_g = e^{Q_g}$,

$$\delta\Omega = U_g \Omega U_g^{-1} - \Omega = [Q_g, \Omega]. \quad (3.5)$$

With this we can compute the variations of all dynamical variables. One obtains

$$\delta p = -gM, \quad \delta q = -g, \quad \delta Q_\alpha = gR_\alpha$$
$$\delta p_\alpha = -gN_\alpha, \quad \delta\lambda = -\dot{g}, \quad \delta p_\lambda = 0 \quad (3.6)$$

where (with subscript f meaning $\frac{d}{df}$)

$$M = f'^2 \Gamma_{\alpha\beta,f} Q_\beta p_\alpha, \quad R_\alpha = f'\Gamma_{\alpha\beta}Q_\beta, \quad N_\alpha = f'\Gamma_{\beta\alpha}p_\beta. \quad (3.7)$$

We also have

$$[\phi, q] = -i, \quad [\phi, p] = -iM$$
$$[\phi, Q_\alpha] = iR_\alpha, \quad [\phi, p_\alpha] = -iN_\alpha \quad (3.8)$$

and, since $V(R_i(f(q)))$ is a scalar depending only on $R_i^2 = Q_\alpha^2$ also $\delta V = 0$. It is now straightforward to verify that

$$\delta L = [Q_g, L] = \frac{d}{dt}[g(t)\phi], \quad (3.9)$$

i. e. the action is invariant under the gauge transformation (3.6). However,

$$\delta H_\lambda = [Q_g, H_\lambda] = -\dot{g}(t)\phi. \quad (3.10)$$

Thus the Hamiltonian is only invariant for time-independent gauge transformations. Of course, since the physical states must be gauge invariant we must also have a Hamiltonian which is gauge invariant. This is exactly what we shall achieve with the BRST transformation below.

4 BRST and anti–BRST transformations and gauge fixing

The BRST charge is taken to be (following Ref. [2])

$$Q_B := -ic(t)\phi - i\dot{c}(t)p_\lambda + i(\pi_{\bar{c}} - \dot{c})b \tag{4.1}$$

where $c(t)$ and $\bar{c}(t)$ are ghost and antighost variables respectively and $b(t)$ is the bosonic Nakanishi–Lautrup variable. The anti–BRST charge $\overline{Q}_B$ is the adjoint of Q_B. The momenta canonical to c and $\bar{c}$ are π_c and $\pi_{\bar{c}}$ which we define by directional derivatives, i. e.

$$\pi_c = L\,\frac{\overleftarrow{\partial}}{\partial\dot{c}}, \quad \pi_{\bar{c}} = \frac{\overrightarrow{\partial}}{\partial\dot{\bar{c}}}\,L \tag{4.2}$$

with $\{c, \dot{c}\} = 0$.We also assume that all bosonic operators have canonical commutation relations and all fermionic operators canonical anticommutation relations. Thus, in particular we assume

$$[p, q] = -i, \quad [p_\alpha, Q_\beta] = -i\delta_{\alpha\beta} \tag{4.3}$$

for $\alpha, \beta = 1, \ldots, N$ and

$$\{\pi_c, c\} = i, \quad \{\pi_{\bar{c}}, \bar{c}\} = -i. \tag{4.4}$$

Since we do not (here) choose c to be Hermitian and $\bar{c}$ to be anti–Hermitian as is often customary in the literature, the operators c and $\bar{c}$ here are not independent though they are independent canonical variables. For any Heisenberg operator Ω now

$$\delta\Omega = [Q_B, \Omega]_\mp, \tag{4.5}$$

$\mp$ depending on whether Ω is bosonic or fermionic respectively. We then find (with $[,]_+ \equiv \{,\}$)

$$\begin{aligned}
&\delta q = -c, \quad \delta p = -cM, \quad \delta Q_\alpha = cR_\alpha, \\
&\delta p_\alpha = -cN_\alpha, \quad \delta\lambda = -\dot{c}, \quad \delta p_\lambda = 0, \\
&\delta c = 0, \quad \delta\bar{c} = b - i\{\dot{c}, \bar{c}\}(p_\lambda + b), \quad \delta b = 0, \\
&\delta\pi_c = \phi, \quad \delta\pi_{\bar{c}} = 0, \quad \delta p_b = -(\pi_{\bar{c}} - \dot{c}).
\end{aligned} \tag{4.6}$$

The anti–BRST transformations are obtained by replacing in the above $-c$ by $+\bar{c}$, except for $\bar{\delta}c = b$. It can be seen that with the use of the equations of motion below $\delta\bar{c} = b$ and $\delta p_b = 0$.

We now add to the first–order Lagrangian L of (3.1) a trivially BRST and anti–BRST invariant gauge fixing contribution which we choose as

$$\begin{aligned}
L_{gf} &= -\delta[\bar{c}(\dot{\lambda} + hq + \frac{1}{2}b)] \\
&= -b(\dot{\lambda} + hq) - \frac{1}{2}b^2 + \dot{\bar{c}}\dot{c} - h\bar{c}c
\end{aligned} \tag{4.7}$$

where h is a constant with appropriate dimension, and in the last line we dropped a total time derivative after a partial integration in the action integral. Of course, one could make a different choice, e. g. $L_{gf} = -\delta[\bar{c}(\dot{\lambda}-p+\frac{1}{2}b)]$, but then, in view of $\delta p = -Mc$, one would arrive at a much more complicated Euler–Lagrange equation for c, i. e. $\ddot{c} - cM = 0$ with further complications in the arguments below. In fact, the choice (4.7) which corresponds to the covariant gauge in QED, results in free equations of motion for the ghost fields c and $\bar{c}$ which then allow to build the physical states on a particular state of the free ghost sector.

It is now a simple matter to verify that under the BRST transformation (4.6) $\delta L_B = 0$. We see from (4.7) that the ghost sector completely decouples and its fields satisfy free field equations

$$\ddot{c} + hc = 0, \qquad \ddot{\bar{c}} + h\bar{c} = 0. \tag{4.8}$$

We can therefore write the Heisenberg operators c, $\bar{c}$ as (with $h \equiv \omega^2$)

$$c(t) = e^{i\omega t}B + e^{-i\omega t}D, \qquad \bar{c}(t) = e^{-i\omega t}B^\dagger + e^{i\omega t}D^\dagger. \tag{4.9}$$

5 The BRST Hamiltonian and its diagonalisation

We pass from L_B to the Hamiltonian H_B by defining the latter as the complete Legendre transform of L_B, i. e.

$$\begin{aligned} H_B &= p\dot{q} + p_\alpha \dot{Q}_\alpha + p_\lambda \dot{\lambda} + \pi_c \dot{c} + \dot{\bar{c}}\pi_{\bar{c}} - L_B \\ &= \frac{1}{2}p_\alpha p_\alpha + V + \lambda\phi - hp_\lambda q + \frac{1}{2}p_\lambda^2 + \dot{\bar{c}}\dot{c} + h\bar{c}c. \end{aligned} \tag{5.1}$$

With (4.9) we can check that H_B is Hermitian. In fact

$$\dot{\bar{c}}\dot{c} + h\bar{c}c = 2\omega^2(B^\dagger B + D^\dagger D), \quad h \equiv \omega^2 > 0. \tag{5.2}$$

Here

$$\{c, \bar{c}\} = 0 = \{\pi_c, \pi_{\bar{c}}\} = \{\dot{\bar{c}}, \dot{c}\}. \tag{5.3}$$

Since $c, \bar{c}, \dot{c}, \dot{\bar{c}}$ obey a number of relations, we can use these in order to obtain relations between the operators $B, D, B^\dagger, D^\dagger$. There are six different conditions, i. e. $c^2 = \dot{c}^2 = \{c, \dot{c}\} = \{c, \bar{c}\} = \{\dot{c}, \dot{\bar{c}}\} = 0$, $\{c, \dot{\bar{c}}\} = i$ which give

$$B^2 = D^2 = 0, \quad \{B, D^\dagger\} = 0, \quad \{B, B^\dagger\} = -\frac{1}{2\omega}, \quad \{D, D^\dagger\} = \frac{1}{2\omega}. \tag{5.4}$$

Taking $\omega > 0$ and defining $\mid 0\rangle$ as the vacuum state for which $c(0) \mid 0\rangle = 0$ and $\dot{c}(0) \mid 0\rangle = 0$, i. e. $B \mid 0\rangle = D \mid 0\rangle = 0$, we have

$$\langle 0 \mid DD^\dagger \mid 0\rangle = \frac{1}{2\omega}\langle 0 \mid 0\rangle, \quad \langle 0 \mid BB^\dagger \mid 0\rangle = -\frac{1}{2\omega}\langle 0 \mid 0\rangle. \tag{5.5}$$

For $\langle 0 \mid 0 \rangle$ positive $D^\dagger \mid 0\rangle$ is a state with positive norm and $B^\dagger \mid 0\rangle$ one with negative norm. As a matter of convenience we could take $\langle 0 \mid 0\rangle$ negative in which case $D^\dagger \mid 0\rangle$ is the lowest (negative norm) state. From (5.5) we obtain

$$0 = \langle 0 \mid BD^\dagger \mid 0\rangle. \tag{5.6}$$

Thus $B^\dagger \mid 0\rangle$ is orthogonal to $D^\dagger \mid 0\rangle$ and

$$\langle 0 \mid BH_{ghost}D^\dagger \mid 0\rangle = 0. \tag{5.7}$$

Thus the ghost part of the Hamiltonian does not lead to transitions between the states $D^\dagger \mid 0\rangle$ and $B^\dagger \mid 0\rangle$.

The occurence of the negative norm states here is quite similar to their occurence in QED when a gauge fixing term of the form $(\partial_\mu A^\mu)^2$ is added. It is clear that since the ghost sector is free and completely decouples from the rest of the system its negative norm states lie in that part of Hilbert space which is orthogonal to the subspace of physical states. Thus, in view of (5.6) and (5.7) if we choose $D^\dagger \mid 0\rangle$ at $t = 0$ we completely exclude the $B^\dagger \mid 0\rangle$ states for all time t.

We can verify that under BRST and anti–BRST transformations

$$\delta H_B = 0 = \overline{\delta} H_B, \quad \text{i. e.} \quad [Q_B, H_B] = 0 = [\overline{Q}_B, H_B]. \tag{5.8}$$

We require physical states $\mid \psi\rangle$ to be BRST and anti–BRST invariant, i. e. $Q_B \mid \psi\rangle = 0 = \overline{Q}_B \mid \psi\rangle$ or

$$\begin{aligned} \{-ie^{i\omega t}B(\phi + i\omega p_\lambda) - ie^{-i\omega t}D(\phi - i\omega p_\lambda)\} \mid \psi\rangle = 0, \\ \{ie^{-i\omega t}B^\dagger(\phi - i\omega p_\lambda) + ie^{i\omega t}D^\dagger(\phi + i\omega p_\lambda)\} \mid \psi\rangle = 0. \end{aligned} \tag{5.9}$$

The only way both equations can be satisfied is by states projected out by the constraints, i. e. those satisfying

$$\phi \mid \psi\rangle = 0 \quad \text{and} \quad p_\lambda \mid \psi\rangle = 0. \tag{5.10}$$

Hence the additional anti–BRST symmetry is needed here in order to recover only the physical states projected out by the constraints.

6 Other gauge fixing conditions

For perturbation theory the gauge fixing condition $Q_1 = 0$ is not the most convenient. In that case the calculations become easier and more transparent if we use the gauge fixing condition

$$\chi := Q_1 - r_1 = \eta_1 = 0 \tag{6.1}$$

where $r_1 = \frac{\vec{r}\cdot\vec{r}_f}{\sqrt{\vec{r}_f^2}}$. We set $\vec{Q} = \vec{r} + \vec{\eta}$ and treat $\vec{\eta}$ as a small fluctuation. The derivative $\vec{r}_f \equiv \frac{d\vec{r}}{df}$ is the zero mode of the problem as can be seen by differentiating the classical equation $\vec{r}_{ff} = -\vec{\nabla}V(\vec{r})$ which gives

$$\left[\frac{d^2}{df^2}\delta_{ij} + \left(\frac{\partial^2 V}{\partial R_i \partial R_j}\right)_{\vec{r}}\right](\vec{r}_f)_j = 0. \tag{6.2}$$

We choose the normalisation $f'\vec{r}_f^2 = 1$. For the Poisson bracket $\{\phi, \chi\}_P$ which determines the mass of the ghosts c and $\bar{c}$ we now obtain

$$\{\phi, \chi\}_P = -f'(\vec{r}_{ff}\cdot\vec{n}_a)(\vec{n}_a\cdot\vec{r}) + f'\frac{d}{df}(\vec{r}\cdot\vec{r}_f) + \mathcal{O}(\eta). \tag{6.3}$$

Using $\vec{r}_f \cdot \vec{r}_{ff} = 0$ and the completeness relation $\mid \vec{n}_a\rangle\langle\vec{n}_a \mid = 1 - \mid \vec{r}_f\rangle\langle\vec{r}_f \mid$ we obtain

$$\{\phi, \chi\}_P = f'\vec{r}_f^2 + \mathcal{O}(\eta) = 1 + \mathcal{O}(\eta) \tag{6.4}$$

with appropriate normalisation. Thus the effective mass of the ghosts is determined by the lowest order approximation of the Poisson bracket $\{\phi, \chi\}_P$, and this is, effectively, the Faddeev–Popov determinant. In lowest order this determinant is given by the normalisation of the associated zero mode as is wellknown in the context of soliton considerations. This normalisation, of course, can also be looked at as the finite kinetic energy of the classical particle with zero total energy. The gauge fixing condition (6.1), i. e. $\vec{\eta}\cdot\vec{r}_f = 0$, means that the fluctuations η are orthogonal to the zero mode. This, of course, is precisely the condition for the existence of the Green's function required for the perturbation expansion.

7 Concluding remarks

In the above we considered a quantum mechanical example in order to demonstrate in a relatively simple context how a theory with constraints, specifically a field theory in the neighbourhood of some classical configuration, may be quantised in a way which is very similar to methods applied to theories with gauge fields. The constraints of the theory which result in a singular Lagrangian determine the generators of a gauge group under which the first–order Lagrangian is invariant. This shows clearly that a theory with constraints becomes a theory with a gauge symmetry. The BRST extension of phase space preserves the invariance of the Lagrangian but in addition allows the Hamiltonian to become invariant. One can then diagonalise the Hamiltonian and demonstrate that with the help of the anti–BRST symmetry it is only the constraints which project out the physical states of the theory.

The considerations presented here can be applied to theories with classical configurations with finite energy. Thus in the $(1+1)$ dimensional scalar field theory for a double well potential one has the well known static kink solution $\phi_c(x-x_0)$ where x_0 is the position of the kink. Thus the transformation

$$\phi(x,t) \to \phi(x-x_0(t)) + \eta(x,t) \tag{7.1}$$

where η is the fluctuation is a transformation to a larger number of degrees of freedom. This transformation leads to a constraint very similar to (2.9); for details we refer to Ref. [5]. In the case of a vortex theory in (1+2) dimensions one has three collective coordinates and hence three constraints; for details we refer to Ref. [6]. It is clear that all such theories, including those with topologically unstable configurations can, in principle, be quantised in a way analogous to the method developped here.

Acknowledgements

This research has been supported in part by the European Union under the Human Capital and Mobility programme. One of us (J. - z. Z.) is indebted to the Deutsche Forschungsgemeinschaft (Germany) for financial support. His work has also been supported in part by the National Natural Science Foundation of China under Grant No. 19274017.

References

[1] C. Becchi, A. Rouet and R. Stora (1974), *Phys. Lett*, **52B** 344; V. Tyutin (1975), Lebedev Report No. FIAN–39 (unpublished).

[2] D. Nemeschansky, C. Preitschopf and M. Weinstein (1988), *Ann. Phys. (NY)*, **183** 226.

[3] J. – L. Gervais and B. Sakita (1977), *Phys. Rev.* , **D16** 3507.

[4] H. J. W. Müller–Kirsten and A. Wiedemann (1985), *J. Math. Phys.* , **26** 1680; (1983), *Nuovo Cimento*, **78A** 61 and (1988), **99A** 541.

[5] J. Maharana and H. J. W. Müller–Kirsten (1984), *Nuovo Cimento*, **83A** 229.

[6] H. J. W. Müller–Kirsten and D. H. Tchrakian (1991), *Phys. Rev.* , **D44** 1204.

Geometry of fermionic constraints in superstring theories

Dmitrij P. Sorokin

During recent years there has been an activity in the development of a, so called, twistor-like, doubly supersymmetric approach for describing superparticles and superstrings [1]–[8]. The aim of the approach is to provide with clear geometrical meaning an obscure local fermionic symmetry (κ–symmetry) of superparticles and superstrings [9, 10], which plays an essential role in quantum consistency of the theory. At the same time this local fermionic symmetry causes problems with performing the covariant Hamiltonian analysis and quantization of the th eories. This is due to the fact that the first–class constraints corresponding to the κ–symmetry form an infinit reducible set, and in a conventional formulation of superparticles and superstrings (see [10] and references therein) it turned out impossible to single out an irreducible set of the fermionic first–class constraints in a Lorentz covariant way. So the idea was to replace the κ–symmetry by a local extended supersymmetry on the worldsheet by constructing superparticle and superstring models which would be manifestly supersymmetric in a target superspace and on the worldsheet with the number of local supersymmetries being equal to the number of independet κ–symmetry transformations, that is $n = D - 2$ in a space–time with the dimension D=3, 4, 6 and 10. Note that it is just in these space–time dimensions the classical theory of Green–Schwarz superstrings may be formulated [10], and twistor relations [11] take place.

The doubly supersymmetric formulation provides the ground for natural incorporating twistors into the structure of supersymmetric theories. Twistor components (which are roughly speaking commuting spinors) arise as superpartners of Grassmann spinor coordinates of the target superspace and allow one to solve such problems of superparticles and superstrings as the geometrical nature of the fermionic κ–symmetry, the Lorentz–covariant separation of the first and second class constraints, and finding the way of establishing, at the classical level, the relationship between Green–Schwarz and the Neveu–Schwarz–Ramond formulation of the strings, and the hope is that the twistor–like approach may contain the advantages of the both these formulations.

In this talk I would like to present basic ideas of the twistor–like approach developed so far.

Consider a conventional action describing the dynamics of a massless $N = 1$ superparticle in space-time superspace parametrized by bosonic vector coordinates x^m and fermionic spinor coordinates θ^α (m=0,1,...,D-1; α=1,...,2D-4; D=3,4,6,10). In the first order form the action looks as follows:

$$S = \int d\tau [p_m(\frac{d}{d\tau}x^m - i\frac{d}{d\tau}\bar{\theta}\gamma^m\theta) - \frac{1}{2}e(\tau)p_m p^m], \tag{1}$$

where $e(\tau)$ is an auxiliary (one-dimensional gravity) field which ensures the momentum p_m of the superparticle to be light–like on the mass shell:

$$p_m p^m = 0. \tag{2}$$

The action is invariant under the reparametrization of the time parameter $\tau \to f(\tau)$, N=1 target space supersymmetry:

$$\delta\theta = \epsilon, \qquad \delta x^m = -i\delta\bar{\theta}\gamma^m\theta \tag{3}$$

and local fermionic κ–symmetry:

$$\delta p_m = 0, \qquad \delta\theta_\alpha = i(p_m\gamma)^{m\beta}_\alpha \kappa_\beta(\tau), \qquad \delta x^m = i\delta\bar{\theta}\gamma^m\theta, \qquad \delta e = 4\bar{\kappa}\dot{\theta}, \tag{4}$$

where $\kappa_\alpha(\tau)$ is a 2(D-2)–component parameter of the transformations, while, due to the lightlikeness of p_m (2), only n=D-2 κ–symmetry parameters are independent.

Let us replace the κ–symmetry with a local n=D-2 extended supersymmetry on the worldline of the superparticle. To this end we have to construct a version of the theory being manifestly invariant under the worldline supersymmetry. It means that we shall consider the trajectory of the superparticle to be a worldline superspace parametrized by τ and Grassmann coordinates η^a (a=1,...,D-2), and the target space coordinates $X^m(\tau,\eta) = x^m(\tau) + \eta^a\chi^m_a(\tau) + ...$; $\Theta_\alpha(\tau,\eta) = \theta_\alpha(\tau) + \eta^a\lambda_a(\tau) + ...$ of the superparticle to be superfields in the worldline superspace. Thus, we see that $\theta(\tau)$ acquires commuting spinors λ_a as its superpartners.

The generalization of eq.(1) to the case D=3, N=1, n=1 is straightforward. Instead of the time derivative we use supercovariant derivative $D = \frac{\partial}{\partial\eta} + i\eta\frac{\partial}{\partial\tau}$ and write down a generalized action in the following form [12]:

$$S = \int d\tau d\eta [P_m(\tau,\eta)(DX^m - iD\bar{\Theta}\gamma^m\Theta) - \frac{1}{2}E(\tau,\eta)P_m P^m]. \tag{5}$$

Action (5) is invariant under τ–reparametrization; n=1 local supersymmetry transformations, which, in particular, for $\Theta(\tau,\eta)$ components look as follows:

$$\delta\theta_\alpha = \alpha(\tau)\lambda_\alpha; \tag{6}$$

and fermionic transformations (4), where all variables are replaced by corresponding superfields, denoted by capital letters, and $\frac{d}{d\tau}$ is replaced by the supercovariant derivative D. At the first glance it seems that we have not got rid of the κ–symmetry since it appeared again at the superfield level. But, in addition, action (5) is invariant also under the following bosonic superfield transformations

$$\delta X^m = \Lambda(\tau, \eta) P^m, \qquad \delta E = D\Lambda. \tag{7}$$

And it turns out that the transformations (7) and the superfield generalization of (4) allows one to gauge fix $E(\tau, \eta)$ to be zero *globally* in the worldline superspace [12]. Thus, the last term drops out of the eq.(5), and we get the action originally obtained in [1]:

$$S = \int d\tau d\eta P_m(\tau, \eta)(DX^m - iD\bar{\Theta}\gamma^m\Theta), \tag{8}$$

which possesses only the doubly supersymmetry.

Integrating (8) over η and eliminating auxiliary fields one arrives at a component action

$$S = \int d\tau p_m(\frac{d}{d\tau}x^m - i\frac{d}{d\tau}\bar{\theta}\gamma^m\theta - \bar{\lambda}\gamma^m\lambda). \tag{9}$$

As the solution to the equation of motion of λ we get the twistor representation of the light–like vector in D=3,4,6 and 10 space–time dimensions:

$$p_m\gamma_\alpha^{m\beta}\lambda_\beta = 0 \qquad \rightarrow \qquad p^m \sim \bar{\lambda}\gamma^m\lambda \qquad \rightarrow \qquad p_m p^m = 0. \tag{10}$$

Substituting the twistor representation of p_m into (4) one may convince oneself that the κ transformations coincide (for D=3) with the local supersymmetry transformations (6) with $\alpha(\tau) = \bar{\lambda}\kappa$, and it can be shown that the model is equivalent to the conventional N=1 superparticle [1].

There are different ways of generalizing action (8) to the case of D=4,6,10 and n=D-2 [1, 6, 7, 12]. The most straightforward (and the only known for D=10) generalization is achieved by extending the number of Grassmann coordinates η^a and writing down the action in the form [7]:

$$S = \int d\tau d^{D-2}\eta P_{ma}(\tau, \eta)(D_a X^m - iD_a\bar{\Theta}\gamma^m\Theta). \tag{11}$$

The nontrivial thing is to show that in spite of a rather rich contents of the superfields in this action there are enough local symmetries and equations of motion to kill all auxiliary fields so that the model is classiclally equivalent to the masless Brink–Schwarz superparticle [7].

The next step is to generalize this doubly supersymmetric action to $N = 1$ superstrings. For this we suppose that a hypersurface swept by the string is a worldsheet superspace with heterotic geometry subject to constraints on torsion [13, 4], and the points of this surface are parametrized by (τ, σ, η_a^-). Again, the straightforward generalization of (11) is

$$S = \int d\tau d^{D-2}\eta P_{ma}(\tau, \eta)(D_{-a}X^m - iD_{-a}\bar{\Theta}\gamma^m\Theta), \tag{12}$$

where in a Wess–Zumino gauge $D_{-a} = \frac{\partial}{\partial \eta^a} + i\eta_a^- e^{\mu}_{--}(\xi)\frac{\partial}{\partial \xi^\mu}$ with $e^{\mu}_{--}(\xi)$ being one of the two worldsheet zweinbeins ($\xi^\mu = (\tau, \eta)$).

It can be shown [14] that this action describes so called N=1 null superstring, that is a string with zero tension and a degenerate worldsheet metric [15]. This null superstring is infact a continuous set of massless superparticles moving in such a way that their momenta are orthogonal to the string so that the null string does not fall into pieces. The action (12), as well as the ones for superparticles, was called a geometro–dynamical term [16], since it determines the dynamics of a superstring by specifying the imbedding of superworldsheet into target superspace. This imbedding is characterized by vanishing components of the one form $\Pi^m = dX^m - id\bar{\Theta}\gamma^m\Theta$ along the Grassmann–odd directions of the worldsheet superspace:

$$\frac{\delta S}{\delta P_{ma}} = D_a X^m - iD_a\bar{\Theta}\gamma^m\Theta = 0. \tag{13}$$

Thus, if the dynamics of a superstring is described solely by the geometro–dynamical term (12) it is profitable for the string to propagate as the null superstring.

To get a fully fledged superstring we have to further specify the imbedding in a way which leads to string tension generation [17, 8]. This is achieved by determining the pullback of a Wess–Zumino two form

$$B = i\Pi^m \wedge d\bar{\Theta}\gamma_m\Theta. \tag{14}$$

In the conventional Green–Schwarz approach the pullback of this form on a two–dimensional worldsheet is a closed form as any two form on a two–dimensional manifold.

We would like this property to be valid in our case as well. But now getting the closure of the pullback of the form B is not a trivial problem anymore, since the worldsheet is a supermanifold.

First of all the twistor condition (13) on the one form Π^m must be satisfied.

Secondly, even then we have to modify the form B in the following way [8]:

$$\hat{B} = B - \frac{1}{D-2}(E^{--} \wedge E^{++})D_{-a}(\bar{\Theta}\gamma_m D_{-a}\Theta)E^M_{++}(\partial_M X^m - i\partial_M\bar{\Theta}\gamma^m\Theta), \tag{15}$$

where $E^{\pm\pm}_M$ are components of the worldsheet supervielbeins $(M, N = \mu, a)$.

Thus, porvided the twistor condition (13) is valid, one may check that $\hat{B}$ is a closed form in the worldsheet superspace. It means that external differential of $\hat{B}$ is zero, and locally $\hat{B}$ is an exact form:

$$\hat{B} = dA, \tag{16}$$

where $A_M(\xi, \eta)$ is an "electromagnetic" superfield on the worldsheet [17].

It is desirable to get this condition as one determined by the dynamics of the string. To this end let us add to the string action (12) a term from which this condition can be obtained:

$$S_{wz} = \int d^2\xi d^{D-2}\eta P^{MN}(\hat{B}_{MN} - \partial_M A_N), \tag{17}$$

where $P^{MN} = (-1)^{MN+1}P^{NM}$ is a Grassmann antisymmetric Lagrange multiplier. Action (17) together with (12) is invariant under the following transformations of P^{MN}:

$$\delta P^{MN} = \partial_L \Lambda^{LMN}(\xi, \eta), \tag{18}$$

where Λ^{LMN} is a Grassmann antisymmetric superfield parameter. The equation of motion of A_N gives

$$\partial_M P^{MN} = 0.$$

The solution to this equation, with taking into account gauge fixing for the transformations of P^{MN} (18), reads as follows.

$$p^{\mu\nu} = \varepsilon^{\mu\nu} T \eta^{D-2},$$

while other components of the superfield P^{MN} are zero. T is a constant which is identified with string tension.

When $T = 0$ we again get the null superstring, since then (17) vanishes. When $T \neq 0$, one may eliminate in (17) all auxiliary degrees of freedom and get the conventional $N = 1$ Green–Schwarz superstring action, which is a target–space supersymmetric part of a heterotic string action [10].

To complete the doubly supersymmetric formulation of the heterotic string one has to construct a superfield action for describing chiral fermions which must be taken into account for the quantum consistency of the theory. Two possible ways of how one may try to do this has been proposed [18, 19], but the problem was not completely solved, since either there is a danger that undesirable auxiliary degrees of freedom may become propagative [18], or an internal gauge group associated with the chiral fermions is too small [19]. Recently a modified version of [18] was proposed in [20], where, as the authors argue, both these problems are solved.

The program of doubly supersymmetric twistorization has been fulfilled for superparticles and superstrings in D=2,3,4,6 and 10 space–time dimensions with N=1, and n=D-2, but the generalization of these results to an N=2

Green–Schwarz superstring encount ered problems with not allowing auxiliray fields of the model to propagate. Various versions of the twistor–like N=2 Green-Schwarz superstrings have been studied in [2, 21, 22], and twistor-like supermembrane models were constructed in [23]. A model for describing doubly supersymmetric heterotic string with the both Virasoro constraints solved in the twistor form [24], and the existing versions of the chiral fermion action indicate that one might hope to overcome the problem of propagating Lagrange multipliers. Work in this direction is in progress.

Acknowledgments

I would like to thank Nathan Berkovits, Paul Howe and the members of the theoretical group for their warm hospitality at Maths. Department of King's College, London, and the Royal Society for awarding me a Kapitza Fellowship for visiting London and Cambridge. This work was partially supported by the International Science Foundation under the grant No RY 9000.

References

[1] D. P. Sorokin, V. I. Tkach and D. V. Volkov (1989), *Mod. Phys. Lett.*, **A4** 901;
D. P. Sorokin, V. I. Tkach, D. V. Volkov and A. A. Zheltukhin (1989), *Phys. Lett.*, **216B** 302.

[2] D. V. Volkov and A. A. Zheltukhin (1989), *Lett. Math. Phys*, **17** 141;
D. V. Volkov and A. A. Zheltukhin (1990), *Nucl. Phys.*, **B335** 7.

[3] N. Berkovits (1989), *Phys. Lett.*, **232B** 184; (1990), *ibid.*, **241B** 497; (1991), *Nucl. Phys.*, **B350** 193; **B358** 169; (1992), *ibid.*, **B379** 96; (1993), *ibid.*, **B395** 77.

[4] M. Tonin (1991), *Phys. Lett.*, **266B** 312; (1992), *Int. J. Mod. Phys.*, **A7** 6013;
S. Aoyama, P. Pasti and M. Tonin (1992), *Phys. Lett.*, **283B** 213.

[5] F. Delduc, E. Ivanov and E. Sokatchev (1992), *Nucl. Phys.*, **B384** 334.

[6] F. Delduc and E. Sokatchev (1991), *Class. Quantum Grav.*, **9** 361.

[7] A. Galperin and E. Sokatchev (1992), *Phys. Rev.*, **D46** 714.

[8] F. Delduc, A. Galperin, P. Howe and E. Sokatchev (1992), *Phys. Rev.*, **D47** 578.

[9] J. A. De Azcarraga and J. Lukierski (1982), *Phys. Lett.*, **113B** 170; W. Siegel (1993), *Phys. Lett.*, **128B** 397; *Class. Quant. Grav.* (1985), **2** 170.

[10] M. B. Green, J. H. Schwarz and E. Witten (1987), *Superstring Theory*, Cambridge University Press, Cambridge (and references therein).

[11] R. Penrose (1967), *J. Math. Phys.*, **8** 345;
R. Penrose and M. A. H. MacCallum (1972), *Phys. Rep.*, **6** 241, and references therein.

[12] A. I. Pashnev and D. P. Sorokin (1993), *Class. Quantum Grav.*, **10** 625.

[13] P. Nelson and G. Moore (1986), *Nucl. Phys.*, **B274** 509;
P. S. Howe and G. Papadopulos (1987), *Class. Quant. Grav.*, **4** 51.

[14] I. Bandos, D. Sorokin, M. Tonin and D. Volkov (1993), *Phys. Lett.*, **319B** 445.

[15] A. Shield (1977), *Phys. Rev.*, **D16** 1722;
A. Karlhede and U. Lindström (1986), *Class. Quant. Grav.*, **3** L73;
A. Zheltukhin (1987), *JETP Lett.*, **46** 208.

[16] A. Galperin and E. Sokatchev (1993), *Phys. Rev.*, **D48** 4810.

[17] J. A. De Azcàrraga, J. M. Izquierdo and P. K. Townsend (1992), *Phys. Rev.*, **45** 3321;
P. K. Townsend (1992), *Phys. Lett.*, **B277** 285;
E. Bergshoeff, L. A. J. London and P. K.Townsend (1992), *Class. Quantum Grav.*, **9** 2545.

[18] D. P. Sorokin and M. Tonin (1994), *Phys. Lett.*, **326B** 84.

[19] P. Howe (1994), King's College preprint.

[20] E. Ivanov and E. Sokatchev (1994), preprint BONN–TH–94–10.

[21] V. Chikalov and A. Pashnev (1993), *Mod. Phys. Lett.*, **A8** 285; ICTP preprint, 1993.

[22] P. Pasti and M. Tonin (1994), Preprint DFPD/94/TH/05, Padova.

[23] P. Pasti and M. Tonin (1994), *Nucl. Phys.*, **B418** 337;
E. Bergshoeff and E. Sezgin (1993), preprint CTP TAMO–67/93.

[24] I. Bandos, M. Cederwall, D. Sorokin and D. Volkov (1994), preprint Götenborg-ITP-94-10.

BRST and new superstring states

L. Dolan

1 Introduction

Properties of the fermion vertex in four-dimensional superstring theory are investigated, and the requirements for corresponding massless states to carry non-zero charge under gauge symmetry are analyzed. In particular, such states are generic for non-abelian gauge bosons in the Ramond sector of type II models. We express the Yang-Mills amplitude, derived from a modified vertex, in BRST form and relate the question of BRST invariance to a construction of the Virasoro supercurrent. Possible use of non-unitary representations for internal degrees of freedom of the matter sector and hermiticity properties of the supercurrent are described.

2 Ramond states

Superstring theory may involve higher dimensional Ramond fermions which provide 'copies' or 'families' when viewed as fermions in four space-time dimensions. The matter spin fields are the conformal fields which correspond to the ground states $|\mathcal{A};k\rangle$ in the Ramond sector:

$$|\mathcal{A}\rangle = S_{\mathcal{A}}(0)|0\rangle \quad ; \quad |\mathcal{A};k\rangle = S_{\mathcal{A}}(0)e^{ik\cdot X(0)}|0\rangle \tag{2.1}$$

In ten dimensions, a degenerate ground state is denoted by $|\mathcal{A};k\rangle\chi^{\mathcal{A}}(k)$; here there is a sum over $1 \leq \mathcal{A} \leq 32$. The spinor wavefunction $\chi^{\mathcal{A}}(k)$ is restricted to be Majorana-Weyl, i.e. that $\Gamma^{11}\chi = \chi$; and in a Majorana representation to be real: $\chi^* = \chi$.

The vertex operator for the massless Ramond states $|\mathcal{A};k\rangle\chi^{\mathcal{A}}(k)$ in the canonical $q = -\frac{1}{2}$ superconformal ghost picture[1 – 3] is given for $k\cdot\Gamma\chi = 0$ by

$$V_{-\frac{1}{2}}(k,\zeta) = \chi^{\mathcal{A}}(k)S_{\mathcal{A}}(\zeta)e^{ik\cdot X(\zeta)}c(\zeta)e^{-\frac{1}{2}\Phi(\zeta)} \tag{2.2a}$$

$$= \lim_{z\to\zeta} e^{\Phi(z)}F(z)V_{-\frac{3}{2}}(k,\zeta)\,. \tag{2.2b}$$

As suggested by (2.2b), this can be derived via picture changing from the spin field in the $q = -\frac{3}{2}$ picture given by:

$$V_{-\frac{3}{2}}(k,z) = -v_{\mathcal{A}}(k)S^{\mathcal{A}}(z)e^{ik\cdot X(z)}c(z)e^{-\frac{3}{2}\Phi(z)} \tag{2.3}$$

with $k{\cdot}\Gamma v = \chi$ and a suitable choice of supercurrent given by:

$$F(z) = a_\mu(z)h^\mu(z) + \bar{F}(z)\,. \tag{2.4}$$

Here $0 \le \mu \le 3$, $\bar{F}(z)$ corresponds to internal degrees of freedom, and the conformal dimension of $:e^{-\frac{1}{2}\Phi}:$ and $:e^{-\frac{3}{2}\Phi}:$ is $\frac{3}{8}$ and of $S_{\mathcal{A}}$ and $S^{\dot{\mathcal{A}}}$ is $\frac{5}{8}$.

van der Waerden notation[4 – 6] for spinor indices can be used for Γ matrices in the Weyl representation. In ten dimensions we consider $\{\Gamma^A, \Gamma^B\} = 2\eta^{AB}$; $\eta^{AB} = \mathrm{diag}\{-1, 1, \ldots, 1\}$. In a Weyl representation they are:

$$\Gamma^\mu = \begin{pmatrix} 0 & I_4 \\ I_4 & 0 \end{pmatrix} \otimes \gamma^\mu\,;\; \Gamma^{a+3} = \begin{pmatrix} 0 & \alpha^a \\ -\alpha^a & 0 \end{pmatrix} \otimes I_4\,;\; \Gamma^{a+6} = \begin{pmatrix} 0 & \beta^a \\ \beta^a & 0 \end{pmatrix} \otimes \tilde{\gamma}^5\,. \tag{2.5}$$

Here $0 \le A, B \le 9$, $1 \le a \le 3$, and $\{\gamma^\mu\,,\,\gamma^\nu\} = 2\eta^{\mu\nu}$. The six matrices α^a, β^a are antisymmetric and real and satisfy the following algebra:

$$\{\alpha^a, \alpha^b\} = \{\beta^a, \beta^b\} = -2\delta^{ab}$$

$$[\alpha^a, \beta^b] = 0\,, \quad [\alpha^a, \alpha^b] = -2\epsilon_{abc}\alpha^c\,, \quad [\beta^a, \beta^b] = 2\epsilon_{abc}\beta^c\,. \tag{2.6}$$

An explicit representation[7] of these matrices can be given in terms of the Pauli matrices. We define

$$\Gamma^{11} \equiv \Gamma^0 \ldots \Gamma^9 = \begin{pmatrix} I_4 & 0 \\ 0 & -I_4 \end{pmatrix} \otimes I_4\,, \quad (\Gamma^{11})^2 = 1 \tag{2.7}$$

and $(\tilde{\gamma}^5)^2 = -1$, so $\tilde{\gamma}^5$ is antihermitian and is given by $\gamma^5 = i\gamma^0\gamma^1\gamma^2\gamma = i\tilde{\gamma}^5$.
We define the direct product of matrices as $A \otimes B \equiv \begin{pmatrix} a_{11}B & a_{12}B & \ldots \\ a_{21}B & a_{22}B & \ldots \\ \ldots & & \end{pmatrix}$;
also $(A \otimes B)(C \otimes D) = AC \otimes BD$.

A general degenerate spinor ground state in ten dimensions is written in van der Waerden notation as

$$\chi^{\mathcal{A}} = \begin{pmatrix} \Psi^A \\ \Phi^{\dot{A}} \end{pmatrix} \tag{2.8}$$

Here $1 \le A, \dot{A} \le 16$. We denote the index structure of the Γ matrices as $\Gamma^{\mu\mathcal{A}}{}_{\mathcal{B}}$ and tensors which raise and lower spinor indices are the antisymmetric tensors $C^{-1}_{\mathcal{AB}}$, $C^{\mathcal{AB}}$, the charge conjugation matrices for $SO(9,1)$. For eg. $\chi_{\mathcal{A}} = C^{-1}_{\mathcal{AB}}\chi^{\mathcal{B}}$, $\chi^{\mathcal{A}} = C^{\mathcal{AB}}\chi_{\mathcal{B}}$, $C^{-1}_{\mathcal{AB}}C^{\mathcal{BC}} = \delta^{\mathcal{C}}_{\mathcal{A}}$, $C^{\mathcal{AB}} = -C^{\mathcal{BA}}$, $C^{-1}_{\mathcal{AB}} = -C^{-1}_{\mathcal{BA}}$. Then $A_{\mathcal{A}}B^{\mathcal{A}} = -A^{\mathcal{A}}B_{\mathcal{A}}$. It follows from the definition of the charge conjugation matrix

$$C^{-1}_{\mathcal{AB}}(\Gamma^\mu)^{\mathcal{B}}{}_{\mathcal{E}}C^{\mathcal{ED}} = -(\Gamma^{\mu T})_{\mathcal{A}}{}^{\mathcal{D}} \tag{2.9}$$

that $(\Gamma^\mu)^{\mathcal{D}}{}_{\mathcal{A}} = (\Gamma^\mu)_{\mathcal{A}}{}^{\mathcal{D}}$; and that $\Gamma^{\mu\mathcal{A}\mathcal{B}}$ and $\Gamma^\mu_{\mathcal{A}\mathcal{B}}$ are symmetric in the spinor indices. The exact form of C depends on the representation of the Γ matrices used. From (2.5), we have

$$C = \begin{pmatrix} 0 & I_4 \\ I_4 & 0 \end{pmatrix} \otimes C_4 \quad ; \qquad C^{-1} = \begin{pmatrix} 0 & I_4 \\ I_4 & 0 \end{pmatrix} \otimes C_4^{-1} \tag{2.10}$$

where C_4 is the charge conjugation matrix for the four-dimensional γ matrices: $C_4^{-1}(\gamma^\mu)C_4 = -(\gamma^{\mu T})$ and thus $\gamma^{5*} = C^{-1}\gamma^5 C$. In this representation we also have

$$\Gamma^4\Gamma^5\Gamma^6 = \begin{pmatrix} 0 & -I_4 \\ I_4 & 0 \end{pmatrix} \otimes I_4 \quad \text{and} \quad \Gamma^7\Gamma^8\Gamma^9 = \begin{pmatrix} 0 & I_4 \\ I_4 & 0 \end{pmatrix} \otimes \tilde{\gamma}^5 . \tag{2.11}$$

Note that if $\Gamma^{11\mathcal{A}}{}_{\mathcal{B}}\chi^{\mathcal{B}} = \chi^{\mathcal{A}}$, i.e. if $\chi^{\mathcal{A}}$ is left-handed, then $\chi_{\mathcal{A}} = -\Gamma^{11}_{\mathcal{A}}{}^{\mathcal{B}}\chi_{\mathcal{B}}$, so that $\chi_{\mathcal{A}}$ has opposite chirality from $\chi^{\mathcal{A}}$. Thus the spin field $S_{\mathcal{A}}(z)$ and its conjugate $S^{\mathcal{A}}(z)$ have opposite chirality. We will require that $\chi^{\mathcal{A}}$ be left-handed, so then $S_{\mathcal{A}}$ is left-handed and $S^{\mathcal{A}}$ is right-handed. In the Weyl representation (2.5), only the off-diagonal $\Gamma^{\mu A}{}_{\dot{B}}$ and $\Gamma^{\mu\dot{A}}{}_{B}$ components are non-vanishing. This makes it natural to relate sixteen-component to thirty-two component spinors as in (2.8), and $k_\mu\Gamma^{\mu\mathcal{A}}{}_{\mathcal{B}}\chi^{\mathcal{B}} = 0$ reduces to the two 'Weyl equations' for the 16-component spinors $k_\mu\Gamma^{\mu A}{}_{\dot{B}}\Phi^{\dot{B}} = 0$ and $k_\mu\Gamma^{\mu\dot{A}}{}_{B}\Psi^{B} = 0$.

The Majorana-Weyl spinor wavefunction $\Psi^{\mathcal{A}}$ with $\Gamma^{11}\Psi = \Psi$, $k_\mu\Gamma^\mu\Psi = 0$ reduces to an eight-fold degenerate state: it is written in the Weyl representation as

$$\Psi^{\mathcal{A}} = \begin{matrix} \Psi^A \\ \\ \\ \\ \Phi^{\dot{A}} \\ \\ \\ \\ \end{matrix} = \begin{matrix} \psi^{1\alpha} \\ \psi^{2\alpha} \\ \psi^{3\alpha} \\ \psi^{4\alpha} \\ 0 \\ 0 \\ 0 \\ 0 \end{matrix} \tag{2.12}$$

and thus describes four 4-dimensional Majorana (real) spinors $\psi^{i\alpha}$, $1 \le i \le 4$, satisfying $k_\mu\gamma^{\mu\alpha}{}_\beta\psi^{i\beta}(k) = 0$, which for each i then describes two real propagating degrees of freedom, and each 4-dimensional spinor is a function of the four-momentum k^μ, $\psi^i_\alpha(k)$. Since $1 \le i \le 4$, we have four copies or 'families'of fermions. Here $1 \le \alpha \le 4$.Eq. (2.12) represents a particular ansatz for the ten-dimensional wave function which admits a four-dimensional interpretation.

Since only the off-diagonal components $C^{A\dot{B}}$, $C^{\dot{A}B}$, $C^{-1}_{A\dot{B}}$, $C^{-1}_{\dot{A}B}$ are non-vanishing in (2.10), then from (2.8), we have

$$\chi_{\mathcal{A}} = \begin{pmatrix} \Phi_A \\ \Psi_{\dot{A}} \end{pmatrix} . \tag{2.13}$$

Along with (2.12), it will be useful to define a right-handed Majorana-Weyl spinor $\Phi^{\mathcal{A}}$ satisfying $\Gamma^{11}\Phi = -\Phi$, and written in the Weyl representation as

$$\begin{array}{ccccc} \Phi^{\mathcal{A}} = & \Psi^{A} & = & 0 \\ & & & 0 \\ & & & 0 \\ & & & 0 \\ & \Phi^{\dot{A}} & & \phi^{1\alpha} \\ & & & \phi^{2\alpha} \\ & & & \phi^{3\alpha} \\ & & & \phi^{4\alpha} \end{array} \tag{2.14}$$

An example of a right-handed spinor is given by considering $F^{s.t.}(z) \equiv a_\mu(z)d^\mu(z)$ and defining $\Phi^{\dot{A}}(k)$ such that $|A;k\rangle\Psi^A(k) = F_0^{s.t.}|\dot{A};k\rangle\Phi^{\dot{A}}(k)$, so that in (2.14) we let $\phi^{i\alpha} = v^{i\alpha}$ where $\psi^{i\alpha}(k) = \sqrt{2\alpha'}\,k_\mu\frac{1}{\sqrt{2}}\gamma^{\mu\alpha}_{\ \ \beta}v^{i\beta}(k)$ and $\Psi^A(k) = \sqrt{2\alpha'}\,\frac{1}{\sqrt{2}}k_\mu\Gamma^{\mu A}_{\ \ B}\Phi^{\dot{B}}(k)$. Therefore $\bar{\psi}^i(k) = -\bar{v}^i(k)k_\mu\gamma^\mu\frac{1}{\sqrt{2}}\sqrt{2\alpha'}$; $\bar{\Psi}_G(k)\langle k;G| = -\bar{\Phi}_G(k)\langle k;G|F_0^{s.t.}$; $F_0^{s.t.}|A;k\rangle\Psi_A(k) = 0 = \bar{\Psi}_G(k)\langle k;G|F_0^{s.t.}$. Note $F_0^{s.t.}$ is not hermitian although it provides a unitary representation $(c > \frac{3}{2}, h > 0)$ of the super-Virasoro algebra

$$F_0^{s.t.\dagger} = F_0^{s.t.} + 2a_0^0 d_0^0 . \tag{2.15}$$

Acting on states ψ for which $F_0\psi = 0$, we have $F_0F_0^\dagger = (p^0)^2 + (p^i)^2 +$ nonzeromodes. Thus acting on ground states we have

$$\frac{1}{2(k^0)^2}F_0^\dagger = F_0^{-1} . \tag{2.16a}$$

Therefore, if $F_0u = 0$, then $u = F_0v$ where v is proportional to $F_0^\dagger u$ and is given by

$$v = \frac{1}{2(k^0)^2}F_0^\dagger u . \tag{2.16b}$$

(2.16b) holds since then $F_0v = \frac{1}{2(k^0)^2}F_0F_0^\dagger u = u$ by(2.16a). From (2.16), we have

$$v = \frac{1}{2(k^0)^2}F_0^\dagger u = F_0^{-1}u = \frac{1}{\sqrt{2}k^0}\Gamma^0 u . \tag{2.17}$$

In order to compute amplitudes, we will need the operator products[1, 8] for the conformal fields. For the matter fields X^μ, h^μ, and S_A they are:

$$\begin{aligned} X^\mu(z)X^\nu(\zeta) &= -\eta^{\mu\nu}\ln(z-\zeta) \\ h^\mu(z)h^\nu(\zeta) &= \frac{\eta^{\mu\nu}}{(z-\zeta)} \\ :h^\mu(z)h^\nu(z):h^\lambda(\zeta) &= \frac{1}{(z-\zeta)}(-\eta^{\mu\lambda}h^\nu(\zeta)+\eta^{\nu\lambda}h^\mu(\zeta)) \\ :h^\mu(z)h^\nu(z):S_A(\zeta) &= \frac{1}{(z-\zeta)}\tfrac{1}{4}[\Gamma^\mu,\Gamma^\nu]_A{}^B S_B(\zeta) \end{aligned} \tag{2.18}$$

$$\begin{aligned} \langle 0|S_G(z_1)\psi^\mu(z_2)S_A(z_3)|0\rangle &= \Gamma^\mu_{GA}(z_1-z_2)^{-\frac{1}{2}}(z_2-z_3)^{-\frac{1}{2}}(z_1-z_3)^{-\frac{3}{4}} \\ \langle 0|S_D(z_1)\psi^\mu\psi^a(z_2)S_{\dot{B}}(z_3)|0\rangle &= \tfrac{1}{4}[\Gamma^\mu,\Gamma^a]_{D\dot{B}}(z_1-z_2)^{-1}(z_2-z_3)^{-1}(z_1-z_3)^{-\frac{1}{4}} \end{aligned} \tag{2.19}$$

We will also need the BRST vertex operators for Neveu-Schwarz states. For the massless vector Neveu-Schwarz state $\epsilon{\cdot}b_{-\frac{1}{2}}|k\rangle$, the vertex operator in the canonical $q=-1$ ghost picture is

$$V_{-1}(k,\epsilon,z) = \epsilon{\cdot}h(z)e^{ik\cdot X(z)}c(z)e^{-\Phi(z)} \tag{2.20a}$$

Since $k^2=0$ and the conformal dimension of c is -1 and of $:e^{-\Phi}:$ is $\frac{1}{2}$, the vertex operator $V_{-1}(k,\epsilon,z)$ has zero conformal weight. Its copy in the $q=0$ picture is

$$V_0(k,\epsilon,\zeta) = \lim_{z\to\zeta} e^{\Phi(z)}F(z)V_{-1}(k,\epsilon,\zeta) = (k{\cdot}h(\zeta)\epsilon{\cdot}h(\zeta)+\epsilon{\cdot}a(\zeta))e^{ik\cdot X(\zeta)}c(\zeta) \tag{2.20b}$$

Here ϵ_μ is a four-dimensional polarization wavefunction in the Lorentz gauge: $k{\cdot}\epsilon=0$. In models where internal degrees of freedom are described by free world sheet fermions, similarly forthe scalar Neveu-Schwarz state $b^a_{-\frac{1}{2}}|k\rangle$, the vertex operator in the canonical $q=-1$ ghost picture is

$$V^a_{-1}(k,z) = h^a(z)e^{ik\cdot X(z)}c(z)e^{-\Phi(z)} \tag{2.21a}$$

Its copy in the $q=0$ picture depends on the choice of the internal supercurrent $\bar{F}(z)$ which is determined by the gauge symmetry of the theory:

$$\begin{aligned} V^a_0(k,\zeta) &= \lim_{z\to\zeta} e^{\Phi(z)}F(z)V^a_{-1}(k,\zeta) \\ &= \lim_{z\to\zeta}[\,k{\cdot}h(\zeta)\,h^a(\zeta)+(z-\zeta)\bar{F}(z)h^a(\zeta)\,]e^{ik\cdot X(\zeta)}c(\zeta) \end{aligned} \tag{2.21b}$$

As examples of different choices of the supercurrent, we consider

$$\tilde{V}_0^a(k,\zeta) = [\,k\cdot h(\zeta)\,h^a(\zeta) + T^a(\zeta)\,]e^{ik\cdot X(\zeta)}c(\zeta) \qquad (2.21c)$$

$$V'^a_0(k,\zeta) = [\,k\cdot h(\zeta)\,h^a(\zeta) - \frac{i}{2\sqrt{\frac{c_\psi}{2}}} f_{abc}h^b(\zeta)h^c(\zeta)\,]e^{ik\cdot X(\zeta)}c(\zeta) \qquad (2.21d)$$

where $T^a(\zeta)$ are the currents of an abelian Kac-Moody algebra, and f_{abc} are the structure constants of some non-abelian Lie algebra.

3 BRST tree amplitudes

Using these vertices, we present a non-abelian three gluon coupling for the emission of the massless Neveu-Schwarz vector $\epsilon\cdot b^L_{\frac{1}{2}} b^{aR}_{-\frac{1}{2}}|k\rangle$ from Ramond-Ramond vector bosons via the amplitude of a closed superstring given by $A_3^L A_3^R$ where

$$\begin{aligned}
A_3^L &= \bar{\Psi}_G(-k_1)\langle -k_1; G|\, V_0(k_2,\epsilon_2,1)\,|\,A;k_3\rangle \Psi_A(k_3) \\
&= \langle 0|V_{-\frac{1}{2}}(k_1,z_1)V_{-1}(k_2,\epsilon_2,z_2)V_{-\frac{1}{2}}(k_3,z_3)|0\rangle \\
&= \langle 0|c(z_1)c(z_2)c(z_3)|0\rangle\langle 0|:e^{-\frac{1}{2}\phi(z_1)}::e^{-\phi(z_2)}::e^{-\frac{1}{2}\phi(z_3)}:|0\rangle \\
&\quad \cdot\chi^G(k_1)\chi^A(k_3)\epsilon_{2\mu}\,\langle 0|S_G(z_1)\psi^\mu(z_2)S_A(z_3)|0\rangle \\
&= [\,(z_1-z_2)(z_2-z_3)(z_1-z_3)\,]\,[\,(z_1-z_2)^{-\frac{1}{2}}(z_2-z_3)^{-\frac{1}{2}}(z_1-z_3)^{-\frac{1}{4}}\,] \\
&\quad \cdot\chi^G(k_1)\chi^A(k_3)\epsilon_{2\mu}\,\Gamma^\mu_{GA}\,[\,(z_1-z_2)^{-\frac{1}{2}}(z_2-z_3)^{-\frac{1}{2}}(z_1-z_3)^{-\frac{3}{4}}\,] \\
&= \bar{\chi}_{\dot{G}}(k_1)\Gamma^{\mu\dot{G}}{}_A\chi^A(k_3)\,\epsilon_{2\mu} \qquad (3.1a)
\end{aligned}$$

and

$$\begin{aligned}
&A_3^R \\
&= \bar{\Psi}_D(-k_1)\langle -k_1; D|V_0^a(k_2,1)\bar{F}_0^1|B;k_3\rangle(\Gamma^0\frac{1}{\sqrt{2}\,k_3^0}\Psi(k_3))_B \\
&= \langle 0|V_{-\frac{1}{2}}(k_1,z_1)V'^a_0(k_2,z_2)V_{-\frac{3}{2}}(k_3,z_3)|0\rangle \\
&= \langle 0|c(z_1)c(z_2)c(z_3)|0\rangle\langle 0|:e^{-\frac{1}{2}\phi(z_1)}::e^{-\frac{3}{2}\phi(z_3)}:|0\rangle \\
&\cdot\chi^D(k_1)v^{\dot{B}}(k_3)\langle 0|S_D(z_1)[\,k_2\cdot h(z_2)h^a(z_2) - \frac{i}{2\sqrt{\frac{c_\psi}{2}}}f_{abc}h^b(z_2)h^c(z_2)\,]S_{\dot{B}}(z_3)|0\rangle \\
&= [\,(z_1-z_2)(z_2-z_3)(z_1-z_3)\,]\,[\,(z_1-z_3)^{-\frac{3}{4}}\,] \\
&\quad \cdot\chi^D(k_1)v^{\dot{B}}(k_3)\,[k_{2\mu}(\Gamma^\mu\Gamma^a)_{D\dot{B}} - \frac{i}{2\sqrt{\frac{c_\psi}{2}}}f_{abc}(\Gamma^b\Gamma^c)_{D\dot{B}}] \\
&\quad \cdot[\,(z_1-z_2)^{-1}(z_2-z_3)^{-1}(z_1-z_3)^{-\frac{1}{4}}\,] \\
&= -k_{2\mu}\chi_{\dot{D}}(k_1)(\Gamma^\mu\Gamma^a)^{\dot{D}}{}_{\dot{B}}v^{\dot{B}}(k_3) + \frac{i}{2\sqrt{\frac{c_\psi}{2}}}f_{abc}\chi_{\dot{D}}(k_1)(\Gamma^b\Gamma^c)^{\dot{D}}{}_{\dot{B}}\,v^{\dot{B}}(k_3) \\
&= k_{2\mu}\bar{\chi}_{\dot{D}}(k_1)(\Gamma^\mu\Gamma^a)^{\dot{D}}{}_{\dot{B}}v^{\dot{B}}(k_3) - \frac{i}{2\sqrt{\frac{c_\psi}{2}}}f_{abc}\bar{\chi}_{\dot{D}}(k_1)(\Gamma^b\Gamma^c)^{\dot{D}}{}_{\dot{B}}\,v^{\dot{B}}(k_3). \qquad (3.1b)
\end{aligned}$$

The first lines in (3.1a,b) correspond to a modification[7] of the standard vertex operator expressed in old covariant gauge notation. Eq. (3.1) transcribes these amplitudes into BRST form. The brackets in (3.1) denote the respective z_i dependence of the various correlation functions, as evaluated in (2.19).

In order to describe four-dimensional states in the above analysis of Ramond ground states, we chose a particular ansatz for the ten-dimensional wave function χ^A as in (2.12). Also, because in ten dimensions the six 'internal' Γ matrices are constructed from the α^a, β^a matrices in (2.6), the appropriate structure constants (2.3d) correspond to the symmetry group $SU(2) \times SU(2)$. Combining (3.1a,b), we find following[7], for a suitable chirality of $\psi^{1\alpha}$ in (2.12), and for $1 \leq a \leq 3$ that

$$A_3^L A_3^R = (128i)\sqrt{2\alpha'} f_{IaJ} [\epsilon_2^+ \cdot \epsilon_3^+ \epsilon_1^- \cdot k_2 + \epsilon_1^- \cdot \epsilon_3^+ \epsilon_2^+ \cdot k_3 + \epsilon_1^- \cdot \epsilon_2^+ \epsilon_3^+ \cdot k_1] \quad (3.2)$$

where f_{IaJ} is the sixteen-dimensional representation ($8\mathbf{2}$'s) of one $SU(2)$, so that these gauge bosons and a similar set for the other $SU(2)$ would enhance the $SU(2) \times SU(2)$ group to $SO(8)$. This model provides a variation on a four-dimensional $N = 8$ free fermion abelian model[10].

An internal supercurrent $\bar{F}(z)$ used to derive(2.21d) from (2.21a) is

$$\bar{F}(z) = -\frac{i}{6\sqrt{\frac{c_\psi}{2}}} f_{abc} h^a(z) h^b(z) h^c(z) \quad (3.3)$$

but (3.3) closes a super-Virasoro algebra with anomaly $c = 3$, not the value $c = 9$ required for the complete description of the internal conformal field theory, and consequently for proving BRST invariance of the vertex operators. We note that what is needed is an additional matter system with $c = 6$. Since $c = 6$ is the critical dimension of a conformal field theory with $N = 2$ superconformal world sheet invariance[11], an $N = 1$ hermitian supercurrent could be constructed as $F = F^+ + F^-$ and an $N = 1$ non-hermitian supercurrent would be $F = i(F^+ - F^-)$. The total conformal field theory has to be modular invariant for a consistent string theory. This is similar to the Gepner strategy[12, 13], but now we take for part of the internal space a conformal field theory with $N = 2$ world-sheet supersymmetry (not specifically tensor products of the discrete series representations), while maintaining modular invariance. Constructions of the supercurrent similar to those of the recent novel use[14] of ghost states with wrong sign statistics could also be relevant here.

4 Modified fermion vertex

In analogy with the Neveu-Schwarz vertex defined in (2.21d), we can define a picture changed modified fermion vertex as

$$
\begin{aligned}
&V'_{-\frac{1}{2}}(k,\zeta) \\
&= \lim_{z\to\zeta} e^{\Phi(z)} F(z) V_{-\frac{3}{2}}(k,\zeta) \\
&= e^{-\frac{1}{2}\Phi(\zeta)}[\chi^A(k) - \frac{i}{6\sqrt{\frac{c_\psi}{2}}}\frac{1}{2\sqrt{2}} f_{abc} v^B(k)(\Gamma^a\Gamma^b\Gamma^c)^A_B] S_A(\zeta) e^{ik\cdot X(\zeta)} c(\zeta) \\
&= e^{-\frac{1}{2}\Phi(\zeta)}[\chi^A(k) - \frac{i}{2\sqrt{2}} v^B(k)(\Gamma^4\Gamma^5\Gamma^6 - \Gamma^7\Gamma^8\Gamma^9)^A_B] S_A(\zeta) e^{ik\cdot X(\zeta)} c(\zeta) \\
&= e^{-\frac{1}{2}\Phi(\zeta)}[\psi^{i\alpha(k)} + \frac{i}{2\sqrt{2}}((1-\tilde{\gamma}^5) v^i(k))^\alpha] S_\alpha(\zeta) e^{ik\cdot X(\zeta)} c(\zeta)\,. \qquad (4.1)
\end{aligned}
$$

This vertex describes massless states since $k^2 v = k\cdot\gamma\psi = 0$. It reflects spontaneous breakdown of global supersymmetry in the internal conformal field theory in the presence of massless fermions. BRST invariance requires a suitable definition of the total supercurrent.

The discussion presented in this lecture is intended to reexpress a possible derivation of the three-gluon vertex in BRST form. It will be interesting to develop these ideas more fully, and give a complete description of BRST invariance properties of these vertices.

References

[1] D. Friedan, E. Martinec and S. Shenker (1986), *Nucl. Phys.*, **B271** 93.

[2] D. Friedan (1986), in *Unified String Theories*, Gross, D. and Green, M. (eds.), Singapore: World Scientific, p. 162.

[3] D. Lust and S. Theisen (1989), *Lectures on String Theory*, Springer-Verlag, New York.

[4] B. L. van der Waerden (1974), *Group Theory and Quantum Mechanics*, Springer-Verlag, New York.

[5] J. Wess and J. Bagger (1992), *Supersymmetry and Supergravity*, Princeton University Press, Princeton; Appendices A, B.

[6] P. West (1986), *Introduction to Supersymmetry and Supergravity*, World Scientific, Singapore; Appendix A, p.270.

[7] L. Dolan and S. Horvath (1994), *Nuclear Physics*, **B416**.

[8] J. Cohn, D. Friedan, Z. Qui and S. Shenker (1986), *Nucl. Phys.*, **B278** 577.

[9] V. Knizhnik and A.B. Zamolodchikov (1984), *Nucl. Phys.*, **B247** 83.

[10] R. Bluhm, L. Dolan and P. Goddard (1987), *Nucl. Phys.*, **B289** 364.

[11] H. Ooguri and C. Vafa (1991), *Nucl. Phys.*, **B361** 469.

[12] D. Gepner (1988), *Nucl. Phys.*, **B296** 757.

[13] D. Gepner (1987), *Phys. Lett.*, **B199** 87.

[14] C. Vafa and N. Berkovits (1993), preprint hep-th/9310170.

Generalized canonical quantization of gauge theories with polarized second–class constraints

I. A. Batalin, S.L.Lyakhovich and I. V. Tyutin

Abstract

The split involution quantization scheme, proposed previously for pure second–class constraints only, is extended to cover the case of the presence of irreducible first–class constraints. The constraint algebra generating equations are formulated and the Unitarizing Hamiltonian is constructed. Physical operators and states are defined in the sense of a natural counterpart to the Dirac's weak equality.

1 Introduction

The split involution formalism has been initially proposed in Ref[1] for canonical quantization of dynamical systems with pure second–class constraints. The method does not imply to convert original second-class constraints into effective first-class ones by introducing any extra variables. On the other hand, the constraint basis should be chosen in an Sp(2)-polarized form by splitting the total constraint set into two interchangeable subsets, T^a_μ, $a = 1, 2$, to satisfy the split involution relations

$$(\imath\hbar)^{-1}[T^{\{a}_\mu, T^{b\}}_\nu] = U^{\{a\rho}_{\mu\nu} T^{b\}}_\rho, \tag{1.1}$$

where $X^{\{ab\}} \equiv X^{ab} + X^{ba}$, and $[\,,\,]$ denotes the standard supercommutator. The Hamiltonian H is supposed to satisfy the relations

$$(\imath\hbar)^{-1}[H, T^a_\mu] = V^\nu_\mu T^a_\nu. \tag{1.2}$$

In contrast to the conventional involution case, the relations (1.1) allow constraints to be of the second class, being the operator-valued matrix

$$\Delta_{\mu\nu} \equiv \varepsilon_{ab}(\imath\hbar)^{-1}[T^a_\mu, T^b_\nu] \tag{1.3}$$

(where ε_{ab} is an $Sp(2)$–invariant constant tensor, $\varepsilon_{ab} = -\varepsilon_{ba}$, $\varepsilon_{12} = 1$) invertible. Arbitrary second-class constraints (whose Fermionic component number is divisible by 4) and Hamiltonian can be transformed locally to the polarized basis subjected to eqs. (1.1), (1.2). What is not so evident that there exists a valuable set of relativistic dynamical systems such that the Dirac's hamiltonization procedure, being applied to the original Lagrangian, just produces the polarized constraint basis. The split involution "gauge" algebra is generated by imposing the equations

$$[Q^a, Q^b] = 0, \quad [Q^a, \mathcal{H}] = 0, \tag{1.4}$$

on the Fermions Q^a and Boson $\mathcal{H}$ searched in the form of a series expansion in ghost powers

$$Q^a = C^\mu T^a_\mu + \ldots, \quad \mathcal{H} = H + \ldots. \tag{1.5}$$

The complete Unitarizing Hamiltonian reads

$$H_{complete} = \mathcal{H} + \varepsilon_{ab}(\imath\hbar)^{-2}[Q^b, [Q^a, B]] \tag{1.6}$$

where B is a "gauge–fixing" Bosonic operator. The generating equations (1.4) as well as the Hamiltonian (1.6) possess the $Sp(2)$–covariant form which is characteristic to the formalism developed in Ref [2,3] to quantize gauge–invariant theories formulated in a ghost–antighost symmetric fashion. However, the number of ghosts (and antighosts) introduced in the formalism [2,3] is twice as compared with the corresponding number in the split involution theory. Moreover, the ghost numbers of the generating operators (Q^1, Q^2) are (+1, +1) in the split involution scheme, while in the ghost–antighost symmetric theory these numbers are (+1, −1).

In this work we intend to include original first-class constraints into the split involution formalism. When doing this we retain the explicit $Sp(2)$–symmetry property of the method to hold. The first–class constraints and the corresponding Fermion generating operator are $Sp(2)$–scalars, whereas their second–class counterparts are $Sp(2)$–vectors. We assign ghost canonical pairs to constraints of the both classes and introduce the ghost number operators G' and G'' of the first and second class respectively. A pair of the ghost number values, denoted by gh$'$ and gh$''$, is thereby assigned to each admitted operator of the theory. The operators G' and G'' themselves are required to be conserved separately. We formulate the extended version of the constraint gauge algebra generating equations and then construct the complete Unitarizing Hamiltonian. The Existence Theorem and automorphism group of the generating equations will be considered elsewhere.

We use the standard conventions and abbreviations of the generalized canonical quantization method. The usual hermiticity properties are implied for all the canonical variables. The only difference in notations is that the indices of first (second)–class constraints are taken from the first (second) half of the Greek alphabet α, β, …, λ (μ, ν, …, ω).

2 Constraint algebra

Let $(q^i,\, p_i)$, $i = 1,\ldots,n = n_+ + n_-$, $\varepsilon(q^i) = \varepsilon(p_i) \equiv \varepsilon_i$, be a set of the original phase variable operators subjected to the canonical equal–time (super)commutation relations. Let us also suppose the Hamiltonian H and the constraint operators $T^a_\mu(q,p)$, $T_\alpha(q,p)$ to satisfy the eqs. (1.1), (1.2) as well as the following relations:

$$(\imath\hbar)^{-1}[T^a_\mu, T_\alpha] = \tilde{U}^{a\beta}_{\mu\alpha} T_\beta + U^{\nu}_{\mu\alpha} T^a_\nu, \tag{2.1}$$

$$(\imath\hbar)^{-1}[T_\alpha, T_\beta] = \tilde{U}^{\gamma}_{\alpha\beta} T_\gamma + \frac{1}{2}\varepsilon_{ab} W^{\mu\nu}_{\alpha\beta}(T^b_\nu \delta^\rho_\mu - T^b_\mu \delta^\rho_\nu (-1)^{\varepsilon_\mu \varepsilon_\nu} - \imath\hbar U^{b\rho}_{\nu\mu}) T^a_\rho, \tag{2.2}$$

$$(\imath\hbar)^{-1}[H, T_\alpha] = \tilde{V}^{\beta}_{\alpha} T_\beta + \frac{1}{2}\varepsilon_{ab} W^{\mu\nu}_{\alpha}(T^b_\nu \delta^\rho_\mu - T^b_\mu \delta^\rho_\nu (-1)^{\varepsilon_\mu \varepsilon_\nu} - \imath\hbar U^{b\rho}_{\nu\mu}) T^a_\rho. \tag{2.3}$$

where the structure coefficient operators are some functions of the original phase variables q, p, and the following antisymmetry properties are supposed to hold:

$$\begin{gathered} U^{a\rho}_{\mu\nu} = -U^{a\rho}_{\nu\mu}(-1)^{\varepsilon_\mu \varepsilon_\nu}, \quad \tilde{U}^{\gamma}_{\alpha\beta} = -\tilde{U}^{\gamma}_{\beta\alpha}(-1)^{\varepsilon_\alpha \varepsilon_\beta}, \\ W^{\mu\nu}_{\alpha\beta} = -W^{\nu\mu}_{\alpha\beta}(-1)^{\varepsilon_\mu \varepsilon_\nu} = -W^{\mu\nu}_{\beta\alpha}(-1)^{\bar{\varepsilon}_\alpha \bar{\varepsilon}_\beta}, \quad W^{\mu\nu}_{\alpha} = -W^{\nu\mu}_{\alpha}(-1)^{\varepsilon_\mu \varepsilon_\nu}. \end{gathered} \tag{2.4}$$

Let us also require the matrix $\Delta_{\mu\nu}$ (1.3) to be an invertible one:

$$\Delta_{\mu\nu} \quad \Rightarrow \quad \exists \;\; \Delta^{-1\mu\nu}: \quad \Delta^{-1\mu\nu}\Delta_{\nu\rho} = \delta^\mu_\rho. \tag{2.5}$$

This condition implies constraints the T^a_μ to be of the second–class. Eqs (1.1),(1.2), (2.1)-(2.3), (2.5) constitute the split involution conjecture for gauge-invariant theories subject to second–class constraints. (For the sake of simplicity the first–class constraints T_α are implied here to be irreducible. The detailed analysis of the irreducibility conditions will be given elsewhere). It would be just desirable to avoid imposing further restrictions on the constraint algebra (1.1),(2.1)-(2.5). Unfortunately, we are unable to prevent such restrictions for the present. Therefore we have to impose the following extra condition on the structure coefficients $\tilde{U}^{a\beta}_{\mu\alpha}$ entering the cross–sector relation (2.1) that involves constraints of the both classes :

$$\begin{gathered} (\imath\hbar)^{-1}[T^{\{a}_\mu, \tilde{U}^{b\}\beta}_{\nu\alpha}] - (\imath\hbar)^{-1}[T^{\{a}_\nu, \tilde{U}^{b\}\beta}_{\mu\alpha}](-1)^{\varepsilon_\mu\varepsilon_\nu} - \tilde{U}^{\{a\gamma}_{\mu\alpha}\tilde{U}^{b\}\beta}_{\nu\gamma}(-1)^{\varepsilon_\nu(\bar{\varepsilon}_\alpha+\bar{\varepsilon}_\gamma)} + \\ +\tilde{U}^{\{a\gamma}_{\nu\alpha}\tilde{U}^{b\}\beta}_{\mu\gamma}(-1)^{\varepsilon_\mu(\bar{\varepsilon}_\alpha+\bar{\varepsilon}_\gamma+\varepsilon_\nu)} - U^{\{a\rho}_{\mu\nu}\tilde{U}^{b\}\beta}_{\rho\alpha} = \tilde{U}^{\{a\rho\gamma}_{\mu\nu\alpha}(T^{b\}}_\rho\delta^\beta_\gamma - \imath\hbar\tilde{U}^{b\}\beta}_{\rho\gamma})(-1)^{\varepsilon_\mu\bar{\varepsilon}_\alpha}, \end{gathered} \tag{2.6}$$

where the new structure coefficient operators $\tilde{U}^{a\gamma\rho}_{\mu\nu\alpha}$ are supposed to possess the antisymmetry property

$$\tilde{U}^{a\gamma\rho}_{\mu\nu\alpha} = -\tilde{U}^{a\gamma\rho}_{\nu\mu\alpha}(-1)^{\varepsilon_\mu\varepsilon_\nu+\varepsilon_\nu\bar{\varepsilon}_\alpha+\bar{\varepsilon}_\alpha\varepsilon_\mu}. \tag{2.7}$$

Let us consider the status of the restriction (2.6). By applying the Jacoby identity to the constraint algebra (1.1), (2.1), (2.2) one can show the operators $\tilde{U}^{a\beta}_{\mu\alpha}$ to satisfy the relation that differs from (2.6) by the extra contribution

$$\tilde{\tilde{U}}^{ab\gamma\lambda}_{\mu\alpha}(T_\lambda\delta^\beta_\gamma - T_\gamma\delta^\beta_\lambda(-1)^{\bar{\varepsilon}_\gamma\bar{\varepsilon}_\lambda} - i\hbar\tilde{U}^\beta_{\gamma\lambda}) \tag{2.8}$$

to r.h.s. Thus in fact the condition (2.6) is equivalent to requirement for the contribution (2.8) to vanish. On the other hand one can consider the cross–sector relation (2.12) to be the covariant constancy property of the constraints, being the structure coefficients $\tilde{U}^{a\beta}_{\mu\alpha}$, $U^{a\rho}_{\mu\nu}$ treated to serve as the connection components. From this viewpoint, l.h.s. of (2.6) is nothing else but the corresponding curvature components. The condition (2.6) being treated classically requires for the curvature to vanish on the second–class constraint surface, while the algebra (1.1), (2.1), (2.2) itself implies a weaker condition to be satisfied that the curvature components should vanish on the surface of all the constraints.

Now let us comment in brief the most characteristic features of the involution relations (1.1), (1.2), (2.1)-(2.3). First of all we observe that the split involution relations (1.1), (1.2) retain their original form [1] specific to the pure second–class constraint case. Further, the cross–sector constraint supercommutators (2.1) are actually restricted in two respects: the operators $\tilde{U}^{a\beta}_{\mu\nu}$ are subordinated to the relations (2.6), and the operators $U^\nu_{\mu\alpha}$ do not possess their own $Sp(2)$–indices. Finally, let us turn to the first–class constraint involution relations (2.2), (2.3). Being these relations treated classically, second–class constraints are allowed to contribute only quadratically, which assertion is a consequence of the Jacoby identity. Such quadratic contributions are just represented by the second and third terms in r.h.s. of (2.2), (2.3), and these terms possess the specific structure characterized by the antisymmetry property of the coefficients $\varepsilon_{ab}W^{\nu\mu}$ in their indices a, b and μ, ν. The fourth terms in r.h.s. of (2.2), (2.3) represent the quantum contributions which are necessary in order to provide the operator compatibility of the formal constraint algebra.

The original constraints and Hamiltonian involution relations (1.1), (1.2), (2.1) – (2.3) determine the structure coefficient operators up to a natural arbitrariness. By making use of the Jacoby identity together with the irreducibility property of the constraints, one can derive the necessary compatibility conditions to the initial involution relations (1.1), (1.2), (2.1) – (2.3). These new conditions, including the one (2.6), contain new structure coefficient operators to be determined at this level. On the other hand, these relations reduce to an admissible extent the arbitrariness in the preceding–level structure coefficient operators. Continuing this procedure, one generates, step by step, an infinite gauge algebra initiated by the original constraints and Hamiltonian.

3 Constraint algebra generating equations

As a next step let us introduce a ghost extension for the original phase space. We assign a ghost canonical pair to each first–class constraint operator:

$$T_\alpha \quad \rightarrow \quad (C'^\alpha, \bar{\mathcal{P}}'_\alpha), \quad \varepsilon(C'^\alpha) = \varepsilon(\bar{\mathcal{P}}'_\alpha) = \tilde{\varepsilon}_\alpha + 1, \tag{3.1}$$

$$\mathrm{gh}'(C'^\alpha) = -\mathrm{gh}'(\bar{\mathcal{P}}'_\alpha) = 1, \quad \mathrm{gh}''(C'^\alpha) = \mathrm{gh}''(\bar{\mathcal{P}}'_\alpha) = 0. \tag{3.2}$$

In the same way we assign a ghost canonical pair to each $(a = 1,2)$–pair of the second–class constraint operators:

$$T^a_\mu \quad \rightarrow \quad (C''^\mu, \bar{\mathcal{P}}''_\mu), \quad \varepsilon(C''^\mu) = \varepsilon(\bar{\mathcal{P}}''_\mu) = \varepsilon_\mu + 1, \tag{3.3}$$

$$\mathrm{gh}'(C''^\mu) = \mathrm{gh}'(\bar{\mathcal{P}}''_\mu) = 0, \quad \mathrm{gh}''(C''^\mu) = -\mathrm{gh}''(\bar{\mathcal{P}}''_\mu) = 1 \tag{3.4}$$

Further, let us introduce the generating operators

$$\Omega^a(q,p,C',\bar{\mathcal{P}}',C'',\bar{\mathcal{P}}''), \quad \varepsilon(\Omega^a) = 1, \quad \mathrm{gh}'(\Omega^a) = 0, \quad \mathrm{gh}''(\Omega^a) = 1, \tag{3.5}$$

$$\Omega(q,p,C',\bar{\mathcal{P}}',C'',\bar{\mathcal{P}}''), \quad \varepsilon(\Omega) = \mathrm{gh}'(\Omega^a) = 1, \quad \mathrm{gh}''(\Omega^a) = 0, \tag{3.6}$$

$$K(q,p,C',\bar{\mathcal{P}}',C'',\bar{\mathcal{P}}''), \quad \varepsilon(K) = 0, \quad \mathrm{gh}'(K) = 2, \quad \mathrm{gh}''(K) = -2, \tag{3.7}$$

$$\mathcal{H}(q,p,C',\bar{\mathcal{P}}',C'',\bar{\mathcal{P}}''), \quad \varepsilon(\mathcal{H}) = \quad \mathrm{gh}'(\mathcal{H}) = \mathrm{gh}''(\mathcal{H}) = 0, \tag{3.8}$$

$$\Lambda(q,p,C',\bar{\mathcal{P}}',C'',\bar{\mathcal{P}}''), \quad \varepsilon(\Lambda) = 1, \quad \mathrm{gh}'(\Lambda) = 1, \quad \mathrm{gh}''(\Lambda) = -2, \tag{3.9}$$

to subordinate them to the following generating equations:

$$[\Omega^a, \Omega^b] = 0, \tag{3.10}$$

$$[\Omega^a, \Omega] = 0, \tag{3.11}$$

$$[\Omega, \Omega] = \varepsilon_{ab}(\imath\hbar)^{-1}[\Omega^b, [\Omega^a, K]]. \tag{3.12}$$

$$[\Omega^a, \mathcal{H}] = 0, \tag{3.13}$$

$$[\Omega, \mathcal{H}] = \varepsilon_{ab}(\imath\hbar)^{-1}[\Omega^b, [\Omega^a, \Lambda]]. \tag{3.14}$$

Let us seek for a solution to these equations in the form of $C\bar{\mathcal{P}}$–ordered series expansion in ghost powers:

$$\Omega^a = C''^{\mu}T^a_{\mu} + \frac{1}{2}C''^{\nu}C''^{\mu}U^{a\rho}_{\mu\nu}\bar{\mathcal{P}}''_{\rho}(-1)^{\varepsilon_{\nu}+\varepsilon_{\rho}} + C'^{\alpha}C''^{\mu}\tilde{U}^{a\beta}_{\mu\alpha}\bar{\mathcal{P}}'_{\beta}(-1)^{\bar{\varepsilon}_{\alpha}+\bar{\varepsilon}_{\beta}} + \ldots, \tag{3.15}$$

$$\Omega = C'^{\alpha}T_{\alpha} + \frac{1}{2}C'^{\beta}C'^{\alpha}\tilde{U}^{\gamma}_{\alpha\beta}\bar{\mathcal{P}}'_{\gamma}(-1)^{\bar{\varepsilon}_{\beta}+\bar{\varepsilon}_{\gamma}} + C'^{\alpha}C''^{\mu}U^{\nu}_{\mu\alpha}\bar{\mathcal{P}}''_{\nu}(-1)^{\bar{\varepsilon}_{\alpha}+\varepsilon_{\nu}} + \ldots, \tag{3.16}$$

$$K = \frac{1}{2}C'^{\beta}C'^{\alpha}W^{\mu\nu}_{\alpha\beta}\bar{\mathcal{P}}''_{\nu}\bar{\mathcal{P}}''_{\mu}(-1)^{\bar{\varepsilon}_{\beta}+\varepsilon_{\nu}} + \ldots, \tag{3.17}$$

$$\mathcal{H} = H - C''^{\mu}V^{\nu}_{\mu}\bar{\mathcal{P}}''_{\nu}(-1)^{\varepsilon_{\nu}} - C'^{\alpha}\tilde{V}^{\beta}_{\alpha}\bar{\mathcal{P}}'_{\beta}(-1)^{\bar{\varepsilon}_{\beta}} + \ldots, \tag{3.18}$$

$$\Lambda = \frac{1}{2}C'^{\alpha}W^{\mu\nu}_{\alpha}\bar{\mathcal{P}}''_{\nu}\bar{\mathcal{P}}''_{\mu}(-1)^{\varepsilon_{\nu}} + \ldots. \tag{3.19}$$

By inserting the expansions (3.15)–(3.19) into the generating equations (3.10)–(3.14), one obtains to the second order in ghosts just the constraint involution relations (1.1), (1.2), (2.1)–(2.3), whereas to higher orders in ghosts we obtain all the higher structure relations of the gauge algebra initiated by the given constraint and Hamiltonian operators. Thus the equations (3.10)–(3.14) describe the gauge algebra generating mechanism comprehensively.

The following Existence Theorem holds for the proposed generating equations (3.10)–(3.14): if the constraint involution relations (1.1), (1.2), (2.1)–(2.3) are satisfied together with the conditions (2.5), (2.6) and the ones requiring irreducibility for the first-class constraints T_{α}, then there also exist all the higher structure coefficients in the expansions (3.15)–(3.19) and, thus, there exists a formal solution of the algebra generating equations.

The extended generating equations admit a more wide group of automorphisms as compared with the pure second–class constraint case [1]. These transformations will be considered in details in the forthcoming publication.

4 Unitarizing Hamiltonian

Introduce now the following new canonical variable operators which are the antighosts:

$$(\mathcal{P}'^{\alpha}, \bar{C}'_{\alpha}), \quad \varepsilon(\mathcal{P}'^{\alpha}) = \varepsilon(\bar{C}'_{\alpha}) = \tilde{\varepsilon}_{\alpha} + 1,$$
$$\mathrm{gh}'(\mathcal{P}'^{\alpha}) = -\mathrm{gh}'(\bar{C}'_{\alpha}) = 1, \quad \mathrm{gh}''(\mathcal{P}'^{\alpha}) = \mathrm{gh}''(\bar{C}'_{\alpha}) = 0, \tag{4.1}$$

$$(\mathcal{P}''^{\mu}, \bar{C}''_{\mu}), \quad \varepsilon(\mathcal{P}''^{\mu}) = \varepsilon(\bar{C}''_{\mu}) = \varepsilon_{\mu} + 1,$$
$$\mathrm{gh}'(\mathcal{P}''^{\mu}) = \mathrm{gh}'(\bar{C}''_{\mu}) = 0, \quad \mathrm{gh}''(\mathcal{P}''^{\mu}) = -\mathrm{gh}''(\bar{C}''_{\mu}) = 1, \tag{4.2}$$

and dynamically-active Lagrange multipliers:

$$(\lambda^{\alpha}, \pi_{\alpha}), \quad \varepsilon(\lambda_{\alpha}) = \varepsilon(\pi_{\alpha}) = \tilde{\varepsilon}_{\alpha},$$
$$\mathrm{gh}'(\lambda^{\alpha}) = \mathrm{gh}'(\pi_{\alpha}) = \mathrm{gh}''(\lambda^{\alpha}) = \mathrm{gh}''(\pi_{\alpha}) = 0, \tag{4.3}$$

$$(\lambda^{a}_{\mu}), \quad a = 1, 2, \quad \varepsilon(\lambda^{a}_{\mu}) = \varepsilon_{\mu}, \quad \mathrm{gh}'(\lambda^{a}_{\mu}) = \mathrm{gh}''(\lambda^{a}_{\mu}) = 0. \tag{4.4}$$

The equal–time (super)commutators of the Lagrangian multipliers (4.4) are the following:

$$(\imath\hbar)^{-1}[\lambda^{a}_{\mu}, \lambda^{b}_{\nu}] = \varepsilon^{ab} d_{\mu\nu}, \quad \varepsilon^{ab}\varepsilon_{bc} = \delta^{a}_{c}, \tag{4.5}$$

where a constant matrix $d_{\mu\nu}$ is supposed to be invertible and possesses the following symmetry properties

$$d_{\nu\mu} = d_{\mu\nu}(-1)^{\varepsilon_{\mu}\varepsilon_{\nu}}, \qquad d^{*}_{\nu\mu} = d_{\mu\nu}. \tag{4.6}$$

Let us extend the generating operators (3.5), (3.6) by including the variables (4.1) – (4.4) via the formulae

$$Q = \Omega + \mathcal{P}'^{\alpha}\pi_{\alpha}, \tag{4.7}$$

$$Q^{a} = \Omega^{a} + \mathcal{P}''^{\mu}\lambda^{a}_{\mu}, \quad a = 1, 2, \tag{4.8}$$

The extended operators Q, Q^{a} satisfy the same equations (3.10) – (3.14) as their minimal sector counterparts Ω, Ω^{a} do.

The Unitarizing Hamiltonian is defined as

$$H_{complete} = \mathcal{H} + (\imath\hbar)^{-1}[Q, F] + \varepsilon_{ab}(\imath\hbar)^{-2}[Q^{b}, [Q^{a}, B]], \quad [Q^{a}, B] = 0, \tag{4.9}$$

where

$$\varepsilon(F) = 1, \quad \mathrm{gh}'(F) = -1, \quad \mathrm{gh}''(F) = 0, \tag{4.10}$$

$$\varepsilon(B) = 0, \quad \mathrm{gh}'(B) = 0, \quad \mathrm{gh}''(B) = -2. \tag{4.11}$$

The gauge–fixing operators F and B may depend on the total set of phase variables of the extended phase space. In the simplest case these gauge operators can be chosen in the form

$$F = \lambda^\alpha \bar{\mathcal{P}}'_\alpha + (\chi^\alpha + C''^\mu V^{\nu\alpha}_\mu \bar{\mathcal{P}}''_\nu (-1)^{\varepsilon_\nu + \varepsilon_\alpha})\bar{C}'_\alpha + \ldots, \quad (\imath\hbar)^{-1}[T^a_\mu, \chi^\alpha] = V^{\nu\alpha}_\mu T^a_\nu, \tag{4.12}$$

$$B = \bar{\mathcal{P}}''_\mu \bar{C}''_\nu d^{\nu\mu}, \quad d_{\mu\nu} d^{\nu\rho} = \delta^\rho_\mu. \tag{4.13}$$

Finally we define the physical operators and physical states as follows. An operator $\mathcal{O}$ is called the physical one iff

$$\mathrm{gh}'(\mathcal{O}) = \mathrm{gh}''(\mathcal{O}) = 0, \tag{4.14}$$

$$[Q^a, \mathcal{O}] = 0, \quad [Q, \mathcal{O}] = \varepsilon_{ab}(\imath\hbar)^{-1}[Q^b, [Q^a, anything]]. \tag{4.15}$$

A state $|\Phi>$ is called the physical one iff

$$\mathrm{gh}'(|\Phi>) = \mathrm{gh}''(|\Phi>) = Q^a|\Phi>= 0, \quad Q|\Phi>= \varepsilon_{ab} Q^b Q^a |anything> . \tag{4.16}$$

The physical matrix elements $< \Phi|\mathcal{O}|\Phi_1 >$ depend neither on a particular choice of the gauge-fixing operators B and F nor on the arbitrariness of a solution to the eqs (3.10)–(3.14), (4.14)–(4.16).

5 Conclusion

So, we have extended the split involution formalism to cover the case of the presence of irreducible first–class constraints. Thereby the miraculous supersymmetry yielded by the split involution relations is coupled with the actual gauge symmetry initiated by the original first–class constraints.

The most characteristic feature of the formalism proposed is the appearance of the new equivalence criterion explicitly–quadratic in second–class constraints that is a natural counterpart to the Dirac's weak equality concept as applied to the first–class quantities. It is quite evident from this viewpoint that all the double–supercommutator contributions in (3.12), (3.14), (4.9), as well as the quadratic operator in r.h.s. of (4.15), (4.16) are of the same origin.

Acknowledgement

The work is supported in part by the International Science Foundation under the Grant number M2I000.

References

[1] I.A.Batalin, S.L.Lyakhovich and I.V.Tyutin (1992), *Mod.Phys. Lett.*, **A7** 1931.

[2] I.A.Batalin, P.M.Lavrov and I.V.Tyutin (1990), *J.Math. Phys.*, **31** 6.

[3] I.A.Batalin, P.M.Lavrov and I.V.Tyutin (1990), *J.Math. Phys.*, **31** 2708.

Radiation field on superspace

Pedro F. González-Díaz

Abstract

We study the dynamics of multiwormhole configurations within the framework of the Euclidean Polyakov approach to string theory, incorporating a modification to the Hamiltonian which leads to a Planckian probability measure for the Coleman parameters α that allows $\frac{1}{2}\alpha^2$ to be interpreted as the energy of the quanta of a radiation field on superspace whose values might still fix the coupling constants.

Multiwormhole configurations in Polyakov string theory have been studied by looking at the wormholes as the handles on a Riemann surface of genus g, with g giving the number of handles or wormholes in the configuration [1]. The Green function that describes the effects of such wormholes on first order tachyonic amplitudes was calculated by Lyons and Hawking [2] as a path integral over all space-time coordinates x_μ on the Riemann surface. It was assumed that all the fields have the same values at the points on the two circles which result after cutting the handles in such a way that they become divided in two topologically separated discs. In this case, the points were identified by the projective transformations of the Schottky group on each pair of circles, and the Green function can be written as [1-3]

$$< x_\mu(z_1).x_\nu(z_2)... >= \int d[x_\mu]x_\mu(z_1)x_\nu(z_2)...\prod_r\prod_n \delta(x_n - x'_n)e^{-I}, \quad (1)$$

where I is the Euclidean action, r runs from 1 to g,

$$x = \sum x_n e^{i\zeta n}, \quad (2)$$

and the delta function ensures that ζ on one circle is identified with ζ' on another. On can express the Green function (1) in terms of the handle quantum state on the circles using the Fourier transform of the delta function for the zero mode, and expanding the delta function for the nonzero modes in terms of the complete set of orthonormal harmonic-oscillator eigenstates which are the solutions of the string analogue of the Wheeler DeWitt equation [2]

$$H_{WDW}\Psi_{nm_n^{(i)}} = [-\frac{\partial^2}{\partial x_0^2} + \frac{1}{2}\sum_{n>0,i}(-\frac{\partial^2}{\partial (Y_n^{(i)})^2} + n^2(Y_n^{(i)})^2)]\Psi_{nm_n^{(i)}} = 0, \quad (3)$$

with solution ($i = 1, 2$)

$$\Psi_{nm_n^{(i)}} \propto e^{-\frac{1}{2}n(Y_n^{(i)})^2} H_{m_n^{(i)}}(n^{\frac{1}{2}} Y_n^{(i)}) \Psi_K(x_0), \Psi_K(x_0) = e^{iK.x_0}. \tag{4}$$

Eq. (3) is the canonically quantised version of the Hamiltonian constraint derived [2] from the string Euclidean action for the field x on the region of the complex plane outside a disc of radius $r = -\frac{\ln|z|}{t}$ (t is some Euclidean time), K is the momentum of the zero mode $n = 0$, and

$$Y_n^{(1)} = \frac{1}{2}(x_n + x_{-n}), Y_n^{(2)} = \frac{1}{2i}(x_n - x_{-n}).$$

Now, as in the 4-dimensional case [4], one can calculate the effect on Green functions in the fundamental region, by doing a path integral over all fields x^μ on the complex plane with the boundary conditions on the circles given by a set of values $Y_n^{(i)}$, weighting with the wave function $\Psi_{nm_n}^{(i)}$. Again as in the space-time wormhole case, the effect will be given by a vertex operator located approximately at the center of the circles. One can see that, after integrating over the fields, the resulting path integral contains a factor $\kappa = (K^2 + \sum | n | m_n^{(i)} - 2)^{-1}$. We obtain thus a bi-local effective action [1]

$$-\int d\sigma_1 \int d\sigma_2 \sum \int d^4 K \kappa V_p(\sigma_1) V_p(\sigma_2),$$

where the V_p are the handle vertex operators for the fields on the two circles. One can again convert the bi-local action to a local action, by introducing α parameters, to finally obtain a probability measure with the same general form as for the 4-dimensional case [5]. Nevertheless, since now the α parameters are labelled by momentum K, and also the occupation numbers $m_n^{(i)}$ for modes $n > 0$, these parameters should be interpreted [1,3] as a quantum field on the infinite dimensional superspace of all coordinate field values on a single circle. Such a quantum field would be regarded as an infinite tower of fields on the usual space-time minisuperspace; each stage along this tower is labelled by the quantum number $m_n^{(i)}$ that defines the excited states of the basis set of solutions (4). In light of this interpretation, it was concluded [1,3] that the α's could not be regarded as a set of coupling constants to be fixed by quantum measurement, as required by the Coleman mechanism to fix the coupling constants [6].

However, it is not quite clear that this conclusion can be maintained because in string theory the Hamiltonian should be modified [7] by the addition of an (infinite) constant, in such a way that just the ground state of the harmonic oscillators, multiplied by the wave function of the zero mode, will obey the Hamiltonian constraint.

We shall follow here a modified approach where this shortcoming is avoided. The idea consists in considering the more general case allowing for handles on the Riemann surface which, under cutting, give rise to pairs of discs that are no longer disconnected to each other [8]. This would ultimately imply partial or total breakdown of the Schottky group invariance under projective transformations between discs [9]. Such handles need not be on shell in the sense that the analogue of the Wheeler DeWitt operator acting on the excited eigenstates $\Psi_{m_n^{(i)}}(Y_n^{(i)})$ is no longer zero, but gives the corresponding harmonic-oscillator eigenvalues $nm_n^{(i)}$ [8,10]. In this case, the δ function for all field modes in the Green function should be replaced [8] by a path integral which has the x_n fixed at the given values at time $t = 0$ on a circle and $t = t_1$ on another, for some Euclidean time interval t_1 between the two circles. Unlike for the states given by Eqns. (1)-(3), this path integral is not generally separable into a product of wave functions (4), but, after integrating over time t_1, it gives the density matrix for a mixed quantum state, and is equal to the propagator

$$K(x_n, 0; x'_n, t_1) = \prod_{n>0,i} \sum_{m_n^{(i)}>0} \Phi_{m_n^{(i)}}(Y_n^{(i)}) \Phi_{m_n^{(i)}}(Y_n^{(i)\prime}) e^{-nm_n^{(i)} t_1}, \tag{5}$$

where $\Phi_{m_n^{(i)}}(Y_n^{(i)})$ would match the excited eigenstates of the harmonic oscillators ($\Phi_{m_n^{(i)}} \equiv \Psi_{m_n^{(i)}}$) if we allowed an unlimited resolution for the fields, or equal some wave functions which contain only that part of the information contained in the $\Psi_{m_n^{(i)}}$ that associates with the finite eigenenergy sector surviving after the introduction of a given cut off at a finite highest energy scale. If the handles are off shell, then, instead of the quantum constraint equation (3), we must use the "time-independent" wave equation [8,10]

$$(H_{WDW} - \sum_{n>0,i} nm_n^{(i)})\Psi = 0, \tag{6}$$

so that the wave function for the wormhole handles becomes

$$\Psi \equiv \Psi(x_0, Y) = e^{iKx_0} \prod_{n>0,i} e^{-\frac{1}{2}n(Y_n^{(i)})^2}. \tag{7}$$

The K^2 term in (6), resulting from the application of operator $-\frac{\partial^2}{\partial x_0^2}$ to Ψ, gives the energy of the x_0 plane waves, with $K^2 < 0$. This would correspond to a timelike momentum in *Lorentzian* space-time. Wick rotating to Euclidean momentum, $K \to -iK_E$, we have from (6) and (7) $K_E^2 = \sum_{n>0,i}(m_n^{(i)} + \frac{1}{2})n$ and, since we are dealing with harmonic oscillators, we have in general $K_E = \sum_{n>0,i} m_n^{(i)} n$ [11]. Let us then calculate the quantity

$$\sum_{m_n^{(i)}>0} \Psi(x_0, Y)\Psi^*(x'_0, Y')$$

$$= \sum_{m_n^{(i)}>0} \prod_{n>0,i} (e^{-\frac{1}{2}n(Y_n^{(i)})^2} e^{-\frac{1}{2}n(Y_n^{(i)\prime})^2} e^{-m_n^{(i)} n(x'_0 - x_0)}). \tag{8}$$

The relative minus sign between x_0 and x_0' in (8) should be kept anyway in order to ensure an orientable surface when the handles are glued together [2]. Taking $x_0' - x_0 = t_1$, noting that, since each two circles can have any time separations, one should integrate (8) over all possible values of t_1 [8], and denoting the density matrix by ρ, we can see that

$$i \int dt_1 K(x_n, 0; x_n', t_1)$$

$$= \sum_{m_n^{(i)}>0} \int d(x_0' - x_0)\Psi(x_0, Y)\Psi^*(x_0', Y') \equiv i\rho(Y; Y'), \tag{9}$$

whenever we take for the states $\Phi_{m_n^{(i)}}$ in the propagator (5) only that part of the $\Psi_{m_n^{(i)}}$ which corresponds to the harmonic oscillator ground states, surviving after projecting off all the information contained in the Hermite polynomials.

¿From (7) and (9), the density matrix for handles becomes

$$\rho(Y; Y') = \sum_{m_n^{(i)}>0} \prod_{n>0,i} \frac{\Psi_0(Y_n^{(i)})\Psi_0(Y_n^{(i)\prime})}{nm_n^{(i)}}, \tag{10}$$

where $\Psi_0(Y_n^{(i)}) = e^{-\frac{1}{2}n(Y_n^{(i)})^2}$. Thus, for each $m_n^{(i)} > 0$, we should use a Green function for the density matrix of the mixed state case given by

$$< x_\mu(z_1).x_\nu(z_2)... >$$

$$= \int d[x_\mu]x_\mu(z_1)x_\nu(z_2)... \prod_r \prod_{n>0,i} \tilde{\Theta}(x_0 - x_0')\Psi_0(Y_n^{(i)})\Psi_0(Y_n^{(i)\prime})e^{-I}, \tag{11}$$

where we have replaced the full δ function in (1) for the density matrix (7), specialising to a single generic relative probability $(nm_n^{(i)})^{-1}$ for each mode $n > 0$, and the step function $i\tilde{\Theta}(x_0 - x_0') = \Theta(x_0 - x_0')$ arises from integrating the δ function for the zero mode over its argument, as it is done in (9) and (10). In the present approach, if we want to consider a Green function also for the wave function (7), instead of the full δ function, we should use the probability $\mid \Psi \mid^2$ obtained from (7) as the weighting factor. This Green function will then be

$$< x_\mu(z_1).x_\nu(z_2)... >$$

$$= \int d[x_\mu]x_\mu(z_1)x_\nu(z_2)... \prod_r \prod_{n>0,i} \delta(x_0 - x_0')\Psi_0(Y_n^{(i)})\Psi_0(Y_n^{(i)\prime})e^{-I}. \tag{12}$$

If we regard each pair of circles as the ends of a sum of wormholes of different species [1,3], then each species would now be labelled by just the momentum K of the zero mode, but *not* the levels $m_n^{(i)}$ of the other modes.

The quantity K can be interpreted as the conserved scalar charge carried by the wormhole [7].

We can now calculate the effect of wormholes on tachyonic amplitudes for handles whose quantum state is given by both a density matrix and a wave function, using the procedure devised by Lyons and Hawking [2]. For the case of handles in mixed state and tachyons with momenta p_j, unlike the pure-state case considered in Ref. 2, the path integral describing the interaction cannot be factorised into path integrals on each of the two circles [8]. Instead of the wave function (4), one should then introduce as weighting factor the density matrix element $\rho^{(i)}_{m_n}$ which corresponds to each relative probability $\frac{1}{nm_n^{(i)}}$. In the limit of small circle radius $r \to 0$, we then have for each of these density matrix elements

$$D(\rho;p_j) \propto \int dx_0 dx_0' (\prod_{n>0} dY_n^{(i)} dY_n^{(i)'}) \rho_{m_n^{(i)}}$$

$$\times \int dr r^{-3+(\sum_1^M p_j)^2} \int [D\zeta_j] \exp[i(x_0 - x_0') \sum_1^M p_j]$$

$$\times \prod_{n>0} e^{\sum_i [-\frac{1}{2}n((Y_n^{(i)})^2 + (Y_n^{(i)'})^2) + ir^n k_n^{(i)} (Y_n^{(i)} - Y_n^{(i)'})]}, \tag{13}$$

where

$$\rho_{m_n^{(i)}} = \prod_{n>0} \frac{\Psi_0(Y_n^{(i)}) \Psi_0(Y_n^{(i)'})}{nm_n^{(i)}},$$

and we take, as in [2], $r = | z_2 |^{-1}$, $\zeta_j = \frac{z_j}{z_2}$, $j = 3, ..., M$, with M the number of on-shell tachyon vertex operators inserted in the region exterior to the circles, $\zeta_0 = 0$, $\zeta_1 = \infty$, $\zeta_2 = 1$. $\int [D\zeta_j]$ denotes integration over ζ_j with a measure whose explicit form need not be known for our calculation, and $k_n = 2\sum_{j=1}^M p_j \zeta_j^{-n}$, with $k_n^{(1)}$ and $k_n^{(2)}$ the real and imaginary parts of k_n, respectively.

Each integral pair over $Y_n^{(i)}$ and $Y_n^{(i)}$ gives a factor $\frac{e^{-\frac{r^{2n}(k_n^{(i)})^2}{n}}}{n}$ for each n. The Gaussian exponential factor would only contribute higher-order interactions and will be disregarded in our calculation [2]. The other possible contribution would come from integration over each pair of zero-mode fields x_0,x_0'. Since all possible dependence of the density matrix on such field has already been integrated out, unlike for handles in a pure state, we are left with a single delta on $\sum_{j=1}^M p_j$, so in first order approximation the path integral (13) gives essentially a factor $\int \frac{dr}{r^3 m_n^{(i)} n^2}$ for each $m_n^{(i)}$ and n. Note that this factor contains no integration over momentum K. As in the space-time wormhole case [4], the effect of handles will be given again by a vertex operator, located approximately at the center of the circles [1,3]. Thus, for each

$m_n^{(i)}$ and n, the stringy wormholes in mixed state will give rise to a bi-local effective action [1] for each $m_n^{(i)}$

$$-\frac{1}{2}\int d\sigma_1 \int d\sigma_2 \sum_{q,i} \frac{V_q(\sigma_1)V_q(\sigma_2)}{m_n^{(i)}n^2}, \tag{14}$$

where V_q are the vertex functions. Following hereafter the same procedure as for wormholes in space-time [4], this action can be made local by introducing α parameters, i.e.

$$\int d\sigma \sum_{q,i} ((-\frac{1}{2}\alpha_q^2 m_n^{(i)} n^2) + \alpha_q V_q(\sigma)) \tag{15}$$

which leads to a probability measure over the α parameters for each $m_n^{(i)}$

$$Z(\alpha)\prod_{q,i} e^{-\frac{1}{2}\alpha_q^2 m_n^{(i)} n^2}, \tag{16}$$

where again $Z(\alpha)$ is the path integral over all fields x^μ on the two-sphere [1,3], containing the effective interaction $\alpha_q V_q(\sigma)$. The distribution for α-parameters associated with (16) corresponds to just one of the infinite relative probabilities for the state $\Psi_0(Y_n^{(i)})$ of handles. Therefore, one should now sum (16) over all $m_n^{(i)}$ [12,13], to finally obtain a probability measure

$$Z(\alpha)\prod_{q} (e^{\frac{1}{2}n^2\alpha_q^2} - 1)^{-1} \tag{17}$$

for each n and i. Thus, as it was suggested for 4-dimensional space-time [12,13], we obtain a Planckian distribution for α parameters that allows to interpret $\frac{1}{2}\alpha_q^2$ as the energy of the quanta of a radiation field, and n^{-2} as some temperature, on string-theory superspace.

In the case that the quantum state of the handles be given by the wave function (7), the calculation is similar, but with $\rho_{m_n^{(i)}}$ replaced by $\mid \Psi(x_0, Y) \mid^2$ in the path integral (13). In actual calculation, the only difference is in the integration over the field zero modes which now produces $\delta(K - \sum_{j=1}^{M} p_j)$. The essential factor becomes then $-\frac{1}{n(K^2-2)}$ for each n. It follows that each mode n contributes a probability measure

$$Z(\alpha)\prod_{q,i} e^{-\frac{1}{2}\alpha_q^2(1-\frac{K^2}{2})n} = Z(\alpha)\prod_{q,i} e^{-\frac{1}{2}\alpha_q(K)^2 n^2}, \tag{18}$$

where $\alpha_q(K)^2 = \alpha_q^2 \frac{(1-\frac{K^2}{2})}{n}$.

Note that, since the α parameters in both (17) and (18) are labelled by the momentum K, but *not* the levels $m_n^{(i)}$, our results do not allow any interpretation of the α in terms of a quantum field on superspace, which is dimensionally reducible to an infinite tower of fields on space-time. For handles whose state is given by (7), if the initial state is a state with definite values of the α parameters, the final state will be the same as the initial state [14], according to our results, the radiation field α can be dimensionally reduced to just one field, rather than a tower of fields, on the usual space-time, i.e. on the prefered minisuperspace from string-theory superspace, consisting of just the $n = 0$ modes [1,3]. Therefore, there will always exist a set of classical values for α which makes it possible to drive a consistent mechanism that fixes the values of the coupling constants.

Acknowledgements

This work was supported by CAICYT under Research Project N' PB91-0052.

References

[1] S.W. Hawking (1991), *Nucl. Phys.*, **B363** 117.

[2] A. Lyons and S.W. Hawking (1991), *Phys. Rev.*, **D44** 3802.

[3] S.W. Hawking (1991), *Phys. Script.*, **T36** 222.

[4] S.W. Hawking (1988), *Phys. Rev.*, **D37** 904.

[5] S.W. Hawking (1990), *Nucl. Phys.*, **B335** 155.

[6] S. Coleman (1988), *Nucl. Phys.*, **B307** 867.

[7] A. Lyons (1991), Ph. D. Thesis, University of Cambridge, UK.

[8] P.F. González-Díaz (1991), *Nucl. Phys.*, **B351** 767.

[9] S. Mandelstam (1985), in *Unified Field Theories, Proceedings of the 1985 Santa Barbara Workshop*, M. Green and D. Gross (eds.), World Scientific, Singapore, 1986.

[10] S.W. Hawking (1987), in *300 Years of Gravitation*, S.W. Hawking and W. Israel (eds.), Cambridge University Press, Cambridge.

[11] W.H. Louisell (1964), *Radiation and Noise in Quantum Electronics*, McGraw Hill, New York.

[12] P.F. González-Díaz (1993), *Mod. Phys. Lett.*, **A8** 1089.

[13] P.F. González-Díaz (1993), *Class. Quant. Grav.*, **10** 2505.

[14] S. Coleman (1988), *Nucl. Phys.*, **B310** 643.

Participants

Names and E-mail addresses of those attending the Conference

Arlen ANDERSON
Imperial College, London
arley@ic.ac.uk

Abhay ASHTEKAR
Pennsylvania State University
ashtekar@phys.psu.edu

Toby BAILEY
University of Edinburgh
tnbailey@ed.ac.uk

Máximo BAÑADOS
Imperial College, London
mbanados@ic.ac.uk

Igor BARASHENKOV
University of Cape Town
igor@uctvax.uct.ac.za

Julian BARBOUR
Banbury, Oxfordshire

Alessandro BELLINI
Università di Firenze
bellini @fi.infn.it

Friedemann BRANDT
Nikhef-H, Amsterdam
T28@nikhef.nl

José CARIÑENA
Universidad de Zaragoza
jfc@cc.unizar.es

John CHARAP
QMW, London
j.m.charap@qmw.ac.uk

Dariusz CHRUŚCIŃSKI
Torun
darch@phys.uni.torun.pl

Gérard CLÉMENT
Université Paris VI
gecl@ccr.jussieu.fr

Olivier COUSSAERT
Université Libre de Bruxelles
ocouss@IS1.ulb.ac.be

Shanta DE ALWIS
University of Colorado
dealwis@gopika.colorado.edu

Frank De JONGHE
KU Leuven
Frank%tf%fys@cc3.kvleuven.ac.be

Roberto DE PIETRI
Università Degli Studi di Parma
depietri@vaxpr.pr.infn.it

Stefano DE SANTIS
Università di Milano
desantis@vaxmi.mi.infn.it

Louise DOLAN
University of North Carolina
dolan@physics.unc.edu

Fay DOWKER
Fermilab
dowker@fnal.fnal.gov

Michael EASTWOOD
University of Adelaide

Giampiero ESPOSITO
Università di Napoli
esposito@na.infn.it

Jonathan EVANS
CERN
evansjm@suryal1.cern.ch

Héctor FIGUEROA
Universidad de Costa Rica
figo@cariari.ucr.ac.cr

José FIGUEROA-O'FARRILL
QMW, London
j.m.figueroa@qmw.ac.uk

Gerald FISCHER
Universität Regensburg
gerald.fischer
@physik.uni-regensburg.de

César FOSCO
University of Oxford
fosco@thphys.ox.ac.uk

Géza FÜLÖP
Chalmers University of Technology
geza@fy.chalmers.se

Jorge GAMBOA
Universität Siegen
gamboa
@siwaps.fb7.uni-siegen.de

Luis J GARAY
Imperial College, London
l.garay@ic.ac.uk

Jerome GAUNTLETT
University of Chicago

Giorgio GIAVARINI
Università di Parma
giavarini@parma.infn.it

Gary GIBBONS
DAMTP, University of Cambridge
g.w.gibbons@amtp.cam.ac.uk

Gabriele GIONTI
SISSA, Trieste
gionti@tsmi19.sissa.it

Joaquim GOMIS
Universitat de Barcelona
quim@ebubech1.ecm.ub.es

Ennio GOZZI
Università di Trieste
gozzi@trieste.infn.it

Jeffrey GREENSITE
San Francisco State University
greensit@stars.sfsu.edu

Joakim HALLIN
Chalmers University of Technology
& University of Göteborg
tfejh@fy.chalmers.se

Ben HARMS
University of Alabama, Tuscaloosa
harms@ua1vm.ua.edu

Jim HARTLE
Santa Barbara
hartle@cosmic.physics.ucsb.edu

Jeffrey HARVEY
University of Chicago
harvey@yukawa.uchicago.edu

Stephen HAWKING
DAMTP, University of Cambridge
s.w.hawking@phx.cam.ac.uk

Marc HENNEAUX
Université Libre de Bruxelles
henneaux@ulb.ac.be

Gary HOROWITZ
Santa Barbara
gary@cosmic.physics.ucsb.edu

Chris HULL
QMW, London
cmh@qmw.ac.uk

Chris ISHAM
Imperial College, London
cji@rl.ib.ac.uk

Hans KASTRUP
RWTH Aachen
kastrup@thphys.rwth-aachen.de

Richard KERNER
Université Paris VI
rk@ccr.jussieu.fr

Hideo KODAMA
Kyoto University
kodama
@yisun1.yukawa.kyoto-u.ac.jp

Margarita KRAUS
Universität Regensburg
margarita.kraus
@mathematik.uni-regensburg.de

Aldo LAINA
Università di Torino
laina@to.infn.it

Claude LEBRUN
SUNY at Stony Brook
claude@math.sunysb.edu

Steven LEPPARD
King's College London
udah047@bay.cc.kcl.ac.uk

Renate LOLL
Pennsylvania State University
loll@phys.psu.edu

Giorghi LONGHI
Università di Firenze
longhi@fi.infn.it

Luca LUSANNA
Sezione INFN di Firenze
lusanna@fi.infn.it

Simon LYAKHOVICH
Tomsk State University
sll@fftgu.tomsk.su

Nenad MANOJLOVIĆ
Universidade do Algarve, Faro
nmanoj@mozart.si.ualg.pt

Giuseppe MARMO
Università di Napoli
gimarmo@na.infn.it

Lionel MASON
University of Oxford
lmason@maths.ox.ac.uk

Toyoki MATSUYAMA
University of Oxford
matsu@thphys.ox.ac.uk

José MOURÃO
Pennsylvania State University
mourao@phys.psu.edu

Harald MÜLLER-KIRSTEN
University of Kaiserslautern
mueller1@gypsy.physik.uni-kl.de

Joseph MULVEY
University of Durham
j.a.mulvey@durham.ac.uk

Jeanette NELSON
Università di Torino
nelson@to.infn.it

Vladimir NESTERENKO
JINR, Dubna
nestr@theor.jinrc.dubna.su

Naohisa OGAWA
Siegen University
ogawa
@siwaps.physik.uni-siegen.de

George PAPADOPOULOS
University of Hamburg
gpapas@vxdesy.desy.de

Jordi PARIS-MOLLO
Katholieke Universiteit Leuven
jordi=paris%tf%fys
@cc3.kuleuven.ac.be

Roger PENROSE
University of Oxford

Malcolm PERRY
University of Cambridge

Roger PICKEN
Instituto Superior Técnico, Lisboa
picken@ptifm.bitnet

Mikhail PLYUSHCHAY
Universidad de Zaragoza
mikhail@cc.unizar.es

Josep PONS
Universitat de Barcelona
pons@ebubecm1.bitnet

John PRESKILL
Caltech
preskill@theory.caltech.edu

Orlando RAGNISCO
III Università di Roma
ragnisco@roma1.infn.it

Eduardo RAMOS
QMW, London
e.ramos@qmw.ac.uk

Jaume ROCA
QMW, London
j.roca@qmw.ac.uk

Alice ROGERS
King's College London
udah039@uk.ac.kcl.cc.bay

Fernando RUIZ
Nikhef-H, Amsterdam
ruiz@nikhef.nl

Michael SINGER
University of Oxford
singer@maths.ox.ac.uk

Dmitrij SOROKIN
King's College London
dsorokin@mth.kcl.ac.uk
(also at Kharkov Inst. of Physics & Technology)

Kelly STELLE
Imperial College, London

Andy STROMINGER
Santa Barbara
andy@denali.physics.ucsb.edu

Leonard SUSSKIND
Stanford University

Michel TALON
Université Paris VI
talon@lpthe.jussieu.fr

Ranjeet TATE
Pennsylvania State University
rstate@phys.psu.edu

Tigran TCHRAKIAN
St Patrick's College, Maynooth
tchrakian@vax1.may.ie

Thomas THIEMANN
Pennsylvania State University
thiemann@phys.psu.edu

Robin TUCKER
University of Lancaster
r.w.tucker@lancaster.ac.uk

Philip TUCKEY
MPI, München
pht@iws170.mppmu.mpg.de

Paolo VALTANCOLI
INFN, Sezione di Firenze
valtancoli@fi.infn.it

JW van HOLTEN
Nikhef-H, Amsterdam
t32@nikhef.nl

Claude VIALLET
Université Paris VI
viallet@lpthe.jussieu.fr

Richard WARD
Durham University
richard.ward@durham.ac.uk

Shing Tung YAU
Harvard University
yau@math.harvard.edu

Jian-Zu ZHANG
University of Kaiserslautern
jzzhang@de.uni-kl.physik
(also at East China Inst. for Th. Phys., Shanghai)